Mathematical Physics Studies

The series publishes original research monographs on contemporary mathematical physics. The focus is on important recent developments at the interface of Mathematics, and Mathematical and Theoretical Physics. These will include, but are not restricted to: application of algebraic geometry, D-modules and symplectic geometry, category theory, number theory, low-dimensional topology, mirror symmetry, string theory, quantum field theory, noncommutative geometry, operator algebras, functional analysis, spectral theory, and probability theory.

More information about this series at http://www.springer.com/series/6316

Lawrence Horwitz · Yosef Strauss

Unstable Systems

Lawrence Horwitz
School of Physics, Raymond and Beverly
Sackler, Faculty of Exact Sciences
Tel Aviv University
Ramat Aviv, Israel

Department of Physics
Bar-Ilan University
Ramat Gan, Israel

Department of Physics
Ariel University
Samaria, Israel

Yosef Strauss
Department of Mathematics
Ben-Gurion University of the Negev
Beér Sheva, Israel

ISSN 0921-3767 ISSN 2352-3905 (electronic)
Mathematical Physics Studies
ISBN 978-3-030-31572-6 ISBN 978-3-030-31570-2 (eBook)
https://doi.org/10.1007/978-3-030-31570-2

This Springer imprint is published by the registered company Springer Nature Switzerland AG
The registered company address is: Gewerbestrasse 11, 6330 Cham, Switzerland

Acknowledgments

I would like to thank Jean-Paul Marchand for originally posing the central problems of describing unstable quantum systems with mathematical precision and his collaboration, and Josef M. Jauch for his important contributions to the discussions at the University of Geneva. I would like to thank as well Israel Michael Sigal for his collaboration on the construction of the Gel'fand triple for the Lee–Friedrichs model, a work motivated in part by discussions with Peter Lax. I would also like to express my appreciation to Calvin Wilcox, from whom I first learned about the theory of Lax and Phillips, and to Constantin Piron and Eli Eisenberg for their assistance in our understanding of the theory, and for their collaborations. I am also grateful to I. Prigogine for many discussions of this subject, for his hospitality at the Free University of Brussels and a meeting he sponsored in Provence which focused on this subject. I would also like to thank Jacob Levitan, Meir Lewkowicz, Yossi Ben Zion, and Marcelo Schiffer for discussions and collaboration, as well as Asher Yahalom and Gil Elgressy.

I am grateful to Stephen L. Adler for his warm hospitality at the Institute for Advanced Study in Princeton over a period of several visits during which, among other things, the application of our methods for treating the Lee–Friedrichs model was worked out for quaternionic quantum theory. I would also like to thank William C. Schieve for his collaboration over many years, and for the hospitality he and his wife, Florence, offered me at the University of Texas at Austin during many visits where this subject was discussed extensively, both with him and Arno Bohm.

I am, of course, grateful to my co-author Yossi Strauss for his collaboration over several years, for his insight in embedding the Lax–Phillips theory into quantum mechanics and his creativity and mathematical precision in the development of the theory.

Finally, I wish to thank my wife, Ruth, for her patience and support in the many years during which the research was done in developing and exploring our evolving ideas on the subject of unstable systems.

Lawrence Horwitz

I would like to thank Israel Michael Sigal at the University of Toronto and Avi Soffer at Rutgers University for their hospitality during the period in which the modified Lax–Phillips theory began to take shape and to Peter D. Lax for closely following the development of the formalism and for his advice. I would like to thank Jonathan Silman and Shai Machnes and for their collaboration in the development of the subject of Lyapunov operators. I would also like to thank Arno Bohm, Manuel Gadela, Sugeev Wikramaskara, and Helmut Buamgartel for extensive and fruitful discussions of the role of Rigged Hilbert spaces and the Lax–Phillips formalism in the description of resonances and irreversible evolution in quantum mechanics. I would also like to thank Jacob Levitan and Asher Yahalom for discussions and collaboration, as well as Gil Elgressy and Michal Wagman.

I am deeply grateful to my co-author, Larry Horwitz, for introducing me to the fascinating subject of unstable systems and irreversible processes in quantum mechanics and the role of the Lax–Phillips formalism in the description of such irreversible evolution. I would like to thank him for his collaboration over the years and for his many innovative ideas.

Last, but not least, I wish to thank my family; my mother Miriam, my wife Ela, son Roee, and daughter Noa for their patience and support over the years the research work described in the present book has been in development.

Yosef Strauss

Introduction

There is a large literature on classically unstable systems (Guckenheimer 1983) (such as an oscillator with friction (Maassen 1989), a perturbed pendulum, or a particle moving in a double-well potential) ranging from rigorous mathematical studies of phenomena such as strange attractors (e.g., Ruelle (Ruelle 1989), limit cycles, and period doubling to phenomenologically useful measures of instability, some of which is based on geometrical methods involving the Jacobi metric of Hamiltonian mechanics (Pettini 2007). Recent developments have been concerned with new geometrical methods, in many cases more effective than methods using the Jacobi metric, based on the construction of a manifold underlying the usual potential model Hamiltonian mechanics.

The study of the instability and chaotic behavior of quantum mechanical systems has also received much attention, much of it based on the work of Wigner (Wigner 1955), who analyzed the spectrum of systems with random potentials, as well as some conjectured associations of classical with quantum behavior, such as those of Bohigas (Bohigas 1984) and Zaslavsky (Zaslavsky 1985), but the available characterization of quantum unstable systems is far from complete.

It is difficult to overstate the importance of the study of quantum instability. In addition to the everyday phenomena of electromagnetic radiation from excited atomic states, almost all of the elementary particles that are identified as such today are actually resonances; many, like the $\Sigma, \Lambda \ldots$ baryons (10^{-23} s), and the $\pi, K \ldots$ mesons are short-lived ($10^{-8} - 10^{-10}$ s), and some, like the neutron, relatively long-lived (about 10 mins). In general, it has not been possible to rigorously assign these objects to real quantum states, that is, as vectors in a Hilbert space, although there is much evidence (for example, the saturation of the Adler-Weisberger sum rules (Adler 1965)) that they should be. We discuss recent developments by the authors (Strauss 2000) that extends the work of Lax and Phillips (Lax 1967), who succeeded to construct states in a Hilbert space for the resonances of classical wave systems, to apply as well to quantum systems, thus providing a framework for which resonances may be considered as proper states in a quantum mechanical Hilbert space.

One can see, for example, in the work of Maassen (Maassen 1989), in his study of the oscillator with friction, that an application of the ideas of Nagy and Foias (Nagy 1976), allowing the system to be quantized by "dilation", results in a representation of the friction in terms of emission and absorption of radiation. Studying the stability of a classical Hamiltonian system from a geometrical point of view, the authors have recently shown (Strauss 2015) that the geodesic deviation condition for stability, which has the form of an oscillator equation (called the Jacobi equation) can be similarly quantized, providing an understanding of the instability as excitation modes being exchanged with a medium. There is, therefore, an intimate connection between the fundamental quantum mechanical description of unstable systems and the well known and studied classical behavior (Maassen 1989).

It was our feeling that the description of the underlying quantum theory of unstable systems, and the recent developments in the subject permitting the representation of resonances as physical states in a quantum theory should be treated in Part I, providing a microscopic framework for the analysis of classically unstable systems. In Part II we provide a description of approaches to classical stability analysis, and introduce the recently developed geometrical methods, shown to be highly effective in diagnosing instability and, in many cases, chaotic behavior, and explain its mathematical basis as an underlying geometrical structure (involving geodesic motion on a manifold) for standard Hamiltonian potential models. We shall discuss a new and rigorous method of embedding a Hamiltonian system of standard form into a space with curvature by means of a canonical transformation corresponding to a symplectomorphism.

Contents

Chapter 1
Spectral Analysis in Hilbert Space

1.1 Background and Wigner–Weisskopf Theory

After some years of observed radioactivity, Gamow (1928) wrote down a formula in quantum mechanics that was supposed to govern the decay of the nucleus of an atom into a lighter nucleus with the emission of an alpha particle (helium atoms without their electron cloud, i.e. ionized helium). This formula, a Schrödinger equation for a stable particle of energy E, was written with a complex energy (with negative imaginary part), implying that the quantum wave function would go to zero exponentially fast in time, as was observed of a radioactive nucleus. The resulting exponential decay law obeyed an essential property called the "semigroup" law, stating that for an evolution $Z(t)$ on the quantum states,

$$Z(t_1)Z(t_2) = Z(t_1 + t_2), \tag{1.1.1}$$

for both t_1 and t_2 positive. This law, even in more general situations, such as the two-channel K meson decay, has been very precisely measured experimentally. In fact, one can quite generally argue that for an irreversible process, for which the system has no memory, that at every moment the process starts anew, and therefore after t_1, the application of $Z(t_2)$ brings the system to the state that it would have arrived at due to the application of the evolution $Z(t_1 + t_2)$, i.e., for the full interval $t_1 + t_2$, as in Eq. (1.1.1).

Gamow's idea was well received as the first application of quantum mechanics to nuclear physics. It had, however, a serious flaw in that the momentum of a (freely moving) particle is the square root of the energy, and the momentum would then be a complex number. The representation of such a wave function in the quantum theory would then contain a complex variable as the coefficient of the *coordinate* dependence, and cause an exponential divergence of the function at large distances.

In 1930, Weisskopf and Wigner(1930) wrote a more mathematically consistent theory which achieved almost the same results as Gamow's idea, that is, an exponential decay (at least for times not too long or too short), and this very useful theory

L. Horwitz and Y. Strauss, *Unstable Systems*, Mathematical Physics Studies,
https://doi.org/10.1007/978-3-030-31570-2_1

has been widely used used up to the present. We describe this theory briefly in the following, and point out that it contains serious difficulties as well. A solution to this problem, which has led to deep mathematical investigations, was first proposed by Lax and Phillips (1967) in the framework of classical wave theory, such as acoustic or electromagnetic waves, has more recently been adapted by the authors to the quantum theory.

The theory of Wigner and Weisskopf assumed that there is a quantum wave function that represents the system in its initial state, thought of as, e.g., a *particle*, and that an evolution of this state by the standard quantum mechanical unitary evolution $U(t) = e^{-iHt}$ corresponds to a *decay* of that object. This picture is not an idea intrinsic to the quantum theory, but it allowed Wigner and Weisskopf to construct a mathematical model for particle decay. One may then calculate the amplitude for the particle to remain in the initial state by computing the scalar product

$$A(t) = (\psi, e^{-iHt}\psi), \tag{1.1.2}$$

where $A(t)$ is known as the "survival amplitude"; its absolute square is called the "survival probability" at time t. Wigner and Weisskopf showed, using time dependent perturbation theory, that the survival probability decays approximately exponentially for times not too short and not too long. A short time power series analysis, however, shows that the initial decay is just quadratic (for a time that would be difficult, for most unstable systems, to observe experimentally directly), and at very long times, using methods of Laplace transform and complex function analysis, the decay is predicted to be inverse polynomial. Thus, the semigroup property is not represented precisely in this formulation. This defect emerges explicitly, in an experimentally verifiable way, in its prediction for the two channel K-meson decay (Horwitz and Marchand 1971).

The Laplace transform of (1.1.2) provides a useful method (Horwitz and Marchand 1971) for analyzing this so-called "survival amplitude" i.e.

$$A(z) = \int_0^\infty e^{izt} A(t) = i\left(\psi, \frac{1}{z - H}\psi\right) \equiv (\psi, G(z)\psi) \tag{1.1.3}$$

is well-defined (and analytic) for $Imz > 0$, exhibiting explicitly the Green's function (often called the resolvent kernel), now understood in terms of a "propagator", as the Laplace transform of the unitary evolution.

The inverse Laplace transform is an integral to be carried out on a line in the complex z plane just above the real axis (where the function is analytic) from $+\infty$ to $-\infty$,

$$A(t) = \frac{1}{2\pi i}\int_{\infty+i\epsilon}^{-\infty+i\epsilon} dz e^{-izt} A(z). \tag{1.1.4}$$

If the spectrum of H, for the reduced motion of a two body system runs from zero to infinity (i.e., bounded below), then the integration on the negative real axis can be moved to the lower half plane, running along the imaginary axis, where the value is suppressed by the exponential $\exp(-izt)$; the contributions are very small for the interval sufficiently below the branch point, and t sufficiently large. The integral along (and slightly above) the positive real axis can be lowered to the second Riemann sheet by considering the difference

$$(\psi, G(E+i\epsilon)\psi) - (\psi, G(E-i\epsilon)\psi) = 2\pi i\chi(E), \tag{1.1.5}$$

where $\chi(E)$ is the spectral weight factor in the expectation value of the spectral representation of the operator $(z-H)^{-1}$. The term $(\psi, G(E-i\epsilon)\psi)$ is evaluated by analytic continuation around the real axis to a point just below the positive real axis.[1] For the case of a continuous spectrum from $-\infty$ to $+\infty$, two functions, one analytic in the upper half plane, and the other in the lower half plane (Ben Ari and Horwitz 2004), according to forward and backward evolution, may be defined, and a similar method may be applied by moving the integral along the real line in the upper half plane into the lower half plane. Rotating the integral on the positive real axis into the lower half plane (second sheet), one finds that there can be contributions from singularities in the lower half plane. In the case of a pole, which one might understand as the remnant of a bound state, a pole on the real axis, pulled down into the lower half plane by the interaction, the passage of the contour of integration over this pole extracts a residue proportional to $e^{-iz_P t}$. This term may dominate the entire integral (for times not too long and not too short), in agreement with the proposal of Gamow, and corresponds to a *resonance* in the scattering amplitude.

For very short times (as we show below) this expression will not be dominant, and for very long times, the contribution of the branch cut dominates, for which there may be a polynomial type decay law (Horwitz and Marchand 1971; Horwitz et al., 1971a). Horwitz and Katznelson (1983) suggested that the proton decay predicted in some particle physics models could be suppressed in a nuclear environment due to this short time effect.

There are systems with multiplicity on the continuum, such as many body systems or an atom with several electrons in bound states which may be successively excited, for which the continuous spectrum has a sequence of thresholds; the analytic continuation through the cut is then more complicated. Aguilar and Combes (1971) have suggested that one apply a unitary transformation to the resolvent of the form $e^{-i\lambda D}$, where $D = i\frac{\partial}{\partial E}$ (dilation); and analytic continuation in λ then rotates the spectrum about each of the branch points, thus separating the continuous spectrum at each threshold. The analytic continuation of the resolvent is then more easily defined, and the resonances appear as complex poles in the lower half place of each of the sheets defined in this way.

However, the pole contributions do not correspond to any physical states. It has been shown, on the other hand, that there is a vector in a Banach space (an element of a Gel'fand triple constructed in the dual to a subspace of the original Hilbert space (Horwitz and Sigal 1978; Baumgartel 1976) that can be constructed to correspond to this pole with exact exponential decay, as we shall see in the next section, but it is

[1]*.

difficult to interpret such a construction as a physically meaningful state (expectation values would not be generally defined). Nevertheless, this description provides an experimentally useful definition for resonances in terms of the decay law (Bohm and Gadella 1989).

Although the result Wigner–Weisskopf obtained from (1.1.2) may provide an exponential behavior for $|A(t)|^2$ for sufficiently long (but not too long) times, as we have seen, for very short times it generally displays a decay law which is not consistent with the exponential form. For short times,

$$A(t) = 1 - i < H > \tau - \frac{< H^2 >}{2}\tau^2 + \cdots \tag{1.1.6}$$

It then follows that

$$|A(t)|^2 = 1 - \Delta H^2 \tau^2 + \cdots , \tag{1.1.7}$$

with $\Delta H^2 =< H^2 > - < H >^2$, not consistent with the semigroup property.

The model proposed by Gamow, obeys the *semigroup* law of evolution (1.1.1), expected from reasonable arguments to be valid for irreversible processes, such as the decay of an unstable system[2]; however $|A(t)|^2$ does not have the property that it approaches unity at zero linearly, but quadratically, as it would for almost any Hamiltonian (with finite dispersion in state ψ) in the Wigner–Weisskopf model. One can argue that in many cases the very short time before an approach to exponential (Sudarshan and Misra 1977) behavior would not be observable experimentally, and this has justified its use in many cases, but the fact that the evolution is not semigroup has consequences for the application of the idea to two of more dimensions, such as for the neutral K meson decay, where it has been shown to be quantitatively inapplicable (Cohen and Horwitz 2001). In this case, one finds that the poles of the resolvent for the Wigner–Weisskopf evolution of the two channel system results in non-orthogonal residues that generate interference terms, which make the non-semigroup property evident even for times for which the pole approximation is valid (Cohen and Horwitz 2001), a domain in which exponential decay for the single channel system is very accurately described by the Wigner–Weisskopf model.

The (Wu and Yang 1964) parametrization of the K^0 decay processes, based on a Gamow type evolution generated by an effective 2×2 non-Hermitian matrix Hamiltonian, on the other hand, results in an evolution that is an exact semigroup. It appears that the phenomenological parametrization of Yang and Wu, actually a two dimensional form of the Gamow hypothesis, is indeed consistent to a high degree of accuracy with the experimental results on K-meson decay (Patrignani et al., 2016).

We shall discuss below a theory (Strauss and Horwitz 2002) (based on the work of Lax and Phillips (1967)) in which the evolution law is precisely semigroup and identifies the resonance with a quantum state, and discuss how it can be applied to the

---

[2]Based on the argument that one can stop the evolution at any moment and then proceed as if starting from the new initial conditions, with a result equivalent to letting the system develop undisturbed for the entire time, an essentially Markovian hypothesis.

relativistic evolution of the neutral K-meson system, explaining as well the origin of the phenomenological parametrization of Yang and Wu.

1.2 Analytic Continuation and the Gel'fand Triple

We have discussed in the previous section the emergence of resonances as complex poles in the lower half plane of the resolvent kernel, and pointed out that the complex poles do not correspond to a physical state. As we emphasized in the Introduction, there is, however, considerable evidence in the phenomenological discussion of resonances that these phenomena should be assigned in a systematic way to some precise description of states of the system. We discuss in this section the quantum mechanical form given by Friedrichs (1950) of the Lee (1956) model as a simple example of how such an assignment can be made in a Banach space (Horwitz and Sigal 1978; Baumgartel 1976).

The Lee–Friedrichs model, in its simplest form, is defined in terms of a Hamiltonian

$$H = H_0 + V, \tag{1.2.1}$$

where H_0 has a simple absolutely continuous spectrum in $(0, \infty)$ and V is a rank one potential. These operators are defined by[3]

$$< E|H_0 f) = E < E|f), \tag{1.2.2}$$

for any f in the Hilbert space $\mathcal{H}$, and E real in $[0, \infty)$,

$$H_0\phi = E_0\phi \tag{1.2.3}$$

where ϕ is an eigenstate for an eigenvalue of H_0 imbedded in the continuum,

$$< E|V|E' >= 0 \quad \forall E, E', \tag{1.2.4}$$

and

$$< E|V\phi) \equiv v(E) \tag{1.2.5}$$

is some smooth complex valued function of E.

The state represented by ϕ is considered to be the unstable state that decays into the continuum in the model of Wigner and Weisskopf.

The operator V is bounded if $v(E)$ is square integrable and continuous, since $\|V\phi\|^2 = \int dE|v(E)|^2$ and $\|V|E>\|^2 = |v(E)|^2$.

We now use the identity for the Green's function (or propagator)

---

[3]We use angular brackets for generalized vectors and round brackets for proper normalized vectors in the Hilbert space. The set $\{< E|, \phi\}$ is complete in $\mathcal{H}$.

$$G(z) = \frac{1}{z - H} = \frac{1}{z - H_0} + \frac{1}{z - H_0} V \frac{1}{z - H} \tag{1.2.6}$$

or

$$G(z) = G_0(z) + G_0(z) V G(z). \tag{1.2.7}$$

We now study the matrix elements of this equation, and show that this model is completely soluble.

According to (1.2.3) the quantity we are interested in is

$$(\phi|G(z)|\phi) = \frac{1}{z - E_0} + \frac{1}{z - E_0} \int dE (\phi V|E >< E|G(z)|\phi), \tag{1.2.8}$$

where we have used the fact that V connects ϕ only with the continuum. It is this property that makes the model soluble. What we have to do now is to compute (the first term does not contribute)

$$< E|G(z)|\phi) = \frac{1}{z - E} < E|V\phi)(\phi|G(z)|\phi). \tag{1.2.9}$$

Substituting this result into (1.2.8), and multiplying both sides by $z - E_0$, we obtain a closed formula for $(\phi|G(z)|\phi)$,

$$\left(z - E_0 - \int dE \frac{|v(E)|^2}{z - E} \right) G(z) = 1. \tag{1.2.10}$$

This formula makes the cut along the real axis in the analytic function $G(z)$ evident. Let us define

$$h(z) = z - E - \int dE \frac{|v(E)|^2}{z - E}, \tag{1.2.11}$$

so that (1.2.10) reads

$$h(z) G(z) = 1. \tag{1.2.12}$$

To analytically continue across the cut on the real line, we first remark that, since the cut is on $(0, \infty)$, we may analytically continue the function $G(z)$ through the negative half line with no change in its structure, thus defining it on the lower half plane in the first Riemann sheet. Then, using the relation

$$\frac{1}{x + i\epsilon} - \frac{1}{x - i\epsilon} = -2\pi i \delta(x), \tag{1.2.13}$$

we see that (crossing at the value E)

$$h(E + i\epsilon) - h(E - i\epsilon) = 2\pi i |v(E)|^2, \tag{1.2.14}$$

so that we can define the *second sheet function*, which goes continuously across the cut as

$$h^{II}(E - i\epsilon) = h(E - i\epsilon) + 2\pi i|v(E)|^2.$$

We now assume that $|v(E)|^2$ is the boundary value of a function analytic in a sufficient region of the lower half plane, so that for z in the lower half plane,

$$h^{II}(z) = h(z) + 2\pi i\chi(z), \tag{1.2.15}$$

where $\chi(z)$ is the analytic continuation of $|v(E)|^2$, and then

$$Imh^{II}(z) = Imz\left(1 + \int dE\frac{|v(E)|^2}{|z - E|^2}\right) + 2\pi Re\chi(z) \approx Im(h(z)) + 2\pi Re\chi(z) \tag{1.2.16}$$

If the value of Imz in the lower half plane that makes this vanish is sufficiently small, then $\chi(z)$ should be real and approximately equal to $|v(E)|^2|_{E=Rez}$. Since from (1.2.12), under analytic continuation,

$$h^{II}(z)G^{II}(z) = 1. \tag{1.2.17}$$

The imaginary part of $h^{II}(z)$ vanishes, as we see from (1.2.16), for

$$Imz = -\left(1 + \int dE\frac{|v(E)|^2}{|z - E|^2}\right)^{-1} 2\pi Re\chi(z) \approx -|v(E)|^2|_{E=Rez}. \tag{1.2.18}$$

If the real part vanishes also, $G^{II}(z)$ would have a pole at this point.

In quantum mechanical scattering theory (Taylor 1972), the scattering part of the S matrix is expressed as

$$T(z) = V + VG(z)V, \tag{1.2.19}$$

analytic in the same domain as $G(z)$. Since the matrix elements of V connect the continuum only with the state ϕ in the Lee–Friedrichs model, the amplitude $< E|T(z)|E' >$ contains the factor $(\phi|G(z)|\phi)$ and therefore would have a pole in the second sheet, dominating the scattering cross section on the real line, corresponding to the resonance of the Lee–Friedrichs model.

As we have seen, according to (1.2.6), one might think of the singularity in $G(z)$ as corresponding to an "eigenvalue" of H; since H is self-adjoint, however, it cannot have an eigenfunction with complex eigenvalue in the Hilbert space $\mathcal{H}$. Let us, however, study the condition

$$Hf(z) = zf(z), \tag{1.2.20}$$

where $f(z) \in \mathcal{H}$. Once again, using (1.8), we see that this equation is exactly solvable in the Lee–Friedrichs model; for the $< E|$ component of (1.2.20), we have

$$< E|f(z)) = \frac{1}{z-E} < E|V|\phi)(\phi, f(z)). \tag{1.2.21}$$

The remaining (ϕ) component of $f(z)$ is determined by

$$E_0(\phi, f(z)) + \int dE(\phi|V|E >< E|f(z)) = z(\phi, f(z)). \tag{1.2.22}$$

Substituting (1.2.21) into (1.2.22), we obtain

$$\left(z - E_0 - \int dE \frac{|(\phi|V|E>|^2}{z-E}\right)(\phi, f(z)) = 0. \tag{1.2.23}$$

If $(\phi, f(z))$ is not zero, which we shall assume, then the first factor in (1.31) must vanish. However, for the imaginary part, this would require

$$Imz\left(1 + \int dE \frac{|(\phi|V|E>|^2}{|z-E|^2}\right) = 0, \tag{1.2.24}$$

clearly not possible in the upper half plane. Moreover, on the positive real axis, this condition would require, for real $\lambda < 0$ (the expression would diverge for $\lambda > 0$)

$$\lambda - E_0 + \int dE \frac{|(\phi|V|E>|^2}{|\lambda|+E} = 0. \tag{1.2.25}$$

If, then

$$\int dE \frac{|(\phi|V|E>|^2}{|\lambda|+E} < \int \frac{dE}{E} |(\phi|V|E>|^2 \tag{1.2.26}$$

(assuming $|(\phi|V|E>|^2$ vanishes quickly enough at $E \to 0$) is smaller than E_0, then there can be no solution on the negative real axis. However, with sufficiently large coupling, a pole could occur on the negative real axis, corresponding to the creation, through the interaction, of a real bound state by means of this very strong interaction.

If this does not occur, there may be, as discussed above, a solution in the lower half plane. This solution corresponds to an analytic continuation of the Hilbert space vector (defined in (1.2.2)) from the upper half plane to the lower half plane, as we shall see, to an element of a Banach space as part of a *Gel'fand-triple* (Horwitz and Sigal 1978; Baumgartel 1976; Bohm and Gadella 1989). To define this analytic continuation, consider the Hilbert space scalar product, for z in the upper half plane,

$$(g, f(z)) = \int (g|E >< E|f(z)) + (g, \phi)(\phi, f(z)). \tag{1.2.27}$$

The first term can be studied with the help of (1.2.21). Equation (1.2.22) is homogeneous in $(\phi, f(z))$, allowing us to assume that it remains analytic in a region across the cut. In terms of functions defined in the first sheet

$$(g, f(\lambda + i\epsilon)) - (g, f(\lambda - i\epsilon)) = -2\pi i (g|\lambda >< \lambda|V|\phi)(\phi, f(\lambda). \tag{1.2.28}$$

We can then define the second sheet of the scalar product function

$$(g, f(z))^{II} = (g, f(z)) - 2\pi i G(z), \tag{1.2.29}$$

where $G(z)$ is the analytic continuation of $(g|\lambda >< \lambda|V|\phi)$ into a region of the lower half plane large enough to include the zero of (1.2.16).

We see that this construction cannot be carried out if $(g|\lambda >$ is not analytic in this region. Therefore $(g, f(z))^{II}$ is only defined on $g \in \mathcal{D} \subset \mathcal{H}$, where $\mathcal{D}$ is the set of g's for which $(g|\lambda >$ has analytic continuation to a sufficient region of the lower half plane. The functional defined by $(g, f(z))$ defined in this way lies, however, in a space larger than $\mathcal{H}$, say, $\bar{\mathcal{H}}$. The hierarchy

$$\mathcal{D} \subset \mathcal{H} \subset \bar{\mathcal{H}} \tag{1.2.30}$$

is called a Gel'fand triple, or rigged Hilbert space.

This construction provides a "vector" in a Banach space that can be considered a representation of the resonance of the Lee–Friedrichs model, but, although it has a maximum modulus bound, one cannot compute scalar products of such elements, or the expectation value of observables in such a resonant "state".

To achieve a more useful description of a resonance as an element in a Hilbert space, we develop in the next chapter the Lax–Phillips quantum theory (Strauss and Horwitz 2002) as a quantum generalization of the theory originally developed by Lax and Phillips (1967) for the description of resonances in classical wave theory.

Chapter 2
Semigroups and Lax Phillips Theory

2.1 The Lax Phillips Scattering Theory

In this chapter, we discuss the scattering theory of Lax and Phillips (1967), originally developed for the description of resonances in the scattering of classical waves, such as electromagentic or acoustic waves, off compactly supported obstacles. In Appendix B, we give an extensive treatment of the mathematical background enabling a more complete understanding of the mathematical structure underlying a theory of this type, which seems to be categorical for the description of resonances.

In the Lax–Phillips theory, one assumes the existence of a Hilbert space $\mathcal{H}^{LP}$ of physical states in which there are two distinguished subspaces $\mathcal{D}_+$ and $\mathcal{D}_-$ called, respectively, the *outgoing* and *incoming* subspaces, such that $\mathcal{D}_+$ is orthogonal to $\mathcal{D}_-$ and

$$\begin{aligned} U(\tau)\mathcal{D}_+ &\subset \mathcal{D}_+ \qquad \tau \geq 0 \\ U(\tau)\mathcal{D}_- &\subset \mathcal{D}_- \qquad \tau \leq 0 \\ \bigcap_\tau U(\tau)\mathcal{D}_\pm &= \{0\} \\ \overline{\bigcup_\tau U(\tau)\mathcal{D}_\pm} &= \mathcal{H}^{LP}, \end{aligned} \tag{2.1.1}$$

The subspace $\mathcal{D}_+$, called the *outgoing subspace*, is stable under the evolution of the system for positive values of the evolution parameter τ. Similarly, the subspace $\mathcal{D}_-$, called the *incoming subspace*, is stable under the evolution for negative τ. The stability properties of $\mathcal{D}_+$ and $\mathcal{D}_-$ correspond to the fact that in problems in which the velocity of wave propagation is bounded there exists a subspace of waves that do not interact with the scattering target before $\tau = 0$ (i.e., $\mathcal{D}_-$) and there exists a subspace of waves that do not interact with the scattering target after $\tau = 0$ (i.e., $\mathcal{D}_+$; of course, in most cases we have time translation invariance and the reference point $\tau = 0$ is arbitrarily chosen). The third line of (2.1.1) correspond to the fact that

L. Horwitz and Y. Strauss, *Unstable Systems*, Mathematical Physics Studies,
https://doi.org/10.1007/978-3-030-31570-2_2

for $\tau \to \pm\infty$ all scattered waves propagate to spatial infinity. The final condition in (2.1.1) assumes that the evolution group $\{U(\tau)\}_{\tau \in R}$ generates a dense set in $\mathcal{H}^{LP}$ from either $\mathcal{D}_+$ or $\mathcal{D}_-$. This condition is equivalent to the assumption of asymptotic completeness of the scattering problem.

Let P_+ be the orthogonal projection onto the orthogonal complement of $\mathcal{D}_+$ in $\mathcal{H}^{LP}$. Let P_- be the orthogonal projection onto the orthogonal complement of $\mathcal{D}_-$ in $\mathcal{H}^{LP}$. Lax and Phillips define a family $\{Z(\tau)\}_{t \geq 0}$ of operators on $\mathcal{H}^{LP}$ by

$$Z(\tau) \equiv P_+ U(\tau) P_-, \quad t \geq 0. \tag{2.1.2}$$

From the definition (2.1.2) it is evident that $Z(\tau)$ annihilates $\mathcal{D}_-$ and its range is orthogonal to $\mathcal{D}_+$. Using the stability property of $\mathcal{D}_+$ in (2.1.1) we have, for any element $x \in \mathcal{D}_+$ and $\tau \geq 0$,

$$Z(\tau)x = P_+ U(\tau) P_- f = P_+ U(\tau) f = 0$$

hence $Z(\tau)$ annihilates $\mathcal{D}_+$. Furthermore, for any $x \in \mathcal{D}_-$ and $y \in \mathcal{H}^{LP}$ and $\tau \geq 0$, we have,

$$(x, Z(\tau)y)_{\mathcal{H}^{LP}} = (x, P_+ U(\tau) P_- y)_{\mathcal{H}^{LP}} = (P_- U(-\tau)x, y)_{\mathcal{H}^{LP}} = 0,$$

where the last equality is obtained using the stability properties of $\mathcal{D}_-$ in (2.1.1). By the last equation $\mathcal{D}_-$ is orthogonal to the range of $Z(\tau)$, $\forall \tau \geq 0$. We conclude that for each $\tau \geq 0$, $Z(\tau)$ annihilates $\mathcal{D}_\pm$ and maps the supspace $\mathcal{K} = \mathcal{H}^{LP} \ominus (\mathcal{D}_- \oplus \mathcal{D}_+)$ into itself.

It is easily proved that the family of operators $\{Z(\tau)\}_{t \geq 0}$ forms a continuous semigroup. Considering a vector $x \in \mathcal{K}$ we have

$$\begin{aligned} &Z(\tau_1)Z(\tau_2)x = \\ &\quad = P_+ U(\tau_1) P_- Z(\tau_2)x = P_+ U(\tau_1) P_+ U(\tau_2) P_- x = P_+ U(\tau_1) P_+ U(\tau_2)x, \quad \tau_1, \tau_2 \geq 0. \end{aligned}$$

Moreover, by the stability properties of $\mathcal{D}_+$ we have $P_+ U(\tau)(I - P_+) = 0$, for $\tau \geq 0$. Hence

$$\begin{aligned} Z(\tau_1)Z(\tau_2)x =& P_+ U(\tau_1) P_+ U(\tau_2)x = P_+ U(\tau_1)[P_+ + (I - P_+)]U(\tau_2)x = \\ &= P_+ U(\tau_1) U(\tau_2)x = P_+ U(\tau_1 + \tau_2) P_- x = Z(\tau_1 + \tau_2)x, \quad \tau_1, \tau_2 \geq 0, x \in \mathcal{K} \end{aligned}$$

Lax and Phillips prove the following theorem, providing further properties of the semigroup $\{Z(\tau)\}_{\tau \geq 0}$,

It is easily proved that the family of operators $\{Z(\tau)\}_{t \geq 0}$ forms a continuous semigroup. Considering a vector $x \in \mathcal{K}$ we have

$$\begin{aligned} Z(\tau_1)Z(\tau_2)x = P_+ U(\tau_1) P_- Z(\tau_2)x = P_+ U(\tau_1) P_+ U(\tau_2) P_- x = \\ = P_+ U(\tau_1) P_+ U(\tau_2)x, \quad \tau_1, \tau_2 \geq 0 \end{aligned}$$

Moreover, by the stability properties of $\mathcal{D}_+$ we have $P_+U(\tau)(I - P_+) = 0$, for $\tau \geq 0$. Hence

$$Z(\tau_1)Z(\tau_2)x = P_+U(\tau_1)P_+U(\tau_2)x = P_+U(\tau_1)[P_+ + (I - P_+)]U(\tau_2)x = \\ = P_+U(\tau_1)U(\tau_2)x = P_+U(\tau_1 + \tau_2)P_-x = Z(\tau_1 + \tau_2)x, \quad \tau_1, \tau_2 \geq 0, x \in \mathcal{K}.$$

Lax and Phillips prove the following theorem, providing further properties of the semigroup $\{Z(\tau)\}_{\tau\geq 0}$:

Theorem 2.1 *The operators $\{Z(\tau)\}_{\tau\geq 0}$ annihilate $\mathcal{D}_+$ and $\mathcal{D}_-$, map the orthogonal complement $\mathcal{K} = \mathcal{H}^{LP} \ominus (\mathcal{D}_+ \oplus \mathcal{D}_-)$ into itself and form a strongly continuous semigroup of contraction operators on $\mathcal{K}$. Furthermore, $Z(\tau)$ tends strongly to zero for $\tau \to \infty$, i.e., $\lim_{\tau\to\infty} \|Z(\tau)x\|_{\mathcal{H}^{LP}} = 0$, $\forall x \in \mathcal{K}$.* □

The semigroup $\{Z(\tau)\}_{\tau\geq 0}$ is called the *Lax–Phillips semigroup* and is the central object of investigation in the Lax–Phillips scattering theory.

Let $L^2(R; H)$ be the Hilbert space of Lebesgue square integrable functions defined on R and taking their values in a Hilbert space H. Ja. G. Sinai (Cornfeld et al. 1982) proved that if the conditions of (2.1.1) hold for the outgoing subapce $\mathcal{D}_+$ then the following theorem holds:

Theorem 2.2 (Ja. G. Sinai) *If $\mathcal{D}_+$ is an outgoing subspace with respect to an evolution group of unitary operators $\{U(\tau)\}_{\tau\in R}$ then the Hilbert space $\mathcal{H}^{LP}$ can be represented isometrically as a Hilbert space of functions $L^2(R; H)$ for some auxiliary Hilbert space H so that $U(\tau)$ is represented by translation to the right by τ units, and $\mathcal{D}_+$ is mapped onto $L^2([0, \infty); H)$. This representation is unique up to isomorphism on H.* □

A representation of this kind is called *outgoing translation representation* for the group $\{U(\tau)\}_{t\in R}$. An analogous representation theorem holds for the incoming subspace $\mathcal{D}_-$, i.e., there is a representation in which $\mathcal{H}^{LP}$ is mapped isometrically onto the Hilbert space $L^2(R; H)$, the incoming subspace $\mathcal{D}_-$ is mapped onto the subspace $L^2((-\infty, 0]; H)$ and $U(\tau)$ acts as translation to the right by τ units. This representation is called the *incoming translation representation*.

Let W_+ and W_- be the operators mapping elements of $\mathcal{H}^{LP}$ to their outgoing, respectively incoming, translation representers. These operators are referred to, respectively, a the outgoing and incoming *Lax–Phillips wave operators*. We call the operator

$$\mathbf{S}^{LP} := W_+W_-^{-1}$$

the *Lax–Phillips scattering operator* associated with the group $\{U(\tau)\}_{\tau\in R}$ and the pair $\mathcal{D}_\pm$. It was proved by Lax and Phillips that S^{LP} is equivalent to the standard definition of the scattering opertor in scattering theory (see, for example, Taylor 1972). The Lax–Phillips scattering operator possess the following properties:

(a) $\mathbf{S}^{LP}$ is unitary,
(b) $\mathbf{S}^{LP}$ commutes with translations,

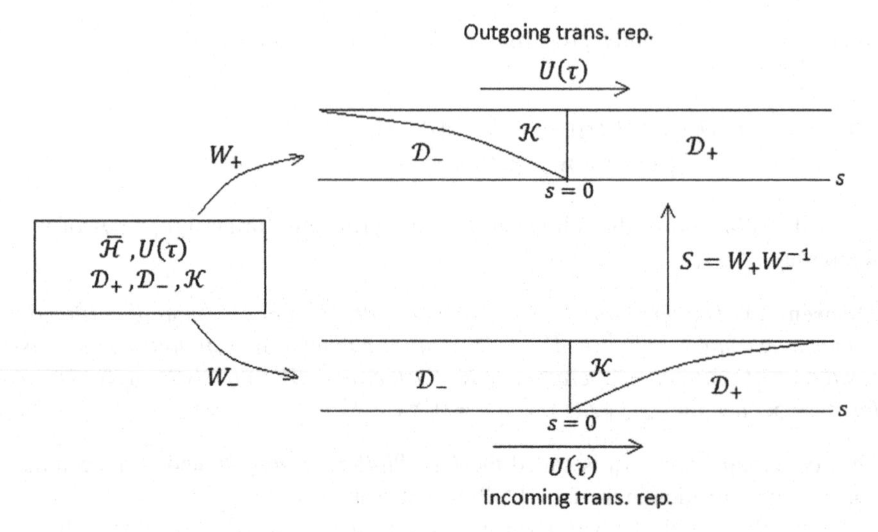

Fig. 2.1 The Lax–Phillips translation representations. The wave operators $W_{\pm}$ map the Lax–Phillips Hilbert space $\overline{\mathcal{H}}$ onto the incoming/outgoing translation representations. In both representations evolution is represented by translation to the right. In the incoming translation representation $\mathcal{D}_-$ is defined by its support property on $(-\infty, 0]$. In the outgoing translation representation $\mathcal{D}_+$ is defined by its support property on $[0, \infty)$

(c) $\mathbf{S}^{LP}$ maps $L^2((-\infty, 0]; H)$ into itself.

Property (b) is due to the fact that $\mathbf{S}^{LP}$ is a mapping between two translation representations. Property (c) is understood by noting that the in the incoming translation representation $\mathcal{D}_-$ is identified with the subspace $L^2((-\infty, 0]; H)$ and in the outgoing translation representation $\mathcal{D}_+$ is represented as $L^2([0, \infty); H)$. The orthogonality of $\mathcal{D}_-$ and $\mathcal{D}_+$ then implies that in the outgoing translation representation $\mathcal{D}_-$ is represented by a subspace of $L^2((-\infty, 0]; H)$ and property (c) above follows. Figure 2.1 shows a schematic representation of the incoming and outgoing translation representations and the corresponding mappings.

Applying Fourier transform to the incoming and outgoing translation representations we obtain, respectively, the *incoming spectral representation* and *outgoing spectral representation* which are, in fact, spectral representations for the generator of the evolution group $\{U(\tau)\}_{\tau\in R}$. We shall denote the mappings of $\mathcal{H}^{LP}$ onto the incoming, respectively outgoing spectral representations by $\hat{W}_-$ and $\hat{W}_+$, i.e.,

$$\hat{W}_- := FW_-, \qquad \hat{W}_+ := FW_+$$

where $F : L^2(R; H) \to L^2(R; H)$ is the Fourier transform operator. Fourier transforming the incoming translation representation and using the Paley–Wiener theorem (Paley and Wiener 1934), we find that in the incoming spectral representation the subspace $\mathcal{D}_-$ is represented by the Hardy space $\mathcal{H}^2_+(R; H)$, consisting of boundary values on R of functions in the Hardy space $\mathcal{H}^2(C^+; H)$ of H valued functions analytic in upper half plane C^+ (for a review of Hardy spaces see Appendix B). In a similar manner, Fourier transform of the outgoing translation representation and another application of the Paley–Wiener theorem, shows that in the outgoing spectral representation $\mathcal{D}_+$ is represented by the Hardy space $\mathcal{H}^2_-(R; H)$, consisting of boundary values on R of functions in the Hardy space $\mathcal{H}^2(C^-; H)$ of H valued functions analytic in the lower half plane C^-. Thus, in the incoming and outgoing spectral representations, respectively, the subspaces representing $\mathcal{D}_-$ and $\mathcal{D}_+$ are distinguished by their analytic properties.

Transforming from translation to spectral representations, the scattering operator transforms into

$$\mathcal{S}^{LP} := F\mathbf{S}^{LP}F^{-1} = \hat{W}_+\hat{W}_-^{-1}$$

where F is the Fourier transform operator, Properties (a)–(c) then imply the corresponding properties of $\mathcal{S}^{LP}$:

(a') $\mathcal{S}^{LP}$ is unitary.
(b') $\mathcal{S}^{LP}$ commutes with multiplication by scalar functions.
(c') $\mathcal{S}^{LP}$ maps $\mathcal{H}^2_+(R; H)$ into itself.

According to a special case of a theorem of Foures and Segal (1955), an operator satisfying properties (a')–(c') can be realized as a multiplicative, operator valued function defined on R such that for each σ, $\mathcal{S}(\sigma) : H \to H$ and

(a") $\mathcal{S}(\sigma)$ is the boundary value of an operator valued function $\mathcal{S}(z)$ analytic in the upper half plane C^+;
(b") $\|\mathcal{S}(z)\| \leq 1$ for $z \in C^+$;
(c") For each $\sigma \in R$, $\mathcal{S}(\sigma)$ is, pointwise, a unitary operator on H.

The operator valued function $\mathcal{S}(\sigma)$ with the properties (a")–(c") is called the *Lax–Phillipa S-matrix*. An operator valued function with the properties (a")–(c") is known as (operator valued) *inner function* (For a discussion of inner functions and, more generally, representation theory of contraction operators and contractive semigroups, see Appendix B or Sz.-Nagy and Foias (1970)). Thus, the Lax–Phillips S-matrix is characterized as being an operator valued inner function. Tha analytic continuation of $\mathcal{S}(z)$ to the lower half-plane is given by

$$\mathcal{S}(z) = [\mathcal{S}^*(\overline{z})]^{-1}, \quad Im\, z < 0 \tag{2.1.3}$$

One of the main results of the Lax–Phillips scattering theory is the follwing theorem (Lax and Phillips 1967):

Theorem 2.3 *Let* $\mathbf{B}$ *denote the generator of the semigroup* $\{Z(\tau)\}_{\tau\geq 0}$. *If* $Im\, \mu < 0$, *then* ν *belongs to the point spectrum of* $\mathbf{B}$ *if and only if* $\mathcal{S}^*(\overline{\mu})$ *has a non-trivial null space.* □

In turn, the null space of $\mathcal{S}^*(\overline{z})$ at the point $z = \mu$ corresponds, by (2.1.3), to a pole of $\mathcal{S}(z)$ at $z = \mu$. Theorem 3 establishes a very important relation between the eigenvalues of the generator **B** of the Lax–Phillips semigroup and poles of the analytic continuation of the Lax–Phillips S-matrix $S(z)$ in the lower half-plane. In particular, this theorem motivates the use of the Lax–Phillips framework for the description of quantum mechanical resonances. It enables the possibility of associating certain well defined vectors in Hilbert space H with the resonance poles of the S-matrix, such that these vectors are eigenvectors of an evolution semigroup related in a natural way to the unitary evolution group on H.

Let $x \in \mathcal{K}$ be an eigenvector of the generator **B** of the Lax–Phillips semigroup $\{Z(\tau)\}_{\tau \geq 0}$ corresponding to an eigenvalue $\mu \in C$, with $Im\,\mu < 0$,

$$\mathbf{B}x = \mu x$$

then

$$i\frac{d}{d\tau}Z(t)x = \mathbf{B}Z(\tau)x = Z(\tau)\mathbf{B}x = \mu Z(\tau)x$$

so that

$$Z(\tau)x = e^{-i\mu\tau}x, \quad t \geq 0. \tag{2.1.4}$$

Since $Z(\tau)$ is a map from $\mathcal{K}$ to $\mathcal{K}$ the domain of **B** consists of vectors in $\mathcal{K}$. In the outgoing translation representation the Hilbert space $\mathcal{H}^{LP}$ is represented by the function space $L^2(R; H)$ which we take to consists of functions of the real varible s. Vectors in $\mathcal{K}$ are then represented by certain functions in $L^2((-\infty, 0]; H)$. Let $\tilde{x}(s)$ represent the eigenvector x of (2.1.3) in the outgoing translation representation. In this case (2.1.3) is transformed into

$$\Theta(-s)\tilde{x}(s-\tau) = e^{-i\mu\tau}\tilde{x}(s), \quad \tau \geq 0.$$

(where Θ is the Heavyside step function). Setting $s = 0$ we obtain

$$\tilde{x}(-\tau) = e^{-i\mu\tau}\tilde{x}(0), \quad \tau \geq 0,$$

Denoting $n = \tilde{x}(0) \in H$ and $s = -\tau$ we find that

$$\tilde{x}(s) = \begin{cases} e^{i\mu s}n, & s \leq 0 \\ 0, & s > 0 \end{cases} \tag{2.1.5}$$

Equation (2.1.5) is the general form of of an outgoing translation representer of an eigenvector x of **B**. Transforming to the outgoing spectral representation (i.e. applying a Fourier transform) we obtain

$$\hat{x}(\sigma) = i\frac{1}{\sigma - \mu}n, \quad n \in H, \; Im\,\mu < 0 \tag{2.1.6}$$

Equation (2.1.6) provides the general form of the outgoing spectral representer of an eigenvector x of **B**.

Given a Lax–Phillips structure with an evolution group $\{U(\tau)\}_{\tau\in R}$ on the Hilbert space $\mathcal{H}^{LP}$, consider the family of operators $\{T(\tau)\}_{\tau\geq 0}$ defined by

$$T(\tau) := P_+U(\tau), \quad \tau \geq 0$$

Each element $T(\tau)$, $\tau \geq 0$ of this family annihilates $\mathcal{D}_+$. Moreover, for each element $x \in \mathcal{D}_- \oplus \mathcal{K}$ we have

$$\begin{aligned} T(\tau_1)T(\tau_2)x = P_+U(\tau_1)P_+U(\tau_2)x = P_+U(\tau_1)[P_+ + (I - P_+)]U(\tau_2)x = \\ = P_+U(\tau_1)U(\tau_2)x = P_+U(\tau_1 + \tau_2)x = T(\tau_1 + \tau_2)x, \quad \tau_1, \tau_2 \geq 0 \end{aligned}$$

and, hence, $\{T(\tau)\}_{\tau\geq 0}$ is a continuous, contractive semigroup. Observe also that for $x \in \mathcal{K}$ we have

$$T(\tau)x = P_+U(\tau)x = P_+U(\tau)P_-x = Z(\tau)x, \quad \tau \geq 0,$$

and hence $Z(\tau)$ is obtained by the restriction of $T(\tau)$ to $\mathcal{K}$,

$$Z(\tau) = T(\tau)|\mathcal{K}, \quad \tau \geq 0. \tag{2.1.7}$$

Let $\hat{E}$ be the opertor of multiplication by the independent variable on $L^2(R; H)$. Let $\{u(\tau)\}_{\tau\in R}$ be the continuous, one parameter, unitary evolution group on $L^2(R; H)$ generated by $\hat{E}$, i.e.

$$[u(\tau)f](\sigma) = [e^{-i\hat{E}\tau}f](\sigma) = e^{-i\sigma\tau}f(\sigma), \quad f \in L^2(R; H), \quad \sigma \in R$$

Then, if $\hat{W}_+ : \mathcal{H}^{LP} \to L^2(R; H)$ is the mapping of $\mathcal{H}^{LP}$ onto the outgoing spectral representation, we have

$$\hat{W}_+U(\tau) = u(\tau)\hat{W}_+$$

Furthermore, if $\hat{P}_+$ and $\hat{P}_-$ are, respectively, the orthogonal projections in $L^2(R; H)$ on the Hardy spaces $\mathcal{H}^2_+(R; H)$ and $\mathcal{H}^2_-(R; H)$ then, since in the outgoing spectral representation $\mathcal{D}_+$ is represented by the Hardy space $\mathcal{H}^2_-(R; H)$, we have

$$\hat{W}_+P_+ = \hat{P}_+\hat{W}_+$$

Hence, we get that

$$\hat{W}_+T(\tau) = \hat{W}_+P_+U(\tau) = \hat{P}_+u(\tau)\hat{W}_+ = \hat{T}(\tau)\hat{W}_+,$$

where $\hat{T}(\tau) := \hat{P}_+u(\tau)$. We conclude that the semigroup $\{T(\tau)\}_{\tau\geq 0}$ is represented in the outgoing spectral representation by the semigroup $\{\hat{T}(\tau)\}_{\tau\geq 0}$. If now apply the mapping $\hat{W}_+$ to (2.1.7) and set $\hat{\mathcal{K}} := \hat{W}_+\mathcal{K}$, we get that

$$\forall x \in \mathcal{K},\ \hat{W}_+ Z(\tau)x = \hat{W}_+ T(\tau)x = \hat{T}(\tau)\hat{W}_+ x \ \Rightarrow\ \hat{Z}(\tau) = \hat{T}(\tau)|\hat{\mathcal{K}},\ \tau \geq 0$$

where $\hat{Z}(\tau) = \hat{W}_+ Z(\tau)\hat{W}_+^{-1}$ is the operator representing the element $Z(\tau)$ of the Lax–Phillips semigroup in the outgoing spectral representation. Finally, by the definition $\mathcal{K} = \mathcal{H}^{LP} \ominus (\mathcal{D}_- \oplus \mathcal{D}_+)$ and the known representations of $\mathcal{D}_-$ and $\mathcal{D}_+$ in the incoming, respectively outgoing spectral representations, we obtain

$$\hat{\mathcal{K}} = \hat{W}_+\mathcal{K} = \mathcal{H}_+^2(R;\ H) \ominus \mathcal{S}^{LP}\mathcal{H}_+^2(R;\ H) = \mathcal{H}_+^2(R;\ H) \ominus \mathcal{S}\mathcal{H}_+^2(R;\ H)$$

where $\mathcal{S}$ is the Lax–Phillips S-matrix. We conclude that: *in the outgoing spectral representation the Lax–Phillips semigroup* $\{Z(\tau)\}_{\tau\geq 0}$ *is represented by the restriction of the semigroup* $\{\hat{T}(\tau)\}_{\tau\geq 0}$ *to a subspace of the form* $\mathcal{H}_+^2(R;\ H) \ominus \mathcal{S}\mathcal{H}_+^2(R;\ H)$ *where* $\mathcal{S}$ *is an inner function.*

We note that the general mathemtical context for the mathematical structure just described (i.e. the outging epesentation of the Lax–Phillips semigroup), is the Sz.-Nagy–Foias theory of contraction operators and contractive semigroups on Hilbert space (Sz.-Nagy and Foias 1970). In particular, the outgoing spectral represnetation of the Lax–Phillips semigroup is directly related to the existence of invariant subspaces for the shift operator and canonical model operatos for contractions and contractive semigroups on function spaces. For the convenience of the reader we include in Appendix B a review of the mathematical constructions of the Sz.-Nagy–Foias theory which are relevant for subjects discussed in this book.

2.2 Quantum Lax Phillips Theory

The quantum Lax–Phillips theory (Strauss and Horwitz 2000, Strauss et al. Strauss, Horwitz and Eisneberg 2000) is constructed essentially by embedding the original Lax–Phillips framework into quantum theory, and describes a resonance as a state in a quantum mechanical Hilbert space. It is therefore possible, in principle, to calculate all measurable properties of the system in this state. Moreover, the quantum Lax–Phillips theory provides a framework for understanding the decay of an unstable system as an irreversible process.

Assume that $\mathcal{D}_+$ is an outgoing subspace with respect to a unitary evolution group $\{U(\tau)\}_{\tau\geq 0}$ on a Hilbert space $\overline{\mathcal{H}}$. Then, according to Sinai's theorem (Theorem 2 above), there exists an outgoing translation representation in which $\overline{\mathcal{H}}$ is represented by a space $\overline{H} = L^2(R;\ H)$ of Lebesgue square-integrable functions with values in a Hilbert space H (the auxiliary space), $\mathcal{D}_+$ is represented by the subspace $L^2([0, \infty);\ H)$ and the evolution $U(\tau)$ is represented by translation to the right by τ units. We think of this representation as a family of Hilbert spaces obtained by foliating $\overline{\mathcal{H}}$ along the real line, in the form of a direct integral

$$\overline{\mathcal{H}} = \int_{\oplus} \mathcal{H}_s \, ds, \tag{2.2.1}$$

where the set of auxiliary Hilbert spaces $\mathcal{H}_s$ are all isomorphic. The foliation variable s may be considered as the *observed* time, as recorded by the laboratory clock (and is therefore considered as an *observable*), associated with an event. In the nonrelativistic framework, s and τ are on the same level numerically, but we still preserve the interpretation of Newton (1687) that τ is the universal parameter of dynamical evolution (as is done as well in Floquet theory (Floquet 1883)). Of course, for an incoming subspace $\mathcal{D}_-$ there exists also an incoming translation representation in which $\overline{\mathcal{H}}$ is represented by $\overline{H} = L^2(R; H)$, $\mathcal{D}_-$ is represented by the subspace $L^2((-\infty, 0]; H)$ and $U(\tau)$ is represented by translation to the right by τ units. The norm in $\overline{H}$ is defined to be

$$\| f \|^2_{\overline{H}} = \int_{-\infty}^{\infty} \| f_s \|^2_H ds, \tag{2.2.2}$$

where $f \in \overline{H}$ is the representer of a vector in $\overline{\mathcal{H}}$, and, for every $s \in R$, $f_s \in H$. It is clear that s has the interpretation of an observable; it is conjugate to the "energy" σ in the specral representations defined by Fourier transform of the translation representations.

We see from the construction of the incoming and outgoing translation representations that the outgoing subspace $\mathcal{D}_+$ is given, in the outgoing translation representation, in terms of support properties (as is also true for the incoming subspace in the incoming representation). One can then easily understand that the fundamental difference between the Lax–Phillips theory and standard quantum theory lies in this property; the projection operators $P_\pm$ are associated with *subspaces defined by their support on the time observable* s. The subspace defining the unstable system in standard quantum theory is usually defined as an eigenstate of an unperturbed Hamiltonian, and cannot be associated with an interval of time. The subspaces of the Lax–Phillips theory are associated with time intervals (e.g., the positive and negative half-lines in the outgoing and incoming free representations).

We now show that the generator of this semigroup $\{T(\tau)\}_{\tau \geq 0}$ is symmetric but not self-adjoint (Horwitz and Piron 1993), and therefore does not generate a group. In the outgoing translation representation,

$$(P_+ U(\tau) f)(s) = \theta(-s) f(s - \tau), \tag{2.2.3}$$

and therefore

$$(P_+ K f)(s) = i\theta(-s) \frac{\partial f}{\partial s}(s - \tau)|_{\tau \to 0_+}, \tag{2.2.4}$$

where $f(s)$ is a vector-valued function, and K is the self-adjoint generator associated with $U(\tau)$. If we then compute the scalar product of the vector given in (2.2.4) with a vector $g \in \overline{H}$, we find that

$$\int_{-\infty}^{\infty} ds\, g^*(s)(P_+Kf)(s) = i\delta(s)g^*(0)f(0) + \int_{-\infty}^{\infty} ds(P_+Kg)^*(s)f(s). \quad (2.2.5)$$

The generator is therefore not self-adjoint. It is through this mechanism that the Lax–Phillips theory provides a description that has the semigroup property for the evolution of an unstable system (Horwitz and Piron 1993). In fact, every point $\mu \in C^-$ is an eigenvalue of the generator of $\{T(\tau)\}_{\tau\geq 0}$ with corresponding eigenfunctions given by

$$f_\mu(s) = \begin{cases} e^{-i\mu s}n, & s \leq 0; \\ 0, & s > 0, \end{cases} \quad (2.2.6)$$

where $n = f_\mu(0)$ is a vector in the auxiliary space H. The semigroup property of the Lax–Phillips semigroup $\{Z(\tau)\}_{\tau\geq 0}$ follows directly from that of $\{T(\tau)\}_{\tau\geq 0}$.

If we identify elements in the space $\overline{\mathcal{H}}$ with *physical states*, and identify the subspace $\mathcal{K}$ on which the Lax–Phillips semigroup acts non-trivially with the unstable system, we see that the quantum Lax–Phillips theory provides a framework for the description of an unstable system which decays according to a semigroup law. We remark that, taking a vector ψ_0 in $\mathcal{K}$, and evolving it under the action of $U(\tau)$, the projection back into the original state is

$$A(\tau) = (\psi_0, U(\tau)\psi_0) = (\psi_0, P_{\mathcal{K}}U(\tau)P_{\mathcal{K}}\psi_0) = (\psi_0, Z(\tau)\psi_0), \quad (2.2.7)$$

so that the survival amplitude of the Lax–Phillips theory, analogous to that of the Wigner-Weisskopf formula (1.1.2), has the exact exponential behavior.

Functions in the space $\overline{H}$, representing elements of $\overline{\mathcal{H}}$, depend on the foliation variable s as well as the variables of the auxiliary space H. The measure space of this Hilbert space of states is one dimension larger than that of a quantum theory represented in the auxiliary Hilbert space alone. The absolute square of the wave function corresponds to the probability that the system would be found, as a result of measurement, at time s in a particular configuration in the auxiliary space (decreasing according to the semigroup property). For example, the expectation value of the position variable x at a given s is, in the standard interpretation of the auxiliary space as a space of quantum states,

$$\langle x \rangle_s = \frac{(\psi_s, x\psi_s)}{\|\psi_s\|^2}. \quad (2.2.8)$$

The full expectation value in the physical Lax–Phillips state, according to (2.2.2) and (2.2.8), is then

$$\int (\psi_s, x\psi_s)\, ds = \int \|\psi_s\|^2 \langle x \rangle_s\, ds, \quad (2.2.9)$$

so we see that, understanding that s is a dynamical variable, $\|\psi_s\|^2$ corresponds to the probability to find a signal which indicates the presence of the system at the time s (in the same way that x is interpreted as a dynamical variable in the quantum theory).

One may ask, in this framework, which results in a precise semigroup behavior for an unstable system, whether such a theory can support as well the description of stable systems. It is clear that if $\mathcal{D}_\pm$ span the whole space, i.e., there is no unstable subspace $\mathcal{K}$, and one has a scattering theory without the type of resonances that can be associated with an unstable system.

2.3 Explicit Constructions in Quantum Lax–Phillips Theory

In the following, we give a procedure (Strauss and Horwitz 2000, Strauss et al. Strauss, Horwitz and Eisneberg 2000) for the construction of the subspaces $\mathcal{D}_\pm$, and for defining the representations realizing the Lax–Phillips structure as a quatum theory with a Hamiltonian based on *particle dynamics* rather than wave propagation dynamics assumed by Lax and Phillips.

It follows from the existence of the one-parameter unitary evolution group $\{U(\tau\}_{\tau \in R}$ acting on the Hilbert space of particle states $\overline{\mathcal{H}}$ that there is an essentially self-adjoint operator K generating the dynamical evolution of these particle states; we further assume that there exist *wave operators* $\Omega_\pm$ intertwining this dynamical operator with an unperturbed dynamical operator K_0. We shall require that K_0 has absolutely continuous spectrum on $(-\infty, \infty)$, i.e., $\sigma_{ac}(K_0) = R$ (we discuss in Chap. 4 the construction of a more general theory which does not make this requirement, and provides a good approximation to the Lax–Phillips theory).

We begin the development of the quantum Lax–Phillips theory (Strauss and Horwitz 2000, Strauss et al. Strauss, Horwitz and Eisneberg 2000) with the construction of the incoming and outgoing translation representations. In this way, we shall construct explicitly the required foliation. The *free spectral representation* of K_0 is defined by

$$ {}_f\langle\sigma, \beta|K_0|g\rangle = \sigma\, {}_f\langle\sigma, \beta|g\rangle, \tag{2.3.1}$$

where $|g\rangle$ is an element of $\overline{\mathcal{H}}$, and β corresponds to the varibles (measure space) of the auxiliary space associated to each value of σ, which, with σ, comprise a complete spectral set. The functions ${}_f\langle\sigma, \beta|g\rangle$ may be thought of as a set of functions of the variables β indexed on the variable σ in a continuous sequence of auxiliary Hilbert spaces isomorphic to H.

We now proceed to define the incoming and outgoing subspaces $\mathcal{D}_\pm$. To do this, we define the Fourier transform from the free spectral representation to the foliation variable s of (2.2.8), i.e.,

$$ {}_f\langle s, \beta|g\rangle = \int e^{-i\sigma s}\, {}_f\langle\sigma, \beta|g\rangle d\sigma. \tag{2.3.2}$$

Clearly, K_0 acts as the generator of translations in this representation. We shall say that the set of functions ${}_f\langle s, \beta|g\rangle$ are in the *free translation representation*.

Let us denote, respectively, by D_0^+ and D_0^- the two subspaces of $L^2(-\infty, \infty, H)$ consisting of functions with support in $L^2(0, \infty, H)$ and in $L^2(-\infty, 0, H)$. The application of the Fourier transform back to the free spectral representation to these subspaces provides, by the Paley–Wiener theorem, two Hardy spaces of vector valued functions (Titchmarsh 1939)

$$_f\langle\sigma, \beta|g_0^{\pm}\rangle = \int e^{i\sigma s}\, _f\langle s, \beta|g_0^{\pm}\rangle ds \in \mathcal{H}_{\pm}^2(R; H), \tag{2.3.3}$$

for $g_0^{\pm} \in D_0^{\pm}$ (we refere the reader to the discussion of Hardy spaces in Appendix A).

At this point we may define the subspaces $\mathcal{D}_{\pm}$ in $\overline{\mathcal{H}}$. To do this we first map the Hardy space functions in $\overline{H}$ to $\overline{\mathcal{H}}$, i.e., we define subspaces $\mathcal{D}_0^{\pm} \subset \overline{H}$ by

$$\int \sum_{\beta} |\sigma, \beta\rangle_f\, _f\langle\sigma, \beta|g_0^{\pm}\rangle d\sigma \in \mathcal{D}_0^{\pm}. \tag{2.3.4}$$

As mentioned above, we assume the existence of the wave operators $\Omega_{\pm}$, intertwining K_0 with the full evolution K, i.e., that the limits

$$\Omega_{\pm} = \lim_{\tau\to\mp\infty} e^{iK\tau}e^{-iK_0\tau} = \lim_{\epsilon\to 0^+}\int_0^{\pm\infty} e^{\mp\epsilon\tau}e^{iK\tau}e^{-iK\tau}d\tau \tag{2.3.5}$$

exist on a dense set in $\overline{\mathcal{H}}$. We emphasize that the operator K generates evolution of the entire history, i.e., of elements in $\overline{\mathcal{H}}$, and that these wave operators are defined in this larger space. In general, these operators are not the standard wave (intertwining) operators for the perturbed and unperturbed Hamiltonians that act in the auxiliary space. The conditions for their existence are, however, closely related to those of the usual wave operators. For the existence of the limit, it is sufficient that for potential $V = K - K_0$, for $\tau \to \pm\infty$, $\|Ve^{-iK_0\tau}\phi\| \to 0$ for a dense set in $\overline{\mathcal{H}}$. As for the usual scattering theory, it is possible to construct examples for which the wave operator exists if the potential falls off sufficiently rapidly (Taylor 1972).

The construction of $\mathcal{D}_{\pm}$ is then completed with the help of the wave operators $\Omega_{\pm}$. We define these subspaces by

$$\mathcal{D}_+ = \Omega_+\mathcal{D}_0^+ \qquad \mathcal{D}_- = \Omega_-\mathcal{D}_0^-. \tag{2.3.6}$$

We remark that these two subspaces are not produced by the same unitary map. This procedure is necessary, although not sufficient, in order to realize the Lax–Phillips structure non-trivially; if a single unitary map were used, then there would exist a transformation into the space of functions on $L^2(R, H)$ which has the property that all functions with support on the positive half-line represent elements of $\mathcal{D}_+$, and all functions with support on the negative half-line represent elements of $\mathcal{D}_-$ in the same representation; the resulting Lax–Phillips S-matrix would then be trivial. The requirement that $\mathcal{D}_+$ and $\mathcal{D}_-$ be orthogonal is not an immediate consequence of our construction and must be taken into account in each application.

In the following, we construct the Lax–Phillips S-matrix and the Lax–Phillips wave operators $W_\pm$. The wave operators defined by (2.2.5) intertwine K and K_0, i.e.,

$$K\Omega_\pm = \Omega_\pm K_0; \tag{2.3.7}$$

we may therefore construct the outgoing (incoming) spectral representations from the free spectral representation. Since

$$K\Omega_\pm|\sigma,\beta\rangle_f = \Omega_\pm K_0|\sigma,\beta\rangle_f = \sigma\Omega_\pm|\sigma,\beta\rangle_f, \tag{2.3.8}$$

we may identify

$$|\sigma,\beta\rangle_{out} = \Omega_+|\sigma,\beta\rangle_f, \quad |\sigma,\beta\rangle_{in} = \Omega_-|\sigma,\beta\rangle_f. \tag{2.3.9}$$

The Lax–Phillips S-matrix is defined as the operator on $\overline{H}$ which carries the incoming to outgoing translation representations of the evolution operator K. Suppose g is an element of $\overline{\mathcal{H}}$; its incoming and outgoing spectral representers, according to (2.2.3), are

$${}_{in}\langle\sigma,\beta|g) = {}_f\langle\sigma,\beta|\Omega_-^{-1}g), \quad {}_{out}\langle\sigma,\beta|g) = {}_f\langle\sigma,\beta|\Omega_+^{-1}g). \tag{2.3.10}$$

Then the transformation from incoming to outgoing spectral representations, defining the Lax–Phillips S-matrix in the spectral representation, is given by:

$${}_{out}\langle\sigma,\beta|g) = {}_f\langle\sigma,\beta|\Omega_+^{-1}g) = \int d\sigma' \sum_{\beta'} {}_f\langle\sigma,\beta|\mathbf{S}^{LP}|\sigma',\beta'\rangle_f \; {}_{in}\langle\sigma',\beta'|g) \tag{2.3.11}$$

where $\mathbf{S}^{LP}$ is the Lax–Phillips scattering operator (defined on $\overline{\mathcal{H}}$). Transforming the kernel to the free translation representation with the help of (2.3.2), i.e.,

$${}_f\langle s,\beta|\mathbf{S}^{LP}|s',\beta'\rangle_f = \frac{1}{(2\pi)^2}\int d\sigma d\sigma' \, e^{i\sigma s}e^{-i\sigma' s'} \, {}_f\langle\sigma,\beta|\mathbf{S}|\sigma',\beta'\rangle_f, \tag{2.3.12}$$

we see that the relation (2.3.11) becomes, after using Fourier transform in a similar way to transform the *in* and *out* spectral representations to the corresponding *in* and *out* translation representations,

$$\begin{aligned}{}_{out}\langle s,\beta|g) = {}_f\langle s,\beta|\Omega_+^{-1}g) &= \int ds' \sum_{\beta'} {}_f\langle s,\beta|\mathbf{S}^{LP}|s',\beta'\rangle_f \; {}_f\langle s',\beta'|\Omega_-^{-1}g) \\ &= \int ds' \sum_{\beta'} {}_f\langle s,\beta|\mathbf{S}^{LP}|s',\beta'\rangle_{f\,in}\langle s',\beta'|g).\end{aligned} \tag{2.3.13}$$

Hence the matrix elements of the Lax–Phillips S-matrix are given by

$$S = \{{}_f\langle s, \beta|\mathbf{S}^{LP}|s', \beta'\rangle_f\}, \tag{2.3.14}$$

in free translation representation. It follows from the intertwining property (2.2.7) that

$${}_f\langle \sigma, \beta|\mathbf{S}^{LP}|\sigma', \beta'\rangle_f = \delta(\sigma - \sigma')S^{\beta\beta'}(\sigma), \tag{2.3.15}$$

This result can be expressed in terms of operators on $\overline{\mathcal{H}}$. Let

$$W_-^{-1} = \{{}_f\langle s, \beta|\Omega_-^{-1}\} \tag{2.3.16}$$

be a map from $\overline{\mathcal{H}}$ to $\overline{H}$ in the incoming translation representation, and, similarly,

$$W_+^{-1} = \{{}_f\langle s, \beta|\Omega_+^{-1}\} \tag{2.3.17}$$

a map from $\overline{\mathcal{H}}$ to $\overline{H}$ in the outgoing translation representation. It then follows from (2.3.13) that

$$S = W_+^{-1}W_-, \tag{2.3.18}$$

is a kernel on the free translation representation. This kernel is understood to operate on the representer of a vector g in the incoming translation representation and map it to the representer in the outgoing translation representation.

Note that we have made essential use of he fact that the spectral representations have support on the full real line, corresponding to the energy spectrum of the physical problem. In the quantum theory, the Stark Hamiltonian, for example, gives rise to such a model (Ben Ari and Horwitz 2004); relativistic quantum theory (Horwitz 2015) also provides a framework in which the theory can be directly applied. In the next section we work out the soluble model of Lee (1956) and Friedrichs (1950) in relativistic form Strauss and Horwitz (2000), Strauss et al. (2000). In Chap. 4, we study a modification of the Lax Phillips theory which can be applied to semibounded spectrum, as occurs in the usual nonrelativisic quantum theory.

2.4 Relativistic Lee–Friedrichs Model

In a dynamical theory consistent with special relativity, one introduces (Horwitz 2015; Stueckelberg 1941; Fanchi 1993) a universal invariant time τ to parametrize the motion.

The difference between the absolute dynamical parameter τ and the observed time s of the foliation of the Lax–Phillips Hilbert space becomes an essential conceptual difference in relativity, with s playing the role of the observed time t of Einstein (Horwitz 2015), subject to Lorentz transformation. See the book of Horwitz (2015) for a complete development of a consistent relativistic quantum theory based on this idea.

The variable τ of the relativistic theory has been called the "proper time" (not to be confused with $\sqrt{dt^2 - dx^2}$) by Schwinger (1951) and Itzykson and Zuber (1980). The variable s also plays the role of the t in Maxwell's equations where, in the standard $4D$ theory, there is no intrinsic *dynamical* evolution.[1]

In this framework, we study the relativistic analysis of a resonance in the Lee–Friedrichs model from the point of view of the Lax–Phillips theory. The Lee–Friedrichs model (Lee 1956; Friedrichs 1950) is an exactly soluble model for a scattering theory with resonances, originally developed in a nonrelativistic framework. Lee (1956) wrote the model in terms of a nonrelativistic quantum field theory, for which the interaction is such that there are sectors that make the problem equivalent to a quantum mechanical model with finite rank potential, corresponding to the formulation of Friedrichs (1950). The *a priori* formulation of the model for the relativistic quantum theory from the point of view of Friedrichs is not so clear, as we shall see, and we therefore follow the procedure of Lee (Horwitz 1995) but for the relativistic quantum field theory (for spin zero bosons) (Horwitz 2015).

Following Lee, let us define the fields $b(p), a_N(p)$ and $a_\theta(p)$ as annihilation operators for particles which we chall call the V, N and θ particles, and M_V, M_N and M_θ the corresponding mass parameters. Writing $p^2 = p^\mu p_\mu$, and $k^2 = k^\mu k_\mu$, we define

$$K_0 = \int d^4p\{\frac{p^2}{2M_V}b^\dagger(p)b(p) + \frac{p^2}{2M_N}a_N^\dagger(p)a_N(p)\} + \int d^4k\frac{k^2}{2M_\theta}a_\theta^\dagger(k)a_\theta(k) \tag{2.4.1}$$

For the interaction, we take

$$V = \int d^4p \int d^4k(f(k)b^\dagger(p)a_N^\dagger(p-k)a_\theta(k) + f^*(k)b(p)a_N(p-k)a_\theta^\dagger(k)), \tag{2.4.2}$$

describing the process $V \leftrightarrow N + \theta$. This interaction is clearly rank one, enabling, as we shall see, one to achieve an exact solution. The coefficient $f(k)$ is required for the potential to be a bounded operator. As a quantum field theory, if f is a constant coefficient, the theory becomes poorly defined. The operators

$$\begin{aligned} Q_1 &= \int d^4p[b^\dagger(p)b(p) + a_N^\dagger(p)a_N(p)] \\ Q_2 &= \int d^4p[a_N^\dagger(p)a_N(p) - a_\theta^\dagger(p)a_\theta(p)] \end{aligned} \tag{2.4.3}$$

are strictly conserved, enabling us to decompose the Fock space to sectors. We shall study the problem in the lowest sector $Q_1 = 1, Q_2 = 0$, for which there is just one

[1] The $5D$ gauge covariant extensions of the Maxwell theory (Saad 1989; Aharonovich and Horwitz 2010; Wesson 2006) contains dynamical evolution of the radiation fields, as in the original Lax and Phillips (1967) work.

V or one N and one θ. In this sector the generator of evolution can be written in the form

$$K = \int d^4 p K^p = \int d^4 p (K_0^p + V^p), \tag{2.4.4}$$

where

$$K_0^p = \frac{p^2}{2M_V} b^\dagger(p) b(p) + \int d^4 k \left(\frac{(p-k)^2}{2M_N} + \frac{k^2}{2M_\theta} \right) a_N^\dagger(p-k) a_\theta^\dagger(k) a_\theta(k) a_N(p-k) \tag{2.4.5}$$

and

$$V^p = \int d^4 k (f(k) b^\dagger(p) a_N^\dagger(p-k) a_\theta(k) + (f^*(k) b(p) a_N(p-k) a_\theta^\dagger(k). \tag{2.4.6}$$

In the corresponding nonrelativistic theory, one can see at this point the essential algebraic content of the Friedrichs formulation (Friedrichs 1950) as discussed in Chap. 1.

We now continue the development of the covariant relativistic Lax–Phillips theory, following procedures similar to those discussed in Chap. 1.

In the form (2.4.6) it is clear that both K and K_0 have a direct integral structure, and therefore the corresponding wave operators $\Omega_\pm^p$ have as well. In this sense, from the expression for K_0^p we see that $|V(p) >= b^\dagger(p)|0>$ is a discrete eigenstate of K_0^p, and is therefore annihilated by $\Omega_\pm^p$; it can be shown, in fact, that $\Omega_\pm |V(p) >= 0$ for any p.

We now construct the Lax–Phillips incoming and outgoing spectral representations for this problem, and discuss the properties of the resonant states. In accordance with the discussion following (2.4.8), in order to construct the wave operators, we must obtain solutions for the unperturbed problem. The complete set of such states may be decomposed into two subsets corresponding to quantum numbers for states containing N and θ particles, which we denote by α and those containing a V for which the quantum numbers are denoted by β. Then, the spectral representations are

$$\begin{aligned} |\sigma, \alpha >_0 &= \int d^4 p \int d^4 k |N(p), \theta(k) >< N(p), \theta(k)|\sigma, \alpha >_0 \\ |\sigma, \beta >_0 &= \int d^4 p |V(p) >< V(p)|\sigma, \beta >_0, \end{aligned} \tag{2.4.7}$$

where we define $|N(p), \theta(k) >\equiv a_N^\dagger(p) a_\theta^\dagger(k)|0>$, and $|V(p) >\equiv b^\dagger(p)|0>$. Therefore, since

$$\begin{aligned} K_0|\sigma, \alpha >_0 &= \omega_{N(p)} + \omega_{\theta(k)}|\sigma, \alpha >_0 = \sigma|\sigma, \alpha >_0 \\ K_0|\sigma, \beta >_0 &= \omega_{V(p)}|\sigma, \beta >_0 = \sigma|\sigma, \beta >_0, \end{aligned}$$

where $\omega_{N(p)} = \frac{p^2}{2M_N}$, $\omega_{\theta(k)} = \frac{k^2}{2M_\theta}$ and $\omega_{V(p)} = \frac{p^2}{2M_V}$, we must have

$$< N(p)\theta(k)|\sigma, \alpha > \propto \delta(\sigma - \omega_{N(p)} - \omega_{\theta(k)})$$
$$< V(p)|\sigma, \beta >_0 \ \propto \delta(\sigma - \omega_{V(p)}) \tag{2.4.8}$$

These matrix elements satisfy the requirements of orthogonality and completeness since they are unitary maps.

Due to the structure of the dynanamics of this model given in (2.4.5) and (2.4.6), we may solve explicitly for the matrix elements of the wave operators (as for the simple nonrelativistic model treated above)

$$< V(p+k)|\Omega_+|N(p), \theta(k) > \quad < N(p'), \theta(k')|\Omega_+|N(p), \theta(k) > \tag{2.4.9}$$

We may apply the integral formula (2.4.5) for the wave operator to the states $|N(p)\theta(k) >$ to obtain

$$\Omega_+|N(p)\theta(k) >= |N(p)\theta(k) > -i \lim_{\epsilon\to 0} \int_0^{-\infty} e^{i(\omega_N(p)+\omega_\theta(k)-i\epsilon)\tau} U(\tau) f(k) b^\dagger(p+k)|0 > . \tag{2.4.10}$$

To complete the evaluation of this integral, we must now find the evolution of the state $b^\dagger(p)|0 >$ under the evolution $U(\tau)$. The solution for this evolution involves very similar procedures to that outlined above for the nonrelativistic Lee–Friedrichs model, making use of the finite rank property of the interaction. In the sector of the Fock space that we are using, the state ψ_τ at any time τ can be represented as

$$\psi_\tau = \int d^4q A(q,\tau) b^\dagger(q)|0 > + \int d^4p \int d^4k B(p,k,\tau) a^\dagger(p) a^\dagger(k)|0 > . \tag{2.4.11}$$

Substituting this into the Stueckelberg–Schrödinger equation [30][43]

$$i\frac{\partial}{\partial\tau}\psi_\tau = K\psi_\tau \tag{2.4.12}$$

with the full Hamiltonian given by (2.3.4), one obtains

$$i\frac{\partial A(q,\tau)}{\partial\tau} = \frac{q^2}{2M_V} + \int d^4k f(k) B(q-k,k,\tau)$$
$$i\frac{\partial B(p,k,\tau)}{\partial\tau} = \left(\frac{p^2}{2M_V} + \frac{k^2}{2M_\theta}\right) B(p,k,\tau) + f^*(k) A(p+k,\tau). \tag{2.4.13}$$

These equations reflect the solubility of the model, forming a closed system that can be solved by Laplace transform. Defining

$$\bar{A}(q,z) = \int_0^{-\infty} e^{iz\tau} A(q,\tau) \quad Imz < 0$$
$$\bar{B}(p,k,z) = \int_0^{-\infty} e^{iz\tau} B(p,k,\tau) \quad Imz < 0, \tag{2.4.14}$$

one obtains

$$\big(z - \frac{q^2}{2M_V}\big)\bar{A}(q,z) = iA(q,0) + \int d^4k f(k)\bar{B}(q-k,k,z)$$
$$\big(z - \frac{p^2}{2M_N} - \frac{k^2}{2M_\theta}\big) = B(p,k,0) + f^*(k)\bar{A}(p+k,z). \tag{2.4.15}$$

Using the initial conditions

$$B(p,k,0) = 0, \quad A(q,0) = f(k)\delta^4(q-p-k) \tag{2.4.16}$$

the Laplace transformed solutions become

$$\bar{A}(q,z) = i\frac{A(q,0)}{h(q,z)}$$
$$\bar{B}(p,k,z) = i\big(z - \frac{p^2}{2M_N} - \frac{k^2}{2M_\theta}\big)^{-1} f^*(k)\frac{A(p+q,0)}{h(p+k,z)}, \tag{2.4.17}$$

where

$$h(q,z) = z - \frac{q^2}{2M_V} - \int d^4k \frac{|f(k)|^2}{z - \frac{(q-k)^2}{2M_N} - \frac{k^2}{2M_\theta}}. \tag{2.4.18}$$

These results are the analog of (1.2.11). With the Laplace transform of ψ_T, and the formula for the wave operator (2.4.10) we obtain the matrix elements of the wave operator (Strauss 2000a)

$$< V(p')|\Omega_+|N(p),\theta(k) >= \lim_{\epsilon\to 0}\delta^4(p'-p-k)f(k)h^{-1}(p',\omega - i\epsilon) \tag{2.4.19}$$

and

$$< N(p'),\theta(k')|\Omega_+|N(p),\theta(k) >= \delta^4(p'-p)\delta^4(k'-k)$$
$$+ i\lim_{\epsilon\to 0}\big[-i\big(\omega - i\epsilon - \frac{p'^2}{2M_N} - \frac{k'^2}{2M_\theta}\big)^{-1}\cdot\frac{f^*(k')f(k)}{h(p+k,\omega - i\epsilon)}\big]\delta^4(p'+k'-p-k), \tag{2.4.20}$$

where $\omega = \omega_\theta + \omega_N$.

According to our previous discussion, we are now in a position to calculate the transformation to the *outgoing* spectral representation (with quantum numbers α)

$$< V(p)|\Omega_+|\sigma,\alpha >_0 \quad < N(p),\theta(k)|\Omega_+|\sigma,\alpha >_0 . \tag{2.4.21}$$

The results can be expressed as

$$< V(p)|\Omega_+|\sigma,\alpha >_0 = h^{-1}(p,\sigma - i\epsilon)|n >^{\alpha}_{p,\sigma} \tag{2.4.22}$$

and

$$< N(p), \theta(k)|\Omega_+|\sigma, \alpha >_0 \equiv < N(p), \theta(k)|\sigma, \alpha >_0 + i \lim_{\epsilon \to 0} [-i(\sigma - i\epsilon - \frac{p'^2}{2M_N} - \frac{k'^2}{2M_\theta})^{-1} \times f^*(k')h^{-1}(p' + k', \sigma - i\epsilon)|n >^\alpha_{p'+k',\sigma}], \tag{2.4.23}$$

where we have defined the vector valued (on α) function

$$|n >^\alpha_{p,\sigma} \equiv \int d^4k f(k) < N(p - k)\theta(k)|\sigma, \alpha > \tag{2.4.24}$$

Carrying out a similar calculation for Ω_-, we can write an explicit formula for the Lax–Phillips S-matrix.

$${}_0 < \sigma', \alpha'|S|\sigma, \alpha >_0 = \delta(\sigma' - \sigma)\delta_{\alpha',\alpha} - 2\pi i \int d^4p \frac{|n >_{p,\sigma}{}^{\alpha'} < n|_{p,\sigma}{}^{\alpha}}{h(p, \sigma + i\epsilon)}, \tag{2.4.25}$$

or, in matrix notation (on α', α)

$$S(\sigma) = 1 - 2\pi i \int d^4p \frac{|n >_{p,\sigma} < n|_{p,\sigma}}{h(p, \sigma + i\epsilon)} \tag{2.4.26}$$

This completes our expression for the Lax–Phillips S-matrix for the relativistic Lee model. The variables of the auxiliary space are treated as a reduced two body system (Horwitz 2015), in terms of a direct integral over the total energy momentum P^μ; separating out the variables $\alpha = \gamma, P$ for the complete set for the auxiliary space, we may write for the corresponding vector valued functions $|n >^\gamma_{\sigma,P}$ the state corresponding to that of the reduced motion with conservation expressed by $\delta(\sigma - P^2/2M - p_{rel}^2/2m)$. The square of the relative momentum is then determined by P and σ, and we can write the S matrix entering the direct integral over P as

$$S_P(\sigma) = 1 - 2\pi i \int d^4p \frac{|n >_{P,\sigma} < n|_{P,\sigma}}{h(P, \sigma + i\epsilon)} \tag{2.4.27}$$

It is proved in [25] that for $|n >_P$ a normalized basis vector, we must have the form

$$|n >_{\sigma,P} = g_P(\sigma)|n >_P, \tag{2.4.28}$$

and that the S-matrix must commute with the projection operator formed from these states.

Assuming that in the relativistic Lee model there is just one pole in the lower half plane μ_P, corresponding to a single resonance, then the S-matrix takes on the form

$$S_P(\sigma) = S'_P(\sigma)M_P(\sigma), \tag{2.4.29}$$

where the resonant poles are carried by the matrix S'. If the function $M_P(\sigma)$ is of bounded exponential growth, i.e. $\ln |M_P(\sigma)|$ is bounded by $|Im\ \sigma|$, then there is an equivalence transformation that can bring the S-matrix to the form

$$S'_P(\sigma) = 1_{H,\sigma} - |n>_P\ _P<n| + \frac{\sigma - \bar{\mu}_P}{\sigma - \mu_P}|n>_P\ _P<n|. \tag{2.4.30}$$

The residue of the pole is a projection operator into a state in the auxiliary Hilbert space, identified here as the *state of the resonance*. In terms of the definition of the states of the Lee model (2.4.24), we see that the resonance becomes a proper state in the auxiliary Hilbert space of the Lax–Phillips theory.[2] Moreover, if there are two or more poles, it is a consequence of the proof that the evolution law is an exact semigroup that the pole residues are orthogonal, corresponding to orthogonal subspaces contained in $\mathcal{K}$, the complement of $\mathcal{D}_\pm$ in the Lax–Phillips Hilbert space. We remark that the resonance pole, associated with the center of mass momentum of the two body system (in general of the final state) may be finitely spread out according to the construction of the normalizable wave packet[25], and the corresponding bilinear form of the residue would then correspond to a mixed state over this small interval.

This result, not achievable in the Wigner-Weisskopf approach to the description of resonances, is made possible by the foliation admitted by the Lax–Phillips theory imbedded in a natural way into the relativistic quantum theory.

[2]See Strauss and Horwitz (Strauss 2000b) for a detailed discussion of the (N, θ, V) particle content of the resonances. We furthermore remark that the estimates of the values of the complex poles are not very different from those that follow from those of the model used by Wu and Yang (1964).

Chapter 3
Lyapunov Operators

In this chapter we discuss the construction of self-adjoint operators indicating the *direction of time* within the framework of standard quantum mechanics (Strauss 2011). Such operators will be referred to as Lyapunov operators. The particular construction of Lyapunov operators in this section will enable us to develop a formalism providing a good approximation to the Lax–Phillips theory for scattering problems involving Hamiltonians with semibounded spectra, as we do in Chap. 4. The construction of Lyapunov operators, moreover, has an intrinsic interest of its own.

It is a fundamental question in standard quantum mechanics (SQM) whether there exists an observable that evolves in a way that reflects the passage of time in a well-defined way. More precisely, one can ask whether SQM admits the construction of a self-adjoint operator M with the so-called Lyapunov property, that is, an operator such that, given an arbitrary initial state ψ_0 and the corresponding dynamical evolution $\psi(t) = U(t)\psi_0$, $t \geq 0$, of the physical system the time evolution of the expectation value $\langle M \rangle_{\psi(t)}$ is monotonic. Such an operator would clearly indicate the direction of time.

A natural candidate for a Lyapunov operator would be a time operator T canonically conjugate to the Hamiltonian H such that T and H form an imprimitivity system (implying that each generates translations on the spectrum of the other). However, for Hamiltonian whose spectrum is bounded from below, a well-known theorem of Pauli (1926) tells us that this is impossible. Recently, Galapon (2002) attempted to bypass Pauli's arguments and found pairs of T and H satisfying canonical commutation relations, but do not form an imprimitivity system (Mackey 1976). It can furthermore be shown that the T obtained in this way does not have the Lyapunov property. Other authors have not insisted on the conjugacy of T and H. In this context, Unruh and Wald (1989) prove that a "monotonically perfect clock" does not exist; We take note also of the no-go theorem of Misra, Prigigine and Courbage (1974). Other authors (for example, Holevo 1982) have removed the requirement of self-adjointness.

Self-adjoint Lyapunov operators were recently constructed within the framework of SQM using the properties of Hardy spaces and quasi-affine transforms (Strauss 2011). In this chapter we discuss these operators and study some of their properties

L. Horwitz and Y. Strauss, *Unstable Systems*, Mathematical Physics Studies,
https://doi.org/10.1007/978-3-030-31570-2_3

(Strauss 2010), in particular, their spectrum and generalized eigenstates, and give an example of their application to the motion of a free particle. The results of this study provide the basis for the modified Lax Phillips theory to be discussed in Chap. 4.

3.1 Forward and Backward Lyapunov Operators

A step forward in the efforts to approximate the structure of the Lax–Phillips theory within the context of quantum mechanics has been made in Strauss (2010) with the introduction into the framework of quantum mechanics of *forward and backward Lyapunov operators*, based on properties of Hardy spaces, and subsequent investigation of their properties and their applications in Strauss (2011), (Strauss et al. 2011a, b). If $\mathcal{H}$ is the Hilbert space corresponding to a given system and H is the self-adjoint generator of evolution of the system we define the trajectory Φ_φ corresponding to a state $\varphi \in \mathcal{H}$ to be the set of states

$$\Phi_\varphi := \{\varphi(t)\}_{t\in\mathbb{R}} = \{U(t)\varphi\}_{t\in\mathbb{R}},$$

where $U(t) = \exp(-iHt)$. The definition of a *forward Lyapunov operator* is (Strauss 2010):

Definition 3.1 (*forward Lyapunov operator*) Let M be a bounded self-adjoint operator on $\mathcal{H}$. Let Φ_φ be the trajectory corresponding to an arbitrarily chosen normalized state $\varphi \in \mathcal{H}$. Let $M(\Phi_\varphi) := \{(\psi, M\psi) \mid \psi \in \Phi_\varphi\}$ be the collection of all expectation values of M for states in Φ_φ. Then M is a forward Lyapunov operator if the mapping $\tau_{M,\varphi} : \mathbb{R} \mapsto M(\Phi_\varphi)$ defined by

$$\tau_{M,\varphi}(t) = (\varphi(t), M\varphi(t))$$

is one to one and monotonically decreasing. □

Remark 3.1 We assume throughout the present chapter that all generators of evolution are time independent and, therefore, we have symmetry of the evolution with respect to time translations.

Remark 3.2 If in the definition above we require that $\tau_{M,\varphi}$ be monotonically increasing instead of monotonically decreasing we also obtain a valid definition of a forward Lyapunov operator. The requirement that $\tau_{M,\varphi}$ is monotonically decreasing is made purely for the sake of convenience.

If M is a forward Lyapunov operator then we are able to find the time ordering of states in the trajectory Φ_φ according to the ordering of expectation values in $M(\Phi_\varphi)$. Hence, the existence of a Lyapunov operator introduces temporal ordering into the Hilbert space $\mathcal{H}$ of a problem for which such an operator can be constructed. The definition of a backward Lyapunov operator is similar to that of a forward Lyapunov operator, but with respect to the reversed direction of time. The significance of the existence of forward and backward Lyapunov operators can be understood if we consider again the Lax–Phillips formalism, and, in particular, the properties of the projection operators P_+ and P_-. In fact, from the representation of P_+ in the outgoing translation representation as an orthogonal projection on the subspace $L^2(\mathbb{R}_-; \mathcal{K})$ (or, indeed, directly from the definition of P_+ and the properties of $\mathcal{D}_\pm$ in Eq. (2.1.1)) it is evident that P_+ is a forward Lyapunov operator for the evolution in the Lax–Phillips theory. For every $\psi \in \mathcal{H}^{LP}$ we have

$$(\psi(t_2), P_+\psi(t_2)) \leq (\psi(t_1), P_+\psi(t_1)), \quad t_1 \leq t_2, \qquad \lim_{t\to\infty} (\psi(t), P_+\psi(t)) = 0.$$

Likewise, from the representation of P_- in the incoming translation representation as an orthogonal projection on the subspace $L^2(\mathbb{R}_+; \mathcal{K})$ it is evident that P_- is a backward Lyapunov operator for the Lax–Phillips evolution satisfying

$$(\psi(t_2), P_-\psi(t_2)) \leq (\psi(t_1), P_-\psi(t_1)), \quad t_2 \leq t_1, \qquad \lim_{t\to-\infty} (\psi(t), P_-\psi(t)) = 0.$$

Let $\mathcal{K}$ be a separable Hilbert space and let $L^2(\mathbb{R}; \mathcal{K})$ be the Hilbert space of Lebesgue square integrable $\mathcal{K}$ valued functions defined on $\mathbb{R}$. Let $\hat{E}$ be the operator of multiplication by the independent variable on $L^2(\mathbb{R}; \mathcal{K})$. Let $\{u(t)\}_{t\in\mathbb{R}}$ be the continuous, one parameter, unitary evolution group on $L^2(\mathbb{R}; \mathcal{K})$ generated by $\hat{E}$, i.e.,

$$[u(t)f](E) = [e^{-i\hat{E}t}f](E) = e^{-iEt}f(E), \quad f \in L^2(\mathbb{R}; \mathcal{K}), \quad E \in \mathbb{R}. \tag{3.1.1}$$

Let $\mathcal{H}^2(\mathbb{C}^+; \mathcal{K})$ and $\mathcal{H}^2(\mathbb{C}^-; \mathcal{K})$ be, respectively, the Hardy space of $\mathcal{K}$ valued functions analytic in $\mathbb{C}^+$ and $\mathbb{C}^-$. As discussed in Chap. 2, the Hilbert space $\mathcal{H}^2_+(\mathbb{R}; \mathcal{K})$ consisting of nontangential boundary values on the real axis of functions in $\mathcal{H}^2(\mathbb{C}^+; \mathcal{K})$, is isomorphic to $\mathcal{H}^2(\mathbb{C}^+; \mathcal{K})$. Similarly, the Hilbert space $\mathcal{H}^2_-(\mathbb{R}; \mathcal{K})$ of non-tangential boundary value functions of functions in $\mathcal{H}^2(\mathbb{C}^-; \mathcal{K})$ is isomorphic to $\mathcal{H}^2(\mathbb{C}^-; \mathcal{K})$. The spaces $\mathcal{H}^2_\pm(\mathbb{R}; \mathcal{K})$ are orthogonal subspaces of $L^2(\mathbb{R}; \mathcal{K})$ and we have

$$L^2(\mathbb{R}; \mathcal{K}) = \mathcal{H}^2_+(\mathbb{R}; \mathcal{K}) \oplus \mathcal{H}^2_-(\mathbb{R}; \mathcal{K}).$$

We denote the orthogonal projections in $L^2(\mathbb{R}; \mathcal{K})$ on $\mathcal{H}^2_+(\mathbb{R}; \mathcal{K})$ and $\mathcal{H}^2_-(\mathbb{R}; \mathcal{K})$, respectively, by $\hat{P}_+$ and $\hat{P}_-$.

Recall that in the Lax–Phillips theory P_+ is the orthogonal projection on the orthogonal complement of the outgoing subspace $\mathcal{D}_+$. In the outgoing translation representation the Lax–Phillips Hilbert space $\mathcal{H}^{LP}$ is mapped isometrically onto the function space $L^2(\mathbb{R}; \mathcal{K})$ where $\mathcal{K}$ is the auxiliary Hilbert space and $\mathcal{D}_+$ is mapped onto the subspace $L^2(\mathbb{R}_+; \mathcal{K})$. The evolution $U(t)$ is represented by translation to the right by t units. The outgoing spectral representation is a spectral representation of the generator K of the unitary evolution group $\{U(t)\}_{t\in\mathbb{R}} = \{\exp(-iKt)\}_{t\in\mathbb{R}}$ of the Lax–Phillips theory and is obtained by Fourier transform of the outgoing translation representation. In this representation $\mathcal{H}^{\mathrm{LP}}$ is represented by $L^2(\mathbb{R}; \mathcal{K})$, the evolution group $\{U(t)\}_{t\in\mathbb{R}}$ is represented by the group $\{u(t)\}_{t\in\mathbb{R}}$ in Eq. (3.1.1) and, by the Paley–Wiener theorem (see Appendix A), $\mathcal{D}_+$ is represented by the Hardy space $\mathcal{H}^2_-(\mathbb{R}; \mathcal{K})$. Therefore P_+ is represented in this representation by the projection $\hat{P}_+$ on $\mathcal{H}^2_+(\mathbb{R}; \mathcal{K})$. Hence, if $\hat{W}_+^{\mathrm{LP}} : \mathcal{H}^{\mathrm{LP}} \mapsto L^2(\mathbb{R}; \mathcal{K})$ is the mapping of $\mathcal{H}^{\mathrm{LP}}$, onto the outgoing spectral representation we have

$$P_+ = \left(\hat{W}_+^{\mathrm{LP}}\right)^{-1} \hat{P}_+ \hat{W}_+^{\mathrm{LP}}. \tag{3.1.2}$$

In a similar manner, in the incoming translation representation the Lax–Phillips Hilbert space $\mathcal{H}^{\mathrm{LP}}$ is mapped isometrically onto the function space $L^2(\mathbb{R}; \mathcal{K})$ and the incoming subspace $\mathcal{D}_-$ is mapped onto the subspace $L^2(\mathbb{R}_-; \mathcal{K})$. The evolution $U(t)$ is again represented by translation to the right by t units. The incoming spectral representation, obtained by Fourier transform of the incoming translation representation, is a spectral representation of the generator K of the evolution group $\{U(t)\}_{t\in\mathbb{R}}$. In this representation $\mathcal{H}^{\mathrm{LP}}$ is represented by the function space $L^2(\mathbb{R}; \mathcal{K})$, the evolution group $\{U(t)\}_{t\in\mathbb{R}}$ is represented by the unitary group $\{u(t)\}_{t\in\mathbb{R}}$ of Eq. (3.1.1) and, by the Paley–Wiener theorem, $\mathcal{D}_-$ is represented by the Hardy space $\mathcal{H}^2_+(\mathbb{R}; \mathcal{K})$. Therefore P_-, the projection on the orthogonal complement of $\mathcal{D}_-$, is represented in the incoming spectral representation by the projection $\hat{P}_-$ on $\mathcal{H}^2_-(\mathbb{R}; \mathcal{K})$ and hence, if $\hat{W}_-^{\mathrm{LP}} : \mathcal{H}^{\mathrm{LP}} \mapsto L^2(\mathbb{R}; \mathcal{K})$ is the mapping of $\mathcal{H}^{\mathrm{LP}}$ onto the incoming spectral representation, we have

$$P_- = \left(\hat{W}_-^{\mathrm{LP}}\right)^{-1} \hat{P}_- \hat{W}_-^{\mathrm{LP}}. \tag{3.1.3}$$

Observe that Eqs. (3.1.2) and (3.1.3) provides us with an explicit procedure for the construction of the forward and backward Lyapunov operators $P_\pm$ in the Lax–Phillips theory.

We now construct a closely analogous procedure which admits a similar analysis of applications arising in quantum theory, for which the spectrum of the Hamiltonian is not the whole real line. Let $[a, b]$ be a closed interval on the real line and denote by $P_{[a,b]}$ the orthogonal projection in $L^2(\mathbb{R}; \mathcal{K})$ on the subspace $L^2([a, b]; \mathcal{K})$. Let $\hat{E}_+$ be the operator of multiplication by the independent variable on $L^2([a, b]; \mathcal{K})$. Let $\{u_+(t)\}_{t\in\mathbb{R}}$ be the continuous, one parameter, unitary evolution group on $L^2([a, b]; \mathcal{K})$ generated by $\hat{E}_+$, i.e.,

$$[u_+(t)f](E) = [e^{-i\hat{E}_+ t} f](E) = e^{-iEt} f(E), \quad f \in L^2([a, b]; \mathcal{K}), \quad E \in [a, b].$$

Observe that $u_+(t) = P_{[a,b]} u(t) P_{[a,b]} = u(t) P_{[a,b]}$, where $u(t)$ are elements of the unitary group of Eq. (3.1.1). The following theorem (a restricted version of which was first proved in Strauss 2010) forms the basis for the present discussion of Lyapunov operators and their applications in the context of quantum mechanical scattering:

Theorem 3.4 *Let $M_F : L^2([a, b]; \mathcal{K}) \mapsto L^2([a, b]; \mathcal{K})$ be the operator defined by*

$$M_F := (P_{[a,b]}\hat{P}_+ P_{[a,b]})|_{L^2([a,b];\mathcal{K})}. \tag{3.1.4}$$

Then M_F is a positive, contractive, injective operator on $L^2([a,b];\mathcal{K})$, such that Ran M_F is dense in $L^2([a,b];\mathcal{K})$ and M_F is a Lyapunov operator in the forward direction, i.e., for every $\psi \in L^2([a,b];\mathcal{K})$ we have

$$(\psi(t_2), M_F\,\psi(t_2)) \leq (\psi(t_1), M_F\,\psi(t_1)), \quad t_2 \geq t_1 \geq 0, \quad \psi(t) = u_+(t)\psi, \tag{3.1.5}$$

and, moreover,

$$\lim_{t\to\infty} (\psi(t), M_F\,\psi(t)) = 0. \tag{3.1.6}$$

□

Proof The basic mechanism underlying the proof of Theorem 3.4 is a fundamental intertwining relation, via a quasi-affine mapping, between the unitary evolution group $\{u_+(t)\}_{t\in\mathbb{R}}$ on $L^2([a,b];\mathcal{K})$, and semigroup evolution in the Hardy space of the upper half-plane $\mathcal{H}^2(\mathbb{C}^+;\mathcal{K})$ or the isomorphic space $\mathcal{H}^2_+(\mathbb{R};\mathcal{K})$. Let $\{u(t)\}_{t\in\mathbb{R}}$ be the unitary group on $L^2(\mathbb{R};\mathcal{K})$ defined in Eq. (3.1.1) and recall that $\hat{P}_+$ is the orthogonal projection of $L^2(\mathbb{R};\mathcal{K})$ on $\mathcal{H}^2_+(\mathbb{R};\mathcal{K})$. For every $t \geq 0$ we define an operator $T_u(t)$: $\mathcal{H}^2_+(\mathbb{R};\mathcal{K}) \mapsto \mathcal{H}^2_+(\mathbb{R};\mathcal{K})$ by

$$T_u(t)f := \hat{P}_+ u(t)f, \quad f \in \mathcal{H}^2_+(\mathbb{R};\mathcal{K})\,. \tag{3.1.7}$$

(such an operator is known as a *Töplitz operator* with symbol $u(t)$; See Rosenblum and Rovnyak 1985; Nikol'skii 1986, 2002). The family of operators $\{T_u(t)\}_{t\geq 0}$ forms a strongly continuous, contractive, one parameter semigroup on $\mathcal{H}^2_+(\mathbb{R};\mathcal{K})$. Indeed, since $\mathcal{H}^2_+(\mathbb{R};\mathcal{K})$ is invariant under the action of $u(t)$ for all $t \geq 0$, we have

$$\begin{aligned} T_u(t_1)T_u(t_2) &= \hat{P}_+u(t_1)\hat{P}_+u(t_2) = \hat{P}_+u(t_1)(\hat{P}_+ + \hat{P}_-)u(t_2) = \\ &= \hat{P}_+u(t_1)u(t_2) = \hat{P}_+u(t_1+t_2) = T_u(t_1+t_2), \quad \forall t_1, t_2 \geq 0. \end{aligned} \tag{3.1.8}$$

Moreover, the semigroup $\{T_u(t)\}_{t\geq 0}$ satisfies

$$\|T_u(t_2)f\| \leq \|T_u(t_1)f\|, \quad t_2 \geq t_1 \geq 0, \quad f \in \mathcal{H}^2_+(\mathbb{R};\mathcal{K})\,, \tag{3.1.9}$$

and

$$s - \lim_{t\to\infty} T_u(t) = 0\,. \tag{3.1.10}$$

Below we shall make frequent use of quasi-affine mappings. The definition of this class of maps is as follows:

Definition 3.2 (*quasi-affine map*) A quasi-affine map from a Hilbert space $\mathcal{H}_1$ into a Hilbert space $\mathcal{H}_0$ is a linear, injective, continuous mapping of $\mathcal{H}_1$ into a dense linear manifold in $\mathcal{H}_0$. If $A \in \mathcal{B}(\mathcal{H}_1)$ and $B \in \mathcal{B}(\mathcal{H}_0)$ then A is a quasi-affine transform of B if there is a quasi-affine map $\theta : \mathcal{H}_1 \mapsto \mathcal{H}_0$ such that $\theta A = B\theta$. □

Concerning quasi-affine maps we have the following two important facts (see, for example Sz.-Nagy and Foias 1970):

(I) If $\theta : \mathcal{H}_1 \mapsto \mathcal{H}_0$ is a quasi-affine mapping then $\theta^* : \mathcal{H}_0 \mapsto \mathcal{H}_1$ is also quasi-affine, that is, θ^* is one to one, continuous and its range is dense in $\mathcal{H}_1$.
(II) If $\theta_1 : \mathcal{H}_0 \mapsto \mathcal{H}_1$ is quasi-affine and $\theta_2 : \mathcal{H}_1 \mapsto \mathcal{H}_2$ is quasi-affine then $\theta_2\theta_1 : \mathcal{H}_0 \mapsto \mathcal{H}_2$ is quasi-affine.

We can now turn to the proof of Theorem 3.4. The proof of the Lyapunov properly of M_F follows from an adaptation of a theorem first proved in Strauss (2005b) and subsequently used in the study of resonances in Strauss (2005a, b); Strauss et al. (2006). We have the following proposition:

Proposition 3.1 *(a) Let $\hat{P}_+|_{L^2([a,b];\mathcal{K})} : L^2([a,b];\mathcal{K}) \mapsto \mathcal{H}^2_+(\mathbb{R};\mathcal{K})$ be the restriction of the orthogonal projection $\hat{P}_+$ to the subspace $L^2([a,b];\mathcal{K}) \subset L^2(\mathbb{R};\mathcal{K})$. Then $\hat{P}_+|_{L^2([a,b];\mathcal{K})}$ is a contractive quasi-affine mapping of $L^2([a,b];\mathcal{K})$ into $\mathcal{H}^2_+(\mathbb{R};\mathcal{K})$.*
(b) For $t \geq 0$ the evolution $u_+(t)$ is a quasi-affine transform of $T_u(t)$. For every $t \geq 0$ and $g \in L^2([a,b];\mathcal{K})$ we have

$$\hat{P}_+ u_+(t) g = T_u(t) \hat{P}_+ g, \quad g \in L^2([a,b];\mathcal{K}), \quad t \geq 0. \tag{3.1.11}$$

□

Proof of Proposition 3.1: Item (a) may be thought of as an extension, and adaptation to the vector valued case, of a theorem proved by Van Winter (1971) for scalar valued functions. Taking into account the definition above of quasi-affine maps, this extension of the Van Winter theorem to vector valued functions can be stated in a simple way.

Theorem 3.5 (Van Winter theorem for vector valued functions) *Let $P_{[a,b]}|_{\mathcal{H}^2_+(\mathbb{R};\mathcal{K})} : \mathcal{H}^2_+(\mathbb{R};\mathcal{K}) \mapsto L^2([a,b];\mathcal{K})$ be the restriction of the orthogonal projection $P_{[a,b]}$ to the subspace $\mathcal{H}^2_+(\mathbb{R};\mathcal{K}) \subset L^2(\mathbb{R};\mathcal{K})$. Then $P_{[a,b]}|_{\mathcal{H}^2_+(\mathbb{R};\mathcal{K})}$ is a contractive quasi-affine map.* □

Remark Note that, in contrast to the statement made here, the original Van Winter theorem refers only to the interval $\mathbb{R}_+$ and includes an explicit construction of the inverse of the quasi-affine map $P_{\mathbb{R}^+}|_{\mathcal{H}^2_+(\mathbb{R})}$ (appropriate for the case of scalar valued functions) in terms of Mellin transform. We shall not need such explicit constructions in our discussion below.

We provide here a simple proof of Theorem 3.5:

Proof of Theorem 3.5 We prove the following properties of the mapping $P_{[a,b]}|_{\mathcal{H}^2_+(\mathbb{R};\mathcal{K})} : \mathcal{H}^2_+(\mathbb{R};\mathcal{K}) \mapsto L^2([a,b];\mathcal{K})$:

(a) $P_{[a,b]}|_{\mathcal{H}^2_+(\mathbb{R};\mathcal{K})}$ is contractive,
(b) $P_{[a,b]}|_{\mathcal{H}^2_+(\mathbb{R};\mathcal{K})}$ is one to one,
(c) $P_{[a,b]}|_{\mathcal{H}^2_+(\mathbb{R};\mathcal{K})}\mathcal{H}^2_+(\mathbb{R};\mathcal{K}) = P_{[a,b]}\mathcal{H}^2_+(\mathbb{R};\mathcal{K})$ is dense in $L^2([a,b];\mathcal{K})$.

That (a) holds is an immediate consequence of the fact that $P_{[a,b]}$ is a projection. The validity of (b) follows from the fact that if $P_{[a,b]}|_{\mathcal{H}^2_+(\mathbb{R};\mathcal{K})}$ is not one to one then there exists a function $f \in \mathcal{H}^2_+(\mathbb{R};\mathcal{K})$ such that $f \neq 0$ and $P_{[a,b]}f = 0$ so that f vanishes a.e. on $[a,b]$. However, a Hardy space function $f \in \mathcal{H}^2_+(\mathbb{R};\mathcal{K})$ cannot be identically zero on any non-zero measure set in $\mathbb{R}$. Therefore $P_{[a,b]}|_{\mathcal{H}^2_+(\mathbb{R};\mathcal{K})}$ must be one to one.

To prove (c) assume that $P_{[a,b]}\mathcal{H}^2_+(\mathbb{R};\mathcal{K})$ is not dense in $L^2([a,b];\mathcal{K})$. Then there exists a function, say $g \in L^2([a,b];\mathcal{K})$, such that $(g, P_{[a,b]}f)_{L^2([a,b];\mathcal{K})} = 0$ for all $f \in \mathcal{H}^2_+(\mathbb{R};\mathcal{K})$. Considering $L^2([a,b];\mathcal{K})$ as a subspace of $L^2(\mathbb{R};\mathcal{K})$ we decompose g into a sum $g = g^+ + g^-$ with $g^+ \in \mathcal{H}^2_+(\mathbb{R};\mathcal{K})$ and $g^- \in \mathcal{H}^2_-(\mathbb{R};\mathcal{K})$. For any $f \in \mathcal{H}^2_+(\mathbb{R};\mathcal{K})$ we then have

$$0 = (g, P_{[a,b]}|_{\mathcal{H}^2_+(\mathbb{R};\mathcal{K})}f)_{L^2([a,b];\mathcal{K})} = (g, P_{[a,b]}f)_{L^2([a,b];\mathcal{K})} = (g, f)_{L^2(\mathbb{R};\mathcal{K})} =$$
$$= (g^+ + g^-, f)_{L^2(\mathbb{R};\mathcal{K})} = (g^+, f)_{L^2(\mathbb{R};\mathcal{K})} = (g^+, f)_{\mathcal{H}^2_+(\mathbb{R};\mathcal{K})}\,.$$

Since f is arbitrary we must have $g^+ = 0$ and $g = g^-$. However g is supported in $[a,b]$ and is identically zero on $\mathbb{R}\backslash[a,b]$ and g^-, being a Hardy space function, cannot identically vanish on $\mathbb{R}\backslash[a,b]$. We get a contradiction. This completes the proof of Theorem 3.5. ■

To prove (a) of Proposition 3.1 we note that by Theorem 3.5 and property (I) above of quasi-affine maps the adjoint of $P_{[a,b]}|_{\mathcal{H}^2_+(\mathbb{R};\mathcal{K})}$ is also quasi-affine. For any $g \in L^2([a,b];\mathcal{K})$ and $f \in \mathcal{H}^2_+(\mathbb{R};\mathcal{K})$ we have,

$$(g, P_{[a,b]}|_{\mathcal{H}^2_+(\mathbb{R};\mathcal{K})}f)_{L^2([a,b];\mathcal{K})} = (g, P_{[a,b]}f)_{L^2([a,b];\mathcal{K})} = (g, P_{[a,b]}f)_{L^2(\mathbb{R};\mathcal{K})} = (g, f)_{L^2(\mathbb{R};\mathcal{K})} =$$
$$= (\hat{P}_+ g, f)_{L^2(\mathbb{R};\mathcal{K})} = (\hat{P}_+|_{L^2([a,b];\mathcal{K})}g, f)_{\mathcal{H}^2_+(\mathbb{R};\mathcal{K})}\,. \tag{3.1.12}$$

Thus, we find that $\hat{P}_+|_{L^2([a,b];\mathcal{K})}$ is the adjoint of $P_{[a,b]}|_{\mathcal{H}^2_+(\mathbb{R};\mathcal{K})}$ and, hence, it is quasi-affine and contractive.

To prove (b) of Proposition 3.1 we first note that for $t \geq 0$ the Hardy space $\mathcal{H}^2_-(\mathbb{R};\mathcal{K})$ is stable under the action of the evolution $u(t)$, i.e., we have $u(t)\mathcal{H}^2_-(\mathbb{R};\mathcal{K}) \subset \mathcal{H}^2_-(\mathbb{R};\mathcal{K})$ for all $t \geq 0$. We then have

$$\hat{P}_+ u_+(t)g = \hat{P}_+ u(t)P_{[a,b]}g = \hat{P}_+ u(t)g = \hat{P}_+ u(t)(\hat{P}_+ + \hat{P}_-)g =$$
$$= \hat{P}_+ u(t)\hat{P}_+ g + \hat{P}_+ u(t)\hat{P}_- g = \hat{P}_+ u(t)\hat{P}_+ g = T_u(t)\hat{P}_+ g, \quad t \geq 0, \quad g \in L^2([a,b];\mathcal{K})\,.$$

This concludes the proof of Proposition 3.1. ■

Going back to the proof of Theorem 3.4, we note that according to (a) in Proposition 3.1 the restriction $\hat{P}_+|_{L^2([a,b];\mathcal{K})}$ is a contractive and quasi-affine mapping of $L^2([a,b];\mathcal{K})$ into $\mathcal{H}^2_+(\mathbb{R}^+;\mathcal{K})$. Let us denote this restriction by Ω_f. It will be convenient below to first consider the mapping $\hat{P}_+ P_{[a,b]} : L^2(\mathbb{R};\mathcal{K}) \mapsto L^2(\mathbb{R};\mathcal{K})$ and then define $\Omega_f : L^2([a,b];\mathcal{K}) \mapsto \mathcal{H}^2_+(\mathbb{R};\mathcal{K})$ using this operator. Thus we set

$$\Omega_f := \hat{P}_+ P_{[a,b]}|_{L^2([a,b];\mathcal{K})} = \hat{P}_+|_{L^2([a,b];\mathcal{K})} \tag{3.1.13}$$

(here the subscript f in Ω_f designates forward time evolution; see Eq. (3.1.15) below). By Eq. (3.1.12)

$$\Omega_f^* = P_{[a,b]}\hat{P}_+|_{\mathcal{H}^2_+(\mathbb{R};\mathcal{K})} = P_{[a,b]}|_{\mathcal{H}^2_+(\mathbb{R};\mathcal{K})}\,. \tag{3.1.14}$$

Now define the operator $M_F : L^2([a,b];\mathcal{K}) \mapsto L^2([a,b];\mathcal{K})$ by

$$M_F := \Omega_f^*\Omega_f = P_{[a,b]}\hat{P}_+ P_{[a,b]}|_{L^2([a,b];\mathcal{K})}\,.$$

By (II) above and the fact that Ω_f, Ω_f^* are quasi-affine we get that M_F is a quasi-affine mapping of $L^2([a,b];\mathcal{K})$ into $L^2([a,b];\mathcal{K})$. Therefore M_F is continuous and injective and $Ran\ M_F$ is dense in $L^2([a,b];\mathcal{K})$. Obviously M_F is symmetric and, since Ω_f and Ω_f^* are bounded, then $Dom(M_F) = L^2([a,b];\mathcal{K})$ and we conclude that M_F is self-adjoint. Since Ω_f and Ω_f^* are both contractive then M_F is contractive.

Following the definition of the mappings Ω_f and Ω_f^* it is convenient to write Eq. (3.1.11) in the form

$$\hat{P}_+ P_{[a,b]} u_+(t) g = T_u(t)\hat{P}_+ P_{[a,b]} g, \quad \forall t \geq 0, \quad g \in L^2([a,b];\mathcal{K})\,,$$

and hence

$$\Omega_f u_+(t) g = T_u(t)\Omega_f g, \quad \forall t \geq 0, \quad g \in L^2([a,b];\mathcal{K})\,. \tag{3.1.15}$$

Hence, for every $\psi \in L^2([a,b];\mathcal{K})$ and $t \geq 0$ we have

$$\begin{aligned}(\psi(t), M_F\psi(t)) &= (u_+(t)\psi, M_F u_+(t)\psi) = \\ &= (u_+(t)\psi, \Omega_f^*\Omega_f u_+(t)\psi) = \|\Omega_f u_+(t)\psi\|^2 = \|T_u(t)\Omega_f\psi\|^2\end{aligned}$$

The fact that M_F is a Lyapunov variable, i.e., the validity of Eqs. (3.1.5) and (3.1.6) then follows immediately from Eqs. (3.1.9) and (3.1.10). ■

The following corollary to Theorem 3.4 provides a more concrete form for the Lyapunov operator M_F defined in Eq. (3.1.4).

Corollary 3.1 *Let M_F be the forward Lyapunov operator defined in Eq. (3.1.4). Then, for any function $f \in L^2([a,b];\mathcal{K})$ we have*

$$(M_F f)(E) = -\frac{1}{2\pi i}\int_a^b \frac{1}{E - E' + i0^+} f(E')\,dE', \quad E \in [a,b]. \tag{3.1.16}$$

□

Proof We know that each function $g \in \mathcal{H}^2(\mathbb{C}^+; \mathcal{K})$ has a non-tangential boundary value function on the real axis $\mathbb{R}$. Denoting this boundary value function by g_+ we have $g_+ \in \mathcal{H}^2_+(\mathbb{R}; \mathcal{K})$ and (Rosenblum and Rovnyak 1985):

$$-\frac{1}{2\pi i}\int_{-\infty}^{\infty} \frac{1}{z-\sigma} g_+(\sigma)d\sigma = \begin{cases} g(z), & \operatorname{Im} z > 0 \\ 0, & \operatorname{Im} z < 0 \end{cases}$$

In a similar manner, a function $g \in \mathcal{H}^2(\mathbb{C}^-; \mathcal{K})$ has a non-tangential boundary value function $g_- \in \mathcal{H}^2_-(\mathbb{R}; \mathcal{K})$ and we have

$$\frac{1}{2\pi i}\int_{-\infty}^{\infty} \frac{1}{z-\sigma} g_-(\sigma)d\sigma = \begin{cases} 0, & \operatorname{Im} z > 0 \\ g(z), & \operatorname{Im} z < 0 \end{cases}$$

Since $L^2(\mathbb{R}; \mathcal{K}) = \mathcal{H}^2_+(\mathbb{R}; \mathcal{K}) \oplus \mathcal{H}^2_-(\mathbb{R}; \mathcal{K})$ is an orthogonal direct sum we then get that the orthogonal projection operator $\hat{P}_+$ in $L^2(\mathbb{R}; \mathcal{K})$ onto the subspace $\mathcal{H}^2_+(\mathbb{R}; \mathcal{K})$ is given by

$$(\hat{P}_+ f)(\sigma) = -\frac{1}{2\pi i}\int_{-\infty}^{\infty} \frac{1}{\sigma - \sigma' + i0^+} f(\sigma')d\sigma', \quad f \in L^2(\mathbb{R}; \mathcal{K})$$

and hence for every $f \in L^2([a, b]; \mathcal{K})$ we immediately obtain

$$(M_F f)(E) = (P_{[a,b]}\hat{P}_+ P_{[a,b]} f)(E) = -\frac{1}{2\pi i}\int_a^b \frac{1}{E - E' + i0^+} f(E'), \quad E \in [a, b].$$

■

We fix a basis for the function space $L^2([a, b]; \mathcal{K})$ as follows: First fix a (generalized) basis $\{\mathbf{e}_\xi\}_{\xi\in\Xi}$ of $\mathcal{K}$ with Ξ an appropriate index set such that $\langle \mathbf{e}_\xi, \mathbf{e}_{\xi'}\rangle_{\mathcal{K}} = \delta_{\xi\xi'}$, where $\delta_{\xi\xi'}$ stands for Kronecker delta for discrete indices and Dirac delta for continuous indices. Next, denote by $\mathbf{e}_{E_0,\xi}$ the chosen basis at the point $E_0 \in [a, b]$, i.e., set

$$\mathbf{e}_{E_0,\xi}(E) = \begin{cases} \mathbf{e}_\xi, & E = E_0 \\ 0, & E \neq E_0 \end{cases}.$$

The set $\{\mathbf{e}_{E,\xi}\}_{E\in[a,b],\xi\in\Xi}$ forms a (generalized) basis for $L^2([a, b], \mathcal{K})$. We shall use regularly this basis in our discussion below. Given a function $f \in L^2([a, b]; \mathcal{K})$ we have

$$f = \sum_\xi \int_a^b dE\, |\mathbf{e}_{E,\xi}\rangle\langle \mathbf{e}_{E,\xi}, f\rangle,$$

and

$$f(E) = \sum_{\xi} |\mathbf{e}_{E,\xi}\rangle\langle\mathbf{e}_{E,\xi}, f\rangle = \sum_{\xi} |\mathbf{e}_{\xi}\rangle\langle\mathbf{e}_{\xi}, f(E)\rangle\,.$$

Using Eq. (3.1.16) and the basis $\{\mathbf{e}_{E,\xi}\}_{E\in[a,b],\xi\in\Xi}$ we obtain

$$M_F = -\frac{1}{2\pi i}\sum_{\xi}\int_a^b dE\int_a^b dE'\,|\mathbf{e}_{E,\xi}\rangle\frac{1}{E-E'+i0^+}\langle\mathbf{e}_{E',\xi}|\,. \tag{3.1.17}$$

In a manner similar to the construction of a forward Lyapunov operator M_F one is able to construct a backward Lyapunov operator M_B. The theorem analogous to Theorem 3.4 in this case is:

Theorem 3.6 *Let $M_B : L^2([a,b];\mathcal{K}) \mapsto L^2([a,b];\mathcal{K})$ be the operator defined by*

$$M_B := (P_{[a,b]}\hat{P}_- P_{[a,b]})\big|_{L^2([a,b];\mathcal{K})}. \tag{3.1.18}$$

Then M_B is a positive, contractive, injective operator on $L^2([a,b];\mathcal{K})$, such that Ran M_B is dense in $L^2([a,b];\mathcal{K})$ and M_B is a Lyapunov operator in the backward direction, i.e., for every $\psi \in L^2([a,b];\mathcal{K})$ we have

$$(\psi(t_2), M_B\,\psi(t_2)) \le (\psi(t_1), M_B\,\psi(t_1)),\quad t_2 \le t_1 \le 0,\quad \psi(t) = u_+(t)\psi,$$

and, moreover,

$$\lim_{t\to-\infty}(\psi(t), M_B\,\psi(t)) = 0.$$

□

The corresponding corollary in this case is:

Corollary 3.2 *Let M_B be the backward Lyapunov operator defined in Eq. (3.1.18). Then, for any function $f \in L^2([a,b];\mathcal{K})$ we have*

$$(M_B f)(E) = \frac{1}{2\pi i}\int_a^b \frac{1}{E-E'-i0^+}f(E')\,dE',\quad E \ge [a,b]. \tag{3.1.19}$$

□

and its proof is similar to the proof of Corollary 3.1. Using Eq. (3.1.19) we get that

$$M_B = \frac{1}{2\pi i}\sum_{\xi}\int_a^b dE\int_a^b dE'\,|\mathbf{e}_{E,\xi}\rangle\frac{1}{E-E'-i0^+}\langle\mathbf{e}_{E',\xi}|.$$

Note that Theorems 3.4 and 3.6 refer, respectively, to positive and negative times. However, due to the time translation invariance of the evolution, the restriction that t_2, t_1 are non-negative in Theorem 3.4 and that t_2, t_1 are non-positive in Theorem 3.6 can be removed (keeping the time ordering between t_2 and t_1 in both cases) and the Lyapunov property extends to all values of time. It is evident from Theorems 3.4 and 3.6 that both the forward and backward Lyapunov operators are defined on a rather abstract level in terms of certain functions spaces and that no relation to any concrete class of physical problems has been made yet. In the following we shall consider quantum mechanical scattering problems satisfying the following assumptions:

(i) Let $\mathcal{H}$ be a separable Hilbert space corresponding to a given quantum mechanical scattering problem. Assume that a self-adjoint, "free", unperturbed Hamiltonian H_0 and a self-adjoint perturbed Hamiltonian H are defined on $\mathcal{H}$ and form a complete scattering system, i.e., the Møller wave operators $\Omega_\pm \equiv \Omega_\pm(H_0, H)$ exist and are complete.
(ii) We assume that $[a, b] \subseteq \sigma_{ac}(H) = \sigma_{ac}(H_0)$.
(iii) We assume that the multiplicity of the absolutely continuous spectrum in (ii) above is uniform over the interval $[a, b]$.

Under assumptions (i)–(iii) above, there exist two mappings $\hat{W}_\pm^{QM} : \mathcal{H}_{ac} \mapsto L^2(\sigma_{ac}(H); \mathcal{K})$ that map the subspace $\mathcal{H}_{ac} \subseteq \mathcal{H}$ isometrically onto the function space $L^2(\sigma_{ac}(H); \mathcal{K})$ for some Hilbert space $\mathcal{K}$ whose dimension corresponds to the multiplicity of $\sigma_{ac}(H)$ and the Schrödinger evolution $U(t) = \exp(-iHt)$ is represented by the group $\{u_+(t)\}_{t\in\mathbb{R}}$. The representation of the scattering problem in the function space $L^2(\sigma_{ac}(H); \mathcal{K})$ obtained by applying $\hat{W}_+^{QM}$ is called the *outgoing energy representation* and is a spectral representation for H in which the action of H is represented by multiplication by the independent variable. In a similar manner, the representation obtain by applying the mapping $\hat{W}_-^{QM}$ is another spectral representation for H, called the *incoming energy representation* of the problem (Strauss 2011). We note that M_F, M_B, in Theorems 3.4 and 3.6–4.8 are, in fact, defined on the level of such spectral representations of H and their construction is made irrespective of the specific spectral representation in which one is working. However, when applied to scattering problems, we need to distinguish between the corresponding objects defined within the incoming energy representation and the outgoing energy representation.

The mappings $\hat{W}_+^{QM}$ and $\hat{W}_-^{QM}$ correspond, respectively, to incoming and outgoing solutions of the Lippmann–Schwinger equation (Taylor 1972; Newton 1967). If $\{\phi_{E,\xi}^-\}_{E\in\sigma_{ac}(H),\,\xi\in\Xi}$ are outgoing solutions of the Lippmann–Schwinger equation, where ξ corresponds to degeneracy indices of the energy E, and if $\{\phi_{E,\xi}^+\}_{E\in\sigma_{ac}(H),\,\xi\in\Xi}$ are incoming solutions of the Lippmann–Schwinger equation, and $\psi \in \mathcal{H}_{ac}$ is any scattering state, then

$$
(\hat{W}_+^{QM}\psi)(E, \xi) = \langle\phi_{E,\xi}^-|\psi\rangle
$$
$$
(\hat{W}_-^{QM}\psi)(E, \xi) = \langle\phi_{E,\xi}^+|\psi\rangle.
$$

With the help of the two mappings $\hat{W}_{\pm}^{QM}$ which are, in fact, associated with the two Møller wave operators $\Omega_{\pm}$, we define the forward Lyapunov operator for a quantum scattering problem satisfying assumptions (i)–(iii) to be

$$M_{+} := \left(\hat{W}_{+}^{QM}\right)^{-1} M_F \hat{W}_{+}^{QM}. \tag{3.1.20}$$

Let $\hat{P}_{[a,b]} : \mathcal{H}_{ac} \mapsto \mathcal{H}_{ac}$ be the spectral projection operator on the subspace $\mathcal{H}'_{ac} \subseteq \mathcal{H}_{ac}$ corresponding to the interval $[a, b] \subseteq \sigma_{ac}(H)$, i.e., $\mathcal{H}'_{ac} = \hat{P}_{[a,b]}\mathcal{H}_{ac}$. By Theorem 3.4 the operator $M_{+} : \mathcal{H}'_{ac} \mapsto \mathcal{H}'_{ac}$ is a positive, contractive, injective operator, such that Ran M_{+} is dense in $\mathcal{H}'_{ac}$ and M_{+} is a forward Lyapunov operator with respect to the quantum evolution on $\mathcal{H}'_{ac}$, i.e., for any $\psi \in \mathcal{H}'_{ac}$ we have

$$(\psi(t_2), M_{+}\psi(t_2)) \leq (\psi(t_1), M_{+}\psi(t_1)), \quad t_1 \leq t_2, \qquad \lim_{t\to\infty} (\psi(t), M_{+}\psi(t)) = 0$$

where $\psi(t) = U(t)\psi = \exp(-iHt)\psi$. Observe that by Eqs. (3.1.16) and (3.1.20) we obtain an explicit form for M_{+},

$$M_{+} = -\frac{1}{2\pi i}\sum_{\xi}\int_a^b dE \int_a^b dE' \, |\phi^{-}_{E,\xi}\rangle \frac{1}{E - E' + i0^{+}} \langle\phi^{-}_{E',\xi}| \, . \tag{3.1.21}$$

Similarly, the backward Lyapunov operator for the quantum scattering problem is defined to be

$$M_{-} := \left(\hat{W}_{-}^{QM}\right)^{-1} M_B \hat{W}_{-}^{QM}, \tag{3.1.22}$$

and according to Theorem 3.6 the operator $M_{-} : \mathcal{H}'_{ac} \mapsto \mathcal{H}'_{ac}$ is a positive, contractive, injective operator on $\mathcal{H}'_{ac}$, such that Ran M_{-} is dense in $\mathcal{H}'_{ac}$ and M_{-} is a backward Lyapunov operator with respect to the quantum evolution on $\mathcal{H}'_{ac}$, i.e., for any $\psi \in \mathcal{H}'_{ac}$ we have

$$(\psi(t_2), M_{-}\psi(t_2)) \leq (\psi(t_1), M_{-}\psi(t_1)), \quad t_2 \leq t_1, \qquad \lim_{t\to-\infty} (\psi(t), M_{-}\psi(t)) = 0 \, .$$

By Eqs. (3.1.19) and (3.1.22) we obtain an explicit form for M_{-},

$$M_{-} = \frac{1}{2\pi i}\sum_{\xi}\int_a^b dE \int_a^b dE' \, |\phi^{+}_{E,\xi}\rangle \frac{1}{E - E' - i0^{+}} \langle\phi^{+}_{E',\xi}| \, . \tag{3.1.23}$$

With Eqs. (3.1.21) and (3.1.23) we have completed the construction of forward and backward Lyapunov operators on $\mathcal{H}'_{ac}$ for scattering problems satisfying conditions (i)–(iii). Note that by the definition of M_{+} in Eq. (3.1.20) and the definition of M_F in Eq. (3.1.4) we obtain

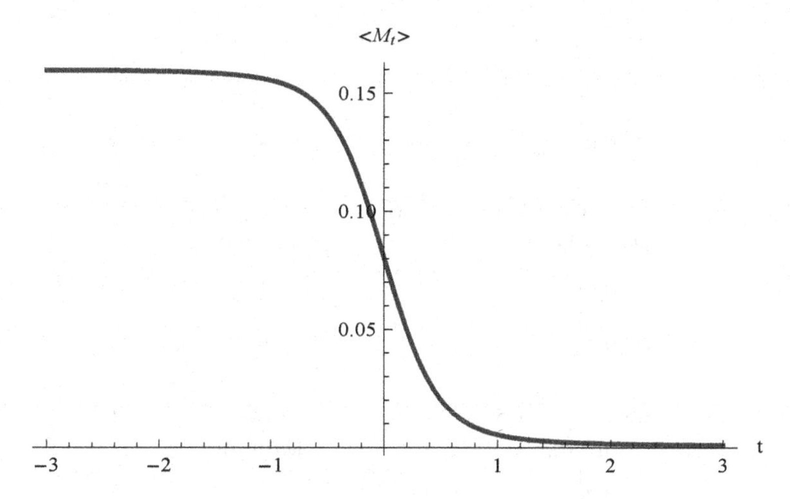

Fig. 3.1 Monotonicity of $\langle M_t \rangle$. The figure depicts the monotonic decrease over time of the expectation value $\langle M_t \rangle$ computed for a free evolution of a Gaussian wave packet with $p_0 = 6.4\eta$ and $\xi_0 = 3\eta$. The time t is given in units of $[\eta]^{-1}$. Reprinted form Strauss et al. (2011b), Copyright 2011, with permission from Elsevier

$$M_+ = \left(\hat{W}_+^{QM}\right)^{-1} M_F \hat{W}_+^{QM} = \left(\hat{W}_+^{QM}\right)^{-1} P_{[a,b]} \hat{P}_+ P_{[a,b]} \hat{W}_+^{QM}$$

and, similarly, by Eqs. (3.1.22) and (3.1.18) we obtain

$$M_- = \left(\hat{W}_-^{QM}\right)^{-1} M_B \hat{W}_-^{QM} = \left(\hat{W}_-^{QM}\right)^{-1} P_{[a,b]} \hat{P}_- P_{[a,b]} \hat{W}_-^{QM} .$$

If we apply these definitions of M_+ and M_- in the case of the original Lax–Phillips theory and set $[a, b] = \mathbb{R}$ so that $P_{[a,b]} = I_{L^2(\mathbb{R};\mathcal{K})}$ (considering the spectral properties of the generator of evolution in the Lax–Phillips case) we get that

$$M_+ = \left(\hat{W}_+^{QM}\right)^{-1} \hat{P}_+ \hat{W}_+^{QM} = P_+, \qquad M_- = \left(\hat{W}_+^{QM}\right)^{-1} \hat{P}_- \hat{W}_+^{QM} = P_- .$$

Hence, in the Lax–Phillips case the general definition of the forward and backward Lyapunov operators $M_\pm$ corresponds to the projection operators $P_\pm$; the natural Lyapunov operators of the Lax–Phillips theory. In this sense the Lax–Phillips case is just a particular instance of large family of Lyapunov operators.

To demonstrate the behavior of Lyapunov operators we consider the free motion of a particle of mass η moving to the right in one dimension, as represented by the evolution of a one-dimensional Gaussian wave packet under the dynamics generated by the free one dimensional Hamiltonian $H_0 = -\frac{\hbar}{2\eta}\frac{d^2}{dx^2}$,

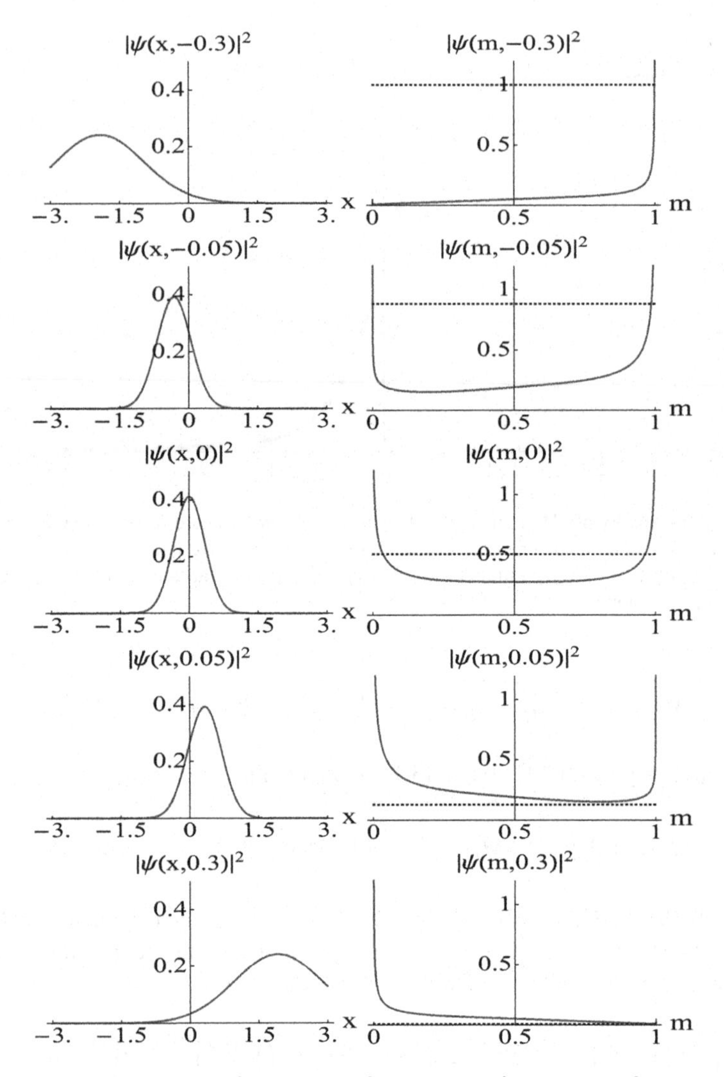

Fig. 3.2 Time frames of $|\psi(x,t)|^2$ and $|\psi(m,t)|^2 = |\langle m,+|\psi\rangle|^2 + |\langle m,-|\psi\rangle|^2$ computed for the Gaussian wave packet of Fig. 3.1. Time frames are taken at $t = -0.3, -0.05, 0, 0.05, 0.3$ with t measured in units of $[\eta]^{-1}$. The spatial variable x is also measured in units of $[\eta]^{-1}$. In each graph in the right column the dotted line correspond to the value of $\langle M_t\rangle$ at the corresponding value of t. Reprinted form Strauss et al. (2011b), Copyright 2011, with permission from Elsevier

$$\psi(x,t) = \left(\frac{\eta^2 {\xi_0}^2}{\pi(\eta + i{\xi_0}^2 t)^2}\right)^{\frac{1}{4}} \exp\left(-\frac{\eta {\xi_0}^2 x^2 + i p_0(p_0 t - 2\eta x)}{2(\eta + i{\xi_0}^2 t)}\right), \tag{3.1.24}$$

where p_0 and ξ_0 are the location and width of the wave packet in momentum space at $t = 0$ (Strauss 2011). The spectrum of the free Hamiltonian is then

$\sigma(H_0) = \sigma_{ac}(H_0) = \overline{\mathbb{R}^+}$ and the forward Lyapunov operator (defined on $\sigma(H_0)$) is given by

$$M_+ = -\frac{1}{i\pi\eta}\sum_{\xi=\pm}\int_0^\infty dp\, p\int_0^\infty dp'\, p'\,|p,\xi\rangle\frac{1}{p^2 - p'^2 + i0^+}\langle p',\xi|, \tag{3.1.25}$$

where $|p,\pm\rangle$ denotes a plane wave state with momentum $\pm p$ (the generalized eigenfunctions of the free Hamiltonian H_0). Calculating the expectation value $\langle M_t\rangle_{\psi(t)} = (\psi(t), M_+\psi(t))$ of M_+ in the state $\psi(t)$ of (3.1.24) one obtains Fig. 3.1, showing the time evolution of this expectation value. The monotonic behavior of $\langle M_t\rangle_{\psi(t)}$ as time increases is clearly evident in Fig. 3.1.

In Sect. 4.3 below we show (see Theorem 4.10) that the spectrum of the Lyapunov operator M_+ is $\sigma(M_+) = \sigma_{ac}(M_F) = [0, 1]$ and calculate the generalized eigenfunctions $|m,\pm\rangle$ of M_F (with $m \in (0, 1]$ the generalized eigenvalues of M_F and $\xi = \pm$ the degeneracy indices as in (3.1.25)). We may then use these eigenfunctions to calculate the probability distribution functions $|\langle m,\pm|\psi(t)\rangle|^2$ of $\psi(t)$ over the spectrum of M_+ at each point of time t and the results are presented in Fig. 3.2. This figure clearly shows the way that the probability distribution of the evolving state $\psi(t)$ over the spectrum of M_+ shifts from higher values of m, close to $m = 1$, for negative values of the time t, to lower values of m, close to $m = 0$, as time progresses. We observe that if the sequence of time frames is shown in reverse order then, with respect to the evolution of the spatial probability distribution of the wave function one is unable to tell whether time is running backwards or whether one is observing a Gaussian wave packet propagating to the left (with time running forward). However, if to each frame we attach the expectation value of M_+, then it is possible to distinguish between these two scenarios. This example clearly illustrates the time-ordering of states introduced into the Hilbert space by the existence of the Lyapunov operator (Strauss 2011).

In this chapter we have focused on the characterization of the projection operators $P_\pm$ of the Lax–Phillips scattering theory as Lyapunov operators and the construction of analogous Lyapunov operators $M_\pm$ for certain classes of quantum mechanical scattering problems. In contrast to $P_\pm$, the Lyapunov operators $M_\pm$ are not projection operators; they do not define an exact resonance subspace as in the Lax–Phillips case and we do not expect to be able to use them for the construction of an exact Lax–Phillips structure. However, as we shall see, we may still use $M_\pm$ for the construction of an approximate Lax–Phillips structure. Such a construction retains some of the advantages of the Lax–Phillips theory such as the natural association of resonances with semigroup evolution and Hilbert space states. This task is taken up in the next chapter.

Chapter 4
Modified Lax–Phillips Scattering Theory

As we have discussed in Chap. 2, several aspects of the Lax–Phillips scattering theory (Lax and Phillips 1967) distinguish it as an appealing abstract formalism for implementation even in situations outside of the strict range of problems for which it has been originally devised. The description of resonances in the framework of the Lax–Phillips theory possesses properties which may be considered as defining properties of an appropriate description of these objects. One such property is a dynamical characterization of resonances via their time evolution given in terms of a continuous, one parameter, strongly contractive semigroup, i.e., the Lax–Phillips semigroup. Specifically, resonances are identified as eigenvalues of the generator of the Lax–Phillips semigroup. This corresponds to another desirable feature of the theory, namely, the fact that each resonance pole is associated with a resonance state (or more generally a subspace) in a Hilbert space. In fact, the Lax–Phillips semigroup is obtained by a projection of the unitary evolution of the full system onto the subspace spanned by resonance states.

The attractive properties of the Lax–Phillips scattering theory, mentioned above, have led to some efforts to adapt the formalism to the framework of quantum mechanics. Early work in this direction can be found, for example, in Pavlov (1977, 1995), Flesia and Piron (1984), Horwitz and Piron (1973, 1993), Eisenberg and Horwitz (1997) (see also Kruglov et al. 2012 for a more recent application of the Lax–Phillips structure to quantum problems). A general formalism was developed in Strauss, Horwitz and Eisneberg (2000) and subsequently applied to several physical models in Strauss and Horwitz (2000, 2002), Ben Ari and Horwitz (2004). However, in general one cannot apply, without modification, the basic structure of the Lax–Phillips scattering theory in the context of standard quantum mechanical scattering problems since incoming and outgoing subspaces $\mathcal{D}_\pm$ having the properties listed in Eq. (2.1.1) cannot be found for large classes of such problems. This can be seen, for example, by noting the fact that in the Lax–Phillips theory the continuous spectrum of the generator of evolution is necessarily unbounded from below as well as from above, a requirement which is not met by most quantum mechanical Hamiltonians. Hence, the range of applications of quantum mechanical adaptations of the Lax–Phillips theory which essentially retain its original

L. Horwitz and Y. Strauss, *Unstable Systems*, Mathematical Physics Studies,
https://doi.org/10.1007/978-3-030-31570-2_4

mathematical structure is rather limited. In this chapter we introduce a modification of the Lax–Phillips theory which enables an implementation of an approximate Lax–Phillips structure within the context of scattering problems for which the original Lax–Phillips formalism cannot be applied.[1]

4.1 The Modified Lax–Phillips Structure

In this section we reexamine the structure of the Lax–Phillips theory and use it as a guideline for the construction of an approximate Lax–Phillips structure applicable to scattering problems which do not satisfy the assumptions of the original Lax–Phillips theory (Strauss 2011, 2015). In essence this implies that we are considering scattering problems for which incoming and outgoing subspaces $\mathcal{D}_\pm$ of the Lax–Phillips theory do not exist. The Lyapunov operators discussed in the previous subsection are a key building block of our construction below.

As a starting point we consider the Lax–Phillips theory and note that, if we define

$$Z(t) := P_+ U(t), \quad t \geq 0 \tag{4.1.1}$$

then, by the stability properties of $\mathcal{D}_+$, we have for $t_1, t_2 \geq 0$

$$Z(t_1)Z(t_2) = P_+U(t_1)P_+U(t_2) = P_+U(t_1)(P_+ + P_+^\perp)U(t_2) = P_+U(t_1+t_2) = Z(t_1+t_2), \tag{4.1.2}$$

where $P_+^\perp = I - P_+$. Hence, the family of operators $\{Z(t)\}_{t\geq 0}$ is a continuous, one parameter, contractive semigroup on $\mathcal{H}^{LP}$. It is easy to show, in addition, that $s - \lim_{t\to\infty} Z(t) = 0$. Moreover, we have

$$P_+U(t) = P_+U(t)(P_+ + P_+^\perp) = P_+U(t)P_+ = Z(t)P_+, \quad t \geq 0, \tag{4.1.3}$$

so that for non-negative times P_+ intertwines the unitary evolution $U(t)$ with the semigroup evolution $Z(t)$. Note that similarity between Eqs. (4.1.1), (4.1.2) and (3.1.7), (3.1.8). In fact, by Eq. (3.1.2) and by the fact that in the outgoing spectral representation of the Lax–Phillips theory we have $U(t) = \left(\hat{W}_+^{LP}\right)^{-1} u(t) \hat{W}_+^{LP}$ we have

$$Z(t) = P_+U(t) = \left(\hat{W}_+^{LP}\right)^{-1} \hat{P}_+ u(t) \hat{W}_+^{LP} = \left(\hat{W}_+^{LP}\right)^{-1} T_u(t) \hat{W}_+^{LP}, \quad t \geq 0.$$

Hence, the semigroups $\{Z(t)\}_{t\geq 0}$ and $\{T_u(t)\}_{t\geq 0}$ are unitarily equivalent. Finally, observe that by the intertwining relation in Eq. (4.1.3) we have

$$Z_{LP}(t) = P_+U(t)P_- = Z(t)P_+P_- = Z(t)P_{\mathcal{K}}, \quad t \geq 0 \tag{4.1.4}$$

[1]This chapter is based on and extends the work reported in Strauss (2011, 2015).

where $P_{\mathcal{K}} = P_+ P_-$. Thus, the Lax–Phillips semigroup is obtained by the projection of the semigroup $\{Z(t)\}_{t \geq 0}$ onto the subspace $\mathcal{K} = P_+ P_- \mathcal{H}$ defined by the Lyapunov operators P_+ and P_-. For the purpose of comparison and, indeed, inclusion of the Lax–Phillips semigroup in the more general case considered below we note also that since $P_\pm$ are orthogonal projection operators then $P_\pm = P_\pm^{1/2}$ so that we may write

$$Z_{\mathrm{LP}} = P_+ U(t) P_- = P_+^{1/2} U(t) P_-^{1/2} = Z(t) P_+^{1/2} P_-^{1/2}, \ t \geq 0.$$

The implementation of an approximate Lax–Phillips structure in scattering problems beyond those satisfying the original Lax–Phillips assumptions, e.g., scattering processes of standard quantum mechanics, is based on the existence of forward and backward Lyapunov operators as in Theorems 3.4 and 3.6 above and on the following two theorems (Strauss 2011), providing direct relation of these Lyapunov operators to semigroup evolution:

Theorem 4.7 *Let $\Lambda_F := M_F^{1/2}$. Then $\Lambda_F : L^2([a, b]; \mathcal{K}) \mapsto L^2([a, b]; \mathcal{K})$ is a positive, contractive, injective operator such that Ran Λ_F is dense in $L^2([a, b]; \mathcal{K})$. Furthermore, there exists a continuous, strongly contractive, one parameter semigroup $\{Z_F(t)\}_{t \in \mathbb{R}_+}$ with $Z_F(t) : L^2([a, b]; \mathcal{K}) \mapsto L^2([a, b]; \mathcal{K})$, such that for every $\psi \in L^2([a, b]; \mathcal{K})$ we have*

$$\|Z_F(t_2)\psi\| \leq \|Z_F(t_1)\psi\|, \quad t_2 \geq t_1 \geq 0 \tag{4.1.5}$$

and

$$s - \lim_{t \to \infty} Z_F(t) = 0 \tag{4.1.6}$$

and the following intertwining relation holds:

$$\Lambda_F u_+(t) = Z_F(t) \Lambda_F, \quad t \geq 0. \tag{4.1.7}$$

□

Proof of Theorem 4.7 Since M_F is a bounded positive operator its positive square root Λ_F is well defined and unique and we may set $\Lambda_F := M_F^{1/2} = (\Omega_f^* \Omega_f)^{1/2}$. Moreover, since

$$M_F L^2([a, b]; \mathcal{K}) = \Lambda_F \Lambda_F L^2([a, b]; \mathcal{K}) \subseteq \Lambda_F L^2([a, b]; \mathcal{K}),$$

and since *Ran* M_F is dense in $L^2([a, b]; \mathcal{K})$ we conclude that *Ran* Λ_F is dense in $L^2([a, b]; \mathcal{K})$. Furthermore, since $M_F = \Lambda_F^2$ is one to one then Λ_F must also be one to one. We can summarize the findings above by stating that the fact that M_F is positive, one to one and quasi-affine implies the same properties for Λ_F. Since M_F is contractive and since for every $\psi \in L^2([a, b]; \mathcal{K})$ we have $(\psi, M_F \psi) = \|\Lambda_F \psi\|^2$ we conclude, by using the Schwartz inequality, that Λ_F is also contractive.

Define a mapping $\tilde{R} : \Lambda_F L^2([a,b];\mathcal{K}) \mapsto \mathcal{H}^2_+(\mathbb{R};\mathcal{K})$ by

$$\tilde{R} := \Omega_f(\Omega_f^*\Omega_f)^{-1/2} = \Omega_f \Lambda_F^{-1} \,. \tag{4.1.8}$$

Obviously $\tilde{R}$ is defined on a dense set in $L^2([a,b];\mathcal{K})$. For any $g \in \Lambda_F L^2([a,b];\mathcal{K})$ we have

$$\begin{aligned}\|\tilde{R}g\|^2 = (\tilde{R}g, \tilde{R}g) =&(\Omega_f\Lambda_F^{-1}g, \Omega_f\Lambda_F^{-1}g) = \\ &= (\Omega_f^*\Omega_f\Lambda_F^{-1}g, \Lambda_F^{-1}g) = (M_F\Lambda_F^{-1}g, \Lambda_F^{-1}g) = \|g\|^2 \,,\end{aligned} \tag{4.1.9}$$

hence $\tilde{R}$ is isometric on a dense set in $L^2([a,b];\mathcal{K})$ and can be extended to an isometric map $R : L^2([a,b];\mathcal{K}) \mapsto \mathcal{H}^2_+(\mathbb{R};\mathcal{K})$ such that

$$R^*R = I_{L^2([a,b];\mathcal{K})} \,.$$

Since R is an isometry $Ran\ R$ is a closed subspace of $\mathcal{H}^2_+(\mathbb{R};\mathcal{K})$. Moreover, $Ran R \supset R\Lambda_F L^2([a,b];\mathcal{K}) = \tilde{R}\Lambda_F L^2([a,b];\mathcal{K}) = \Omega_f L^2([a,b];\mathcal{K})$, so that $Ran\ R$ contains a dense set in $\mathcal{H}^2_+(\mathbb{R};\mathcal{K})$. Therefore $RL^2([a,b];\mathcal{K}) = \mathcal{H}^2_+(\mathbb{R};\mathcal{K})$ and we conclude that $R : L^2([a,b];\mathcal{K}) \mapsto \mathcal{H}^2_+(\mathbb{R};\mathcal{K})$ is, in fact, an isometric isomorphism.

From the definition of $\tilde{R}$ in Eq. (4.1.8), or from Eq. (4.1.9), we see that on the dense set $\Omega_f L^2([a,b];\mathcal{K}) \subset \mathcal{H}^2_+(\mathbb{R};\mathcal{K})$ we have

$$\tilde{R}^* f = (\Omega_f^*\Omega_f)^{-1/2}\Omega_f^* f = \Lambda_F^{-1}\Omega_f^* f, \qquad f \in Ran\ \tilde{R} \tag{4.1.10}$$

and the adjoint R^* of R is an extension of $\tilde{R}^*$ to $\mathcal{H}^2_+(\mathbb{R};\mathcal{K})$. For any $g \in Ran\ \Omega_f$ we have

$$\tilde{R}^* g = (\Omega_f^*\Omega_f)^{-1/2}\Omega_f^* g = (\Omega_f^*\Omega_f)^{-1/2}\Omega_f^*\Omega_f\Omega_f^{-1} g = (\Omega_f^*\Omega_f)^{1/2}\Omega_f^{-1} g = \Lambda_F\Omega_f^{-1} g \,. \tag{4.1.11}$$

Now define (in close analogy to the Lax–Phillips theory)

$$Z_F(t) := \Lambda_F u_+(t)\Lambda_F^{-1}, \qquad t \geq 0 \,. \tag{4.1.12}$$

Obviously, $Z_F(t)$ is well defined on $Ran\ \Lambda_F$ for any $t \geq 0$. Moreover, using the definition of $\tilde{R}$ from Eqs. (4.1.8) and (4.1.11), (3.1.15) we get

$$\begin{aligned}RZ_F(t)R^*g &= \Omega_f\Lambda_F^{-1}Z_F(t)\Lambda_F\Omega_f^{-1}g = [\Omega_f\Lambda_F^{-1}][\Lambda_F u_+(t)\Lambda_F^{-1}][\Lambda_F\Omega_f^{-1}]g = \\ &= \Omega_f u_+(t)\Omega_f^{-1}g = T_u(t)\Omega_f\Omega_f^{-1}g = T_u(t)g, \quad t \geq 0, \quad g \in Ran\ \Omega_f \subset \mathcal{H}^2_+(\mathbb{R};\mathcal{K}) \,.\end{aligned}$$

Thus, on the dense subset $R^*\Omega_F L^2([a,b];\mathcal{K}) \subset L^2([a,b];\mathcal{K})$ we have $Z_F(t) = R^*T_u(t)R$ and since R and $T_u(t)$ are bounded we are able by continuity to extend the domain of definition of $Z_F(t)$ to all of $L^2([a,b];\mathcal{K})$ and obtain

$$RZ_F(t)R^* = T_u(t), \quad Z_F(t) = R^*T_u(t)R, \ \ t \geq 0\,. \tag{4.1.13}$$

From the fact that R is an isometric isomorphism, the fact that $\{T_u(t)\}_{t\in\mathbb{R}^+}$ is a continuous, strongly contractive, one parameter semigroup and Eqs. (3.1.9), (3.1.10) we conclude that $\{Z_F(t)\}_{t\in\mathbb{R}^+}$ is a continuous, strongly contractive, one parameter semigroup as well as Eqs. (4.1.5) and (4.1.6). Finally, from the definition in Eq. (4.1.12) we immediately get that the intertwining relation in Eq. (4.1.7) holds. ■

Theorem 4.8 *Let $\Lambda_B := M_B^{1/2}$. Then $\Lambda_B : L^2([a,b];\mathcal{K}) \mapsto L^2([a,b];\mathcal{K})$ is a positive, contractive, injective operator such that Ran Λ_B is dense in $L^2([a,b];\mathcal{K})$. Furthermore, there exists a continuous, strongly contractive, one-parameter semigroup $\{Z_B(t)\}_{t\in\mathbb{R}_-}$ with $Z_B(t) : L^2([a,b];\mathcal{K}) \mapsto L^2([a,b];\mathcal{K})$, such that for every $\psi \in L^2([a,b];\mathcal{K})$ we have*

$$\|Z_B(t_2)\psi\| \leq \|Z_B(t_1)\psi\|, \quad t_2 \leq t_1 \leq 0$$

and

$$s - \lim_{t\to-\infty} Z_B(t) = 0$$

and the following intertwining relation holds:

$$\Lambda_B u_+(t) = Z_B(t)\Lambda_B, \quad t \leq 0. \tag{4.1.14}$$

□

The proof of Theorem 4.8 is similar to that of Theorem 4.7.

Given a scattering problem satisfying assumptions (i)–(iii) in Sect. 3.1 we regard $L(\sigma_{ac}(H);\mathcal{K})$ in Theorem 4.7 as the function space carrying the outgoing spectral representation of the problem, Similarly, $L^2(\sigma_{ac}(H);\mathcal{K})$ in Theorem 4.8 is considered to be the function space corresponding to the incoming spectral representation. Following Theorems 4.7 and 4.8 and the definitions of M_+ and M_- in Eqs. (3.1.20) and (3.1.22) we then set

$$\Lambda_+ = M_+^{1/2} = \left(\hat{W}_+^{QM}\right)^{-1} M_F^{1/2}\hat{W}_+^{QM}, \qquad \Lambda_- = M_-^{1/2} = \left(\hat{W}_-^{QM}\right)^{-1} M_B^{1/2}\hat{W}_-^{QM}.$$

Note that Theorem 4.7 implies that there exists a continuous, strongly contractive, one parameter semigroup $\{Z_+(t)\}_{t\in\mathbb{R}_+}$ with $Z_+(t) : \mathcal{H}'_{ac} \mapsto \mathcal{H}'_{ac}$ (recall that $\mathcal{H}'_{ac} = \hat{P}_{[a,b]}\mathcal{H})$), such that for every $\psi \in \mathcal{H}'_{ac}$ we have

$$\|Z_+(t_2)\psi\| \leq \|Z_+(t_1)\psi\|, \quad t_2 \geq t_1 \geq 0$$

and

$$s - \lim_{t\to\infty} Z_+(t) = 0$$

and the following intertwining relation holds

$$\Lambda_+ U(t)\psi = Z_+(t)\Lambda_+\psi, \quad \forall t \geq 0, \ \psi \in \mathcal{H}'_{ac}. \tag{4.1.15}$$

In fact, we have

$$Z_+(t) = (\hat{W}_+^{QM})^{-1} Z_F(t) \hat{W}_+^{QM}|_{\mathcal{H}'_{ac}}, \quad t \geq 0, \tag{4.1.16}$$

where $Z_F(t)$ are elements of the semigroup $\{Z_F(t)\}_{t\geq 0}$ in Theorem 4.7. Similarly, Theorem 4.8 implies that there exists a continuous, strongly contractive, one parameter semigroup $\{Z_-(t)\}_{t\in\mathbb{R}_-}$ with $Z_-(t) : \mathcal{H}'_{ac} \mapsto \mathcal{H}'_{ac}$, such that for every $\psi \in \mathcal{H}'_{ac}$ we have

$$\|Z_-(t_2)\psi\| \leq \|Z_-(t_1)\psi\|, \quad t_2 \leq t_1 \leq 0$$

and

$$s - \lim_{t\to-\infty} Z_-(t) = 0$$

and the following intertwining relation holds

$$\Lambda_- U(t)\psi = Z_-(t)\Lambda_-\psi, \quad t \leq 0, \ \psi \in \mathcal{H}'_{ac}. \tag{4.1.17}$$

Upon comparison of the intertwining relations in Eqs. (4.1.15) and (4.1.3) we are led to the following definition:

Definition 4.3 (*Approximate Lax–Phillips semigroup*) Consider a scattering problem satisfying assumptions (i)–(iii) above. Let $\Lambda_+ = M_+^{1/2}$ and $\Lambda_- = M_-^{1/2}$ where M_+ and M_- are, respectively, the forward Lyapunov operator and backward Lyapunov operator defined on $\mathcal{H}'_{ac}$. Then the approximate Lax–Phillips semigroup is defined to be the family of operators $\{Z_{app}(t)\}_{t\in\mathbb{R}_+}$, $Z_{app}(t) : \mathcal{H}'_{ac} \mapsto \mathcal{H}'_{ac}$ defined by

$$Z_{app}(t) := \Lambda_+ U(t)\Lambda_-, \quad t \geq 0, \ \ U(t) = e^{-iHt}. \tag{4.1.18}$$

□

It is precisely due to the central importance of the intertwining relations in Eqs. (4.1.15) and (4.1.17) that the definition of the approximate Lax–Phillips semigroup in Eq. (4.1.18) is made using the square roots $\Lambda_\pm$ of the Lyapunov operators $M_\pm$. As we have already noted above, if we apply similar definitions of $\Lambda_\pm$, as square roots of the Lyapunov operators in the Lax–Phillips case, we obtain $\Lambda_+ = P_+^{1/2} = P_+$ and $\Lambda_- = P_-^{1/2} = P_-$ so that in this case we would have

$$Z_{app}(t) = P_+^{1/2} U(t) P_-^{1/2} = P_+ U(t) P_- = Z_{LP}(t), \quad \forall t \geq 0,$$

and the family of operators $\{Z_{app}(t)\}_{t\geq 0}$ is then identical to the Lax–Phillips semigroup. Moreover, using the intertwining relation in Eq. (4.1.15) we get that

$$Z_{app}(t) := \Lambda_+ U(t)\Lambda_- = Z_+(t)\Lambda_+\Lambda_-, \quad t \geq 0. \tag{4.1.19}$$

Equation (4.1.19) is to be compared with Eq. (4.1.4). Note, however, that in contrast to the original Lax–Phillips case, in general $\{Z_{app}(t)\}_{t\in\mathbb{R}_+}$ is not an exact semigroup.

For a quantum mechanical scattering problem satisfying assumptions (i)–(iii) the scattering operator $S_{QM} = \Omega_-^{-1}\Omega_+$, where Ω_+ and Ω_- are, respectively, the incoming and outgoing Møller wave operators, has a representation as a mapping from the incoming energy representation to the outgoing energy representation in terms of the scattering matrix $\hat{S}_{QM}(\cdot) : \sigma_{ac}(H) \mapsto \mathcal{U}(\mathcal{K})$, where $\mathcal{U}(\mathcal{K})$ is the class of unitary operators on the multiplicity Hilbert space $\mathcal{K}$ (note that $\hat{S}_{QM}(\cdot)$ is, in fact, a representation of the scattering operator S_{QM} in the spectral representation of the unperturbed Hamiltonian H_0). The scattering matrix $\hat{S}_{QM}(\cdot)$ in the quantum case is analogous to the Lax–Phillips scattering matrix $\hat{S}_{LP}(\cdot)$ in the Lax–Phillips case, which is also a mapping between incoming and outgoing spectral representations of the generator of evolution. Adding the correspondence of these two objects to the list of analogous constructions for the quantum machanical scattering theory and the Lax–Phillips scattering theory discussed above we may produce, for a quantum mechanical scattering problem satisfying assumptions (i)–(iii), the following list of correspondences between objects of the Lax–Phillips scattering theory and corresponding objects in the case of quantum mechanical scattering

$$
\begin{aligned}
\underline{\text{LP scattering theory}} &\qquad \underline{\text{QM scattering theory}} \\
U(t) = e^{-iKt} &\Longleftrightarrow U(t) = e^{-iHt} \\
P_\pm &\Longleftrightarrow \Lambda_\pm \\
Z_{\mathrm{LP}}(t) = P_+ U(t) P_-,\ t \ge 0 &\Longleftrightarrow Z_{app}(t) = \Lambda_+ U(t) \Lambda_-,\ t \ge 0 \\
\hat{S}_{LP}(E),\ \ E \in \mathbb{R} &\Longleftrightarrow \hat{S}_{QM}(E),\ \ E \in [a, b]
\end{aligned}
\tag{4.1.20}
$$

We complete the set of relations between constructions of the Lax–Phillips theory and the corresponding constructions for quantum mechanical scattering by considering the incoming and outgoing representations. Observe that the outgoing (spectral or translation) representations of the Lax–Phillips theory are distinguished by the representation of the outgoing subspace $\mathcal{D}_+$. Hence, for example, if $\hat{W}_+^{\mathrm{LP}}$ is the mapping of $\mathcal{H}^{\mathrm{LP}}$ onto the outgoing spectral representation then $\hat{W}_+^{\mathrm{LP}}\mathcal{D}_+ = \mathcal{H}_-^2(\mathbb{R};\mathcal{K})$ and $\hat{W}_+^{\mathrm{LP}}(\mathcal{K} \oplus \mathcal{D}_-) = \mathcal{H}_+^2(\mathbb{R};\mathcal{K})$. Thus the outgoing representations are centered on the separation of the outgoing part of an evolving state $\psi(t) = U(t)\psi$ from the other components of that state which is achieved by the application of the projection P_+. This implies a decomposition of an evolving state $\psi(t) = \exp(-iKt)\psi$, corresponding to an initial state $\psi \in \mathcal{H}^{\mathrm{LP}}$, according to

$$
\psi(t) = P_+\psi(t) + P_+^\perp\psi(t) = \psi_+^b(t) + \psi_+^f(t),
\tag{4.1.21}
$$

where $\psi_+^b(t) := P_+\psi(t)$ and $\psi_+^f(t) := P_+^\perp\psi(t)$. It is readily verified, using the outgoing translation representation, that $\psi_+^b(t)$ is a *backward asymptotic component* of $\psi(t)$, i.e., $\psi_+^b(t)$ vanishes in the forward time asymptote as $t \to \infty$ and is asymptotic to $\psi(t)$ in the backward time asymptote as $t \to -\infty$. Similarly, $\psi_+^f(t)$ is a *forward asymptotic component* of $\psi(t)$, i.e., $\psi_+^f(t)$ vanishes in the backward time asymptote as $t \to -\infty$ and is asymptotic to $\psi(t)$ in the forward time asymptote as $t \to \infty$. The evolution of $\psi(t)$ is then represented as a transition from $\psi_+^b(t)$ to $\psi_+^f(t)$. We call the representation of the evolution obtained by the decomposition in Eq. (4.1.21) a *forward transition representation* and emphasize again its direct relation to the outgoing (translation or spectral) representations in the Lax–Phillips theory. Note that the name given to this representation of the evolution registers both the fact that the representation involves a transition between different components of the evolving state and the fact that the decomposition in Eq. (4.1.21) is obtained using the forward Lyapunov operator P_+.

Following a similar line of argument we may use the backward Lyapunov operator, i.e., the projection P_-, to obtain a decomposition of an evolving state $\psi(t)$ in the form

$$\psi(t) = P_-\psi(t) + P_-^{\perp}\psi(t) = \psi_-^b(t) + \psi_-^f(t), \tag{4.1.22}$$

where $\psi_-^f(t) := P_-\psi(t)$ and $\psi_-^b(t) := P_-^{\perp}\psi(t) = (I - P_-)\psi(t)$. Here $\psi_-^f(t)$ is a forward asymptotic component of $\psi(t)$ and $\psi_-^b(t)$ is a backward asymptotic component of $\psi(t)$ and we obtain another transition representation of the evolution of $\psi(t)$ which we call the *backward transition representation.* In a manner similar to the case of the forward transition representation, the backward transition representation is directly associated with the Lax–Phillips incoming (spectral or translation) representations. It is evident from the structure of the Lax–Phillips theory that such transition representations are useful for the description of transient phenomena in scattering processes, such as resonances.

Turning to the quantum mechanical case (with *e.g.* half line spectrum), we may define forward and backward transition representations analogous to those defined in the Lax–Phillips theory using the following two propositions (proved in Strauss 2011), the first of which concerns the forward transition representation:

Proposition 4.2 *For $\psi(t) = u_+(t)\psi$, $\psi \in L^2([a, b]; \mathcal{K})$, $t \in \mathbb{R}$, apply the following decomposition*

$$\psi(t) = \psi_F^b(t) + \psi_F^f(t) \tag{4.1.23}$$

where

$$\psi_F^b(t) := \Lambda_F\psi(t), \qquad \psi_F^f(t) := (I - \Lambda_F)\psi(t).$$

Then

$$\lim_{t\to-\infty} \|\psi(t) - \psi_F^b(t)\| = 0, \qquad \lim_{t\to\infty} \|\psi_F^b(t)\| = 0,$$

$$\lim_{t\to-\infty} \|\psi_F^f(t)\| = 0, \quad \lim_{t\to\infty} \|\psi(t) - \psi_F^f(t)\| = 0.$$

□

Proof First, by a simple application of the intertwining relation, Eq. (4.1.7), we obtain

$$\lim_{t\to\infty} \|\psi_F^b(t)\|^2 = \lim_{t\to\infty} \|\Lambda_F u_+(t)\psi\|^2 = \lim_{t\to\infty} \|Z_F(t)\psi_{\Lambda_F}\|^2 = 0.$$

Of course, this limit is equivalent also to the limit $\lim_{t\to\infty} \|\psi(t) - \psi_F^f(t)\|^2 = 0$. Next, we note that according to the definitions of M_F and M_B in Eqs. (3.1.4) and (3.1.18) we have

$$M_F + M_B = \left(P_{[a,b]}\hat{P}_+ P_{[a,b]} + P_{[a,b]}\hat{P}_- P_{[a,b]}\right)|_{L^2([a,b];\mathcal{K})} = I_{L^2([a,b];\mathcal{K})}.$$

This identity, together with the Lyapunov property of M_B from Theorem 3.6 leads to the limit

$$\begin{aligned}\lim_{t\to-\infty} \|\Lambda_F\psi(t)\|^2 &= \lim_{t\to-\infty} (\Lambda_F\psi(t), \Lambda_F\psi(t)) = \\ &= \lim_{t\to-\infty} (\psi(t), \Lambda_F^2\psi(t)) = \lim_{t\to-\infty} (\psi(t), M_F\psi(t)) = \lim_{t\to-\infty} (\psi(t), (I - M_B)\psi(t)) = \|\psi\|^2,\end{aligned}$$

hence we get that $\lim_{t\to-\infty} \|\Lambda_F\psi(t)\| = \|\psi\|$. Furthermore, for any $\psi \in L^2([a, b]; \mathcal{K})$ we have

$$\begin{aligned}\|(1 - \Lambda_F)\psi\|^2 &= (\psi, (1 - \Lambda_F)^2\psi) = \|\psi\|^2 + \|\Lambda_F\psi\|^2 - 2(\psi, \Lambda_F\psi) = \\ &= \|\psi\|^2 - \|\Lambda_F\psi\|^2 + 2\big[\|\Lambda_F\psi\|^2 - (\psi, \Lambda_F\psi)\big] = \\ &= \|\psi\|^2 - \|\Lambda_F\psi\|^2 + 2\big[(\Lambda_F\psi, \Lambda_F\psi) - (\psi, \Lambda_F\psi)\big] = \\ &= \|\psi\|^2 - \|\Lambda_F\psi\|^2 - 2(\Lambda_F\psi, (1 - \Lambda_F)\psi)) = \\ &= \|\psi\|^2 - \|\Lambda_F\psi\|^2 - 2(\Lambda_F^{1/2}\psi, (1 - \Lambda_F)\Lambda_F^{1/2}\psi))\end{aligned}$$

where in the last equality we used the fact that Λ_F is positive and its positive square root $\Lambda_F^{1/2}$ is well defined and is self-adjoint. Now, since Λ_F is positive and contractive then so is $(1 - \Lambda_F)$ and so, for any $\psi \in L^2([a, b]; \mathcal{K})$, we have $(\Lambda_F^{1/2}\psi, (1 - \Lambda_F)\Lambda_F^{1/2}\psi) \geq 0$. Thus we obtain the inequality

$$0 \leq \|(1 - \Lambda_F)\psi\|^2 \leq \|\psi\|^2 - \|\Lambda_F\psi\|^2 \tag{4.1.24}$$

Replacing in Eq. (4.1.24) ψ with $\psi(t)$ and taking the limit $t \to -\infty$ we get

$$\lim_{t\to-\infty} \|\psi_F^f(t)\| = \lim_{t\to-\infty} \|(1 - \Lambda_F)\psi(t)\| = 0.$$

The last limit is also equivalent, of course, to the limit $\lim_{t\to-\infty} \|\psi(t) - \psi_F^b(t)\| = 0$ ■

Proposition 4.3 *For $\psi(t) = u_+(t)\psi$, $\psi \in L^2([a, b]; \mathcal{K})$, $t \in \mathbb{R}$, apply the following decomposition*

$$\psi(t) = \psi_B^b(t) + \psi_B^f(t) \tag{4.1.25}$$

where

$$\psi_B^b(t) := (I - \Lambda_B)\psi(t), \qquad \psi_B^f(t) := \Lambda_B \psi(t).$$

Then

$$\lim_{t\to-\infty} \|\psi(t) - \psi_B^b(t)\| = 0, \qquad \lim_{t\to\infty} \|\psi_B^b(t)\| = 0,$$

$$\lim_{t\to-\infty} \|\psi_B^f(t)\| = 0, \quad \lim_{t\to\infty} \|\psi(t) - \psi_B^f(t)\| = 0.$$

□

The proof of Proposition 4.3 is similar to that of Proposition 4.2. Proposition 4.2 states that $\psi(t)$ can be decomposed into a sum of two components, $\psi_F^b(t)$ and $\psi_F^f(t)$ such that $\psi_F^b(t)$ is a backward asymptotic component and $\psi_F^f(t)$ is a forward asymptotic component of $\psi(t)$. Via the decomposition in Eq. (4.1.23) the evolution of $\psi(t)$ is represented as a transition from the backward asymptotic component to the forward asymptotic component and we obtain a transition representation of the evolution which, by the fact that the decomposition is defined using the (square root of the) forward Lyapunov operator, is a forward transition representation. By Proposition 4.3 we have a different decomposition of $\psi(t)$ into a backward asymptotic component $\psi_F^b(t)$ and a forward asymptotic component $\psi_F^f(t)$. The resulting transition representation of the evolution of $\psi(t)$ in this case is a backward transition representation, i.e., the decomposition of the evolving state $\psi(t)$ into the two components in Eq. (4.1.25) is achieved via the use of Λ_B.

Consider a scattering problem satisfying assumptions (i)–(iii) above. Defining the forward Lyapunov operator M_+ for the scattering problem as in Eq. (3.1.20) and noting that $\Lambda_+ = (\hat{W}_+^{QM})^{-1} M_F^{1/2} \hat{W}_+^{QM}$ we immediately obtain, using Proposition 4.2, a forward transition representation for the quantum evolution. The formal definition of this representation is:

Definition 4.4 (*forward transition representation*) Let $\Lambda_+ := M_+^{1/2}$ be the square root of M_+. For any $\psi \in \mathcal{H}'_{ac} = \hat{P}_{[a,b]}\mathcal{H}_{ac}$ the forward transition representation of the evolution of ψ is defined to be the decomposition

$$\psi(t) = \psi_+^b(t) + \psi_+^f(t),$$

where $\psi_+^b(t) := \Lambda_+\psi(t)$, $\psi_+^f(t) := (I - \Lambda_+)\psi(t)$, $\psi(t) = U(t)\psi$ and $U(t) = \exp(-iHt)$ is the Schrödinger evolution in $\mathcal{H}$. □

The asymptotic behavior over time of the two components $\psi_+^b(t)$, $\psi_+^f(t)$ of $\psi(t)$ follows directly from Proposition 4.2. The backward transition representation is defined in a similar manner following the definition of the backward Lyapunov operator M_- for the scattering problem in Eq. (3.1.22) and the fact that $\Lambda_- = (\hat{W}_-^{QM})^{-1} M_B^{1/2} \hat{W}_-^{QM}$:

Definition 4.5 (*backward transition representation*) Let $\Lambda_- := M_-^{1/2}$ be the square root of M_-. For any $\psi \in \mathcal{H}'_{ac}$ the backward transition representation of the evolution of ψ is defined to be the decomposition

$$\psi(t) = \psi_-^b(t) + \psi_-^f(t),$$

where $\psi_-^b(t) := (I - \Lambda_-)\psi(t)$, $\psi_-^f(t) := \Lambda_-\psi(t)$, $\psi(t) = U(t)\psi$ and $U(t) = \exp(-iHt)$ is the Schrödinger evolution in $\mathcal{H}$. □

Of course, the asymptotic behavior over time of the two components $\psi_-^b(t)$, $\psi_-^f(t)$ of the backward transition representation follows directly from Proposition 4.3.

By defining the two transition representations in Definitions 4.4 and 4.5 we complete the construction within the framework of quantum mechanics of objects and representations analogous to the central objects and representations of the Lax–Phillips theory. We may then extend Eq. (4.1.20) as follows:

$$\begin{array}{rcl} \underline{\text{LP scattering theory}} & & \underline{\text{QM scattering theory}} \\ U(t) = e^{-iKt} & \Longleftrightarrow & U(t) = e^{-iHt} \\ P_\pm & \Longleftrightarrow & \Lambda_\pm \\ \psi(t) = P_+\psi(t) + P_+^\perp\psi(t) & \Longleftrightarrow & \psi(t) = \Lambda_+\psi(t) + (I - \Lambda_+)\psi(t) \\ \psi(t) = P_-^\perp\psi(t) + P_-\psi(t) & \Longleftrightarrow & \psi(t) = (I - \Lambda_-)\psi(t) + \Lambda_-\psi(t) \\ Z_{\mathrm{LP}}(t) = P_+U(t)P_-,\ t \geq 0 & \Longleftrightarrow & Z_{app}(t) = \Lambda_+U(t)\Lambda_-,\ t \geq 0 \\ \hat{S}_{\mathrm{LP}}(E),\ \ E \in \mathbb{R} & \Longleftrightarrow & \hat{S}_{\mathrm{QM}}(E),\ \ E \in [a, b] \end{array} \tag{4.1.26}$$

4.2 Resonance States and Resonance Poles in the Modified Lax–Phillips Formalism

Thus far we have considered in the quantum mechanical case a set of objects analogous to the central objects of the Lax–Phillips theory. However, beyond analogy in the construction of certain objects our goal is to construct, in the context of quantum mechanical scattering, a formalism analogous to the Lax–Phillips scattering theory. Thus, upon completion of the set of relations in Eq. (4.1.26), we are left with an important task, i.e., to establish in the context of quantum mechanical scattering a theorem analogous to Theorem 2.3 relating resonance poles of the Lax–Phillips scattering matrix to eigenvalues and eigenvectors of the Lax–Phillips semigroup. Hence, an appropriate definition of resonance states and investigation of their relation to the approximate Lax–Phillips semigroup is a central ingredient in the development of the formalism introduced here. Of course, we do not expect to define resonance states as exact eigenvectors of elements $Z_{app}(t)$ of the approximate Lax–Phillips semigroup since, as its name suggests, it is not an exact semigroup. Instead, we shall prove that to a resonance pole of the quantum mechanical S-matrix $\hat{S}_{QM}(\cdot)$ in the second sheet of the complex energy Riemann surface, at a point μ with Im $\mu < 0$, there corresponds a state ψ_μ which is an approximate eigenfunction of each element of $\{Z_{app}(t)\}_{t\in\mathbb{R}_+}$ in the sense that

$$Z_{app}(t)\psi_\mu = e^{-i\mu t}\psi_\mu + \text{small corrections}, \quad t \geq 0.$$

The state ψ_μ is considered to be a resonance state corresponding to the resonance pole of $\hat{S}_{QM}(\cdot)$ at $z = \mu$. Moreover, ψ_μ is an exact eigenstate of each element of the semigroup $\{Z_+(t)\}_{t\geq 0}$ in the same way that a resonance state in the Lax–Phillips theory is an eigenstate of each element of the semigroup $\{Z(t)\}_{t\geq 0}$. By establishing a result analogous to Theorem 2.3 we obtain, in the context of quantum mechanical resonance scattering (in an appropriately defined approximate sense), a structure analogous to the framework of the Lax–Phillip theory.

The problem of resonances in quantum mechanics and the definition of appropriate resonance states corresponding to resonance poles of the scattering matrix has a long history going back to the early days of quantum mechanics. Several formalisms have been developed for dealing with the problem, notably the methods of complex scaling (Aguilar and Combes 1971; Balslev and Combes 1971; Simon 1973, 1979; Hunziker 1986; Sjöstrand and Zworski 1991; Hislop and Sigal 1996) and rigged Hilbert spaces (Bailey and Schieve 1978; Baumgartel 1975; Bohm and Gadella 1989; Horwitz and Sigal 1978; Parravincini et al. 1980). Here we consider the problem in a simple case which demonstrates the use of basic mathematical constructions of the Lax–Phillips theory and the Sz.-Nagy-Foias theory (Sz.-Nagy and Foias 1970) for the description of resonances in quantum mechanics (the context of the discussion below is essentially within the framework of the semigroup decomposition of resonance evolution Strauss 2005a, b, Strauss et al. 2006). Thus, we shall add to assumptions (i)–(iii) above the assumption:

(iv) Denote by $\mathbb{C}^+$, respectively by $\mathbb{C}^-$ the upper and lower half-planes of the complex plane $\mathbb{C}$. We assume that the S-matrix $\hat{S}_{QM}(\cdot)$, has an extension into a holomorphic operator valued function $\hat{\mathcal{S}}_{QM}(\cdot)$, in some region $\Sigma^+ \subset \mathbb{C}^+$ above the positive real axis $\mathbb{R}$ and having an analytic continuation across the interval $[a, b]$ into a region $\Sigma^- \subset \mathbb{C}^-$ such that the resulting analytically continued function, again denoted by $\hat{\mathcal{S}}_{QM}(\cdot)$, is meromorphic in an open, simply connected region $\Sigma = \Sigma^+ \cup \Sigma^- \cup (\Sigma \cap [a, b])$. We assume that $\hat{\mathcal{S}}_{QM}(\cdot)$ has a finite number of resonance poles at points $z_i = \mu_i \in \Sigma^-$, $1 \leq i \leq m$, and no other singularity in $\overline{\Sigma}$ ($\overline{\Sigma}$ is the closure of Σ). We assume furthermore that there exists an inner operator valued function $\hat{S}^{in}_{QM}$ such that $(\hat{S}^{in}_{QM})^*\hat{S}_{QM}$ has no poles in Σ^-, i.e., $(\hat{S}^{in}_{QM})^*(E)\hat{S}_{QM}(E)$, $E \in [a, b]$ is the boundary value of an operator valued function $(\hat{S}^{in}_{QM})^*(z)\hat{S}_{QM}(z)$, analytic in Σ_-. We shall call $\hat{S}^{in}_{QM}$ the inner part of $\hat{S}_{QM}$ (in Σ).

Observe that by Eqs. (4.1.16) and (4.1.13) we have

$$Z_+(t) = (\hat{W}_+^{QM})^{-1} Z_F(t)\hat{W}_+^{QM} = (\hat{W}_+^{QM})^{-1} R^* T_u(t) R\hat{W}_+^{QM} = R_+^* T_u(t) R_+, \quad t \geq 0,$$

where $R_+ : \mathcal{H}'_{ac} \mapsto \mathcal{H}^2_+(\mathbb{R}, \mathcal{K})$, given by $R_+ := R\hat{W}_+^{QM}$, is an isometric isomorphism and hence the semigroup $\{Z_+(t)\}_{t\geq 0}$ on $\mathcal{H}'_{ac}$ is unitarily equivalent to the semigroup $\{T_u(t)\}_{t\geq 0}$ on $\mathcal{H}^2_+(\mathbb{R}; \mathcal{K})$. Given the inner part $\hat{S}^{in}_{QM}$ of the scattering matrix $\hat{S}_{QM}(\cdot)$

we define subspaces

$$\hat{\mathcal{H}}_{res} := \mathcal{H}^2_+(\mathbb{R};\mathcal{K}) \ominus \hat{S}^{in}_{QM}\mathcal{H}^2_+(\mathbb{R};\mathcal{K}) \subset \mathcal{H}^2_+(\mathbb{R};\mathcal{K})$$

and

$$\mathcal{H}_{res} := R_+^*\hat{\mathcal{H}}_{res} \subset \mathcal{H}'_{ac}.$$

Note that $\{T_u(t)|_{\hat{\mathcal{H}}_{res}}\}_{t\geq 0}$ is a semigroup of Lax–Phillips type and that, by the unitary equivalence induced by R_+, this semigroup is unitarily equivalent to $\{Z_+(t)|_{\mathcal{H}_{res}}\}_{t\geq 0}$. Thus, if $P_{res}:\mathcal{H}'_{ac}\mapsto\mathcal{H}'_{ac}$ is the orthogonal projection on $\mathcal{H}_{res}$ and we define

$$Z_{QM}(t) := Z_+(t)P_{res}, \quad t\geq 0, \tag{4.2.1}$$

we get that $\{Z_{QM}(t)\}_{t\geq 0}$ is unitarily equivalent to a the Lax–Phillips type semigroup $\{\hat{Z}_{LP}(t)\}_{t\geq 0}$, $\hat{Z}_{LP}(t) := T_u(t)\hat{P}_{res}$, where $\hat{P}_{res}$ is the orthogonal projection in $\mathcal{H}^2_+(\mathbb{R},\mathcal{K})$ on $\hat{\mathcal{H}}_{res}$.

Now consider the approximate Lax–Phillips semigroup. Using Eq. (4.1.19) we have

$$\begin{aligned}Z_{app}(t) = \Lambda_+U(t)\Lambda_- = Z_+(t)\Lambda_+\Lambda_- = Z_+(t)P_{res} + Z_+(t)(\Lambda_+\Lambda_- - P_{res}) =\\ = Z_{QM}(t) + Z_+(t)(\Lambda_+\Lambda_- - P_{res}), \quad t\geq 0.\end{aligned}$$

We define a resonance state to be an eigenstate $\psi_\mu\in\mathcal{H}_{res}$ of $Z_{QM}(t)$, $t\geq 0$, such that

$$Z_{QM}(t)\psi_\mu = Z_+(t)\psi_\mu = e^{-i\mu t}\psi_\mu, \quad t\geq 0. \tag{4.2.2}$$

We then get that

$$\begin{aligned}Z_{app}(t)\psi_\mu =& Z_{QM}(t)\psi_\mu + Z_+(t)(\Lambda_+\Lambda_- - P_{res})\psi_\mu =\\ &= e^{-i\mu t}\psi_\mu + Z_+(t)(\Lambda_+\Lambda_- - P_{res})\psi_\mu = e^{-i\mu t}\psi_\mu + Z_+(t)(\Lambda_+\Lambda_-\psi_\mu - \psi_\mu),\end{aligned}$$

so that

$$\|Z_{app}(t)\tilde{\psi}_\mu - e^{-i\mu t}\tilde{\psi}_\mu\| \leq \|\tilde{\psi}_\mu - \Lambda_+\Lambda_-\tilde{\psi}_\mu\|, \tag{4.2.3}$$

where $\tilde{\psi}_\mu = \|\psi_\mu\|^{-1}\psi_\mu$ is normalized resonance state. Note that if $\Lambda_+\Lambda_-$ were an orthogonal projection operator on the resonance subspace, as in the Lax–Phillips case, we would have $\tilde{\psi}_\mu = \Lambda_+\Lambda_-\tilde{\psi}_\mu$, the right hand side of Eq. (4.2.3) would be zero and $\{Z_{app}(t)\}_{t\geq 0}$ would be an exact semigroup. The right hand side of Eq. (4.2.3) therefore measures the deviation from semigroup behavior by estimating the deviation of $\Lambda_+\Lambda_-$ from being an orthogonal projection operator on the resonance subspace.

We can now state our main result for this section:

Theorem 4.9 *Assume a scattering problem for which conditions (i)–(iv) hold. Let $\psi_\mu\in\mathcal{H}'_{ac}$ be the resonance state corresponding to the resonance pole of $\hat{S}_{QM}(\cdot)$ at $z=\mu$, set $\tilde{\psi}_\mu := \|\psi_\mu\|^{-1}\psi_\mu$ and $\psi^{app}_\mu := \Lambda_+\psi_\mu$. Let $\hat{W}^{QM}_+ : \mathcal{H}_{ac}\mapsto L^2(\sigma_{ac}(H);\mathcal{K})$*

be the mapping to the outgoing energy representation and let $\psi^{app}_{\mu,+} \in L^2([a,b];\mathcal{K})$ *be given by* $\psi^{app}_{\mu,+}(E) := [\hat{W}^{QM}_+ \psi^{app}_\mu](E)$. *For* $E, E' \in \mathbb{R}$ *set*

$$G(E,E') := \frac{(\hat{S}^{in}_{QM})(E) - (\hat{S}^{in}_{QM})(E')}{E - E'} (\hat{S}^{in}_{QM})^*(E'),$$

define an operator $\hat{G} : L^2(\mathbb{R};\mathcal{K}) \mapsto L^2(\mathbb{R};\mathcal{K})$ *by*

$$\langle g, \hat{G} f\rangle_{L^2(\mathbb{R};\mathcal{K})} := \int_{-\infty}^{\infty} dE \int_{-\infty}^{\infty} dE' \langle g(E), G(E,E') f(E')\rangle_{\mathcal{K}}, \quad f, g \in L^2(\mathbb{R};\mathcal{K}),$$

and assume that $\hat{G}$ *is bounded. Then there exists a constant* $C > 0$ *such that*

$$\|\tilde{\psi}_\mu - \Lambda_+\Lambda_-\tilde{\psi}_\mu\| \le C\left(1 - \frac{\|\psi^{app}_\mu\|^2}{\|\psi_\mu\|^2}\right)^{1/4} + \left(\int_a^b \|(I - (\hat{S}^{in}_{QM}(E))^* \hat{S}_{QM}(E)\|^2_{\mathcal{B}(\mathcal{K})} \frac{\|\psi^{app}_{\mu,+}(E)\|^2_{\mathcal{K}}}{\|\psi^{app}_\mu\|^2}\, dE\right)^{1/2} \tag{4.2.4}$$

□

Proof Let $\hat{W}^{QM}_+ : \mathcal{H}_{ac} \mapsto L^2(\sigma_{ac}(H),\mathcal{K})$ be the mapping to the outgoing energy representation corresponding to the scattering problem considered in Theorem 4.9. As mentioned in Sect. 3.1, explicit expression for the mapping $\hat{W}^{QM}_+ : \mathcal{H}_{ac} \mapsto L^2(\sigma_{ac}(H),\mathcal{K})$ is obtained by finding a complete set $\{\phi^-_{E,\xi}\}_{E\in\sigma_{ac}(H),\xi\in\Xi}$ of outgoing solutions of the Lippmann–Schwinger equation. Similarly, If $\hat{W}^{QM}_- : \mathcal{H}_{ac} \mapsto L^2(\sigma_{ac}(H),\mathcal{K})$ is the mapping to the incoming energy representation then an explicit expression for this mapping is obtained by finding a complete set $\{\phi^+_{E,\xi}\}_{E\in\sigma_{ac}(H),\xi\in\Xi}$ of incoming solutions of the Lippmann–Schwinger equation. Our starting point for the proof of Theorem 4.9 is following inequality

$$\begin{aligned}\|\tilde{\psi}_\mu - \Lambda_+\Lambda_-\tilde{\psi}_\mu\|_{\mathcal{H}'_{ac}} &= \|\tilde{\psi}_\mu - \Lambda_+\tilde{\psi}_\mu + \Lambda_+\tilde{\psi}_\mu - \Lambda_+\Lambda_-\tilde{\psi}_\mu\|_{\mathcal{H}'_{ac}} = \\ = \|(I-\Lambda_+)\tilde{\psi}_\mu + \Lambda_+(I-\Lambda_-)\tilde{\psi}_\mu\|_{\mathcal{H}'_{ac}} &\le \|(I-\Lambda_+)\tilde{\psi}_\mu\|_{\mathcal{H}'_{ac}} + \|\Lambda_+(I-\Lambda_-)\tilde{\psi}_\mu\|_{\mathcal{H}'_{ac}} \\ &\le \|(I-\Lambda_+)\tilde{\psi}_\mu\|_{\mathcal{H}'_{ac}} + \|(I-\Lambda_-)\tilde{\psi}_\mu\|_{\mathcal{H}'_{ac}}.\end{aligned} \tag{4.2.5}$$

We consider first the term $\|(I-\Lambda_-)\tilde{\psi}_\mu\|_{\mathcal{H}'_{ac}}$ on the right hand side of Eq. (4.2.5). Note that the state we consider during most of the estimates below is the unnormalized state ψ_μ and normalized state $\tilde{\psi}_\mu = \|\psi_\mu\|^{-1}\psi_\mu$ enters only at the end of the proof. Recall that $\Lambda_+ = M_+^{1/2} = (\hat{W}^{QM}_+)^{-1}\Lambda_F \hat{W}^{QM}_+$ and $\Lambda_- = M_-^{1/2} = (\hat{W}^{QM}_-)^{-1}\Lambda_B \hat{W}^{QM}_-$. We then have

$$\|(I-\Lambda_-)\psi_\mu\|_{\mathcal{H}'_{ac}} = \|(I-(\hat{W}_-^{QM})^{-1}\Lambda_B\hat{W}_-^{QM})\psi_\mu\|_{\mathcal{H}'_{ac}} = \|(\hat{W}_-^{QM})^{-1}(I-\Lambda_B)\hat{W}_-^{QM}\psi_\mu\|_{\mathcal{H}'_{ac}}$$
$$\le \|(I-\Lambda_B)\hat{W}_-^{QM}\psi_\mu\|_{L^2([a,b];\mathcal{K})} = \|(I-\Lambda_B)\hat{W}_-^{QM}(\hat{W}_+^{QM})^{-1}\hat{W}_+^{QM}\psi_\mu\|_{L^2([a,b];\mathcal{K})}$$
$$= \|(I-\Lambda_B)\hat{S}^*_{QM}\hat{W}_+^{QM}\psi_\mu\|_{L^2([a,b];\mathcal{K})} = \|(I-\Lambda_B)\hat{S}^*_{QM}\psi_{\mu,+}\|_{L^2([a,b];\mathcal{K})},$$

where we have denoted $\psi_{\mu,+} = \hat{W}_+^{QM}\psi_\mu$ and used the fact that $\hat{S}_{QM} = \hat{W}_+^{QM}(\hat{W}_-^{QM})^{-1}$. We will show that under appropriate conditions the norm $\|(I-\Lambda_B)\hat{S}^*_{QM}\psi_{\mu,+}\|_{L^2([a,b];\mathcal{K})}$ is small. Note first that by the positivity of Λ_B we have the inequality

$$(\varphi,(I+\Lambda_B)\varphi)_{L^2([a,b];\mathcal{K})} \ge \|\varphi\|^2_{L^2([a,b];\mathcal{K})}, \quad \forall\varphi\in L^2([a,b];\mathcal{K}),$$

by which we obtain that

$$\|\varphi\|^2_{L^2([a,b];\mathcal{K})} = (\varphi,(I+\Lambda_B)^{-1}(I+\Lambda_B)\varphi)_{L^2([a,b];\mathcal{K})} =$$
$$=((I+\Lambda_B)^{-1/2}\varphi,(I+\Lambda_B)(I+\Lambda_B)^{-1/2}\varphi)_{L^2([a,b];\mathcal{K})} \ge ((I+\Lambda_B)^{-1/2}\varphi,(I+\Lambda_B)^{-1/2}\varphi)_{L^2([a,b];\mathcal{K})} =$$
$$= (\varphi,(I+\Lambda_B)^{-1}\varphi)_{L^2([a,b];\mathcal{K})}, \quad \forall\varphi\in L^2([a,b];\mathcal{K}).$$

From this inequality we get that $\|(I+\Lambda_B)^{-1}\| \le 1$. Noting that

$$M_F + M_B = (P_{[a,b]}\hat{P}_+P_{[a,b]})\big|_{L^2([a,b];\mathcal{K})} + (P_{[a,b]}\hat{P}_-P_{[a,b]})\big|_{L^2([a,b];\mathcal{K})} =$$
$$= (P_{[a,b]}(\hat{P}_+ + \hat{P}_-)P_{[a,b]})\big|_{L^2([a,b];\mathcal{K})} = I,$$

where $I \equiv I_{L^2([a,b];\mathcal{K})}$) we have

$$M_F = I - M_B = (I+\Lambda_B)(I-\Lambda_B) \quad\Longrightarrow\quad I-\Lambda_B = (I+\Lambda_B)^{-1}M_F.$$

Thus,

$$\|(I-\Lambda_B)\hat{S}^*_{QM}\psi_{\mu,+}\|^2_{L^2([a,b];\mathcal{K})} =$$
$$= \|(I-\Lambda_B)^{1/2}(I-\Lambda_B)^{1/2}\hat{S}^*_{QM}\psi_{\mu,+}\|^2_{L^2([a,b];\mathcal{K})} \le \|(I-\Lambda_B)^{1/2}\hat{S}^*_{QM}\psi_{\mu,+}\|^2_{L^2([a,b];\mathcal{K})} =$$
$$= (\hat{S}^*_{QM}\psi_{\mu,+},(I-\Lambda_B)\hat{S}^*_{QM}\psi_{\mu,+})_{L^2([a,b];\mathcal{K})} = (\hat{S}^*_{QM}\psi_{\mu,+},(I+\Lambda_B)^{-1}M_F\hat{S}^*_{QM}\psi_{\mu,+})_{L^2([a,b];\mathcal{K})} =$$
$$= (\hat{S}^*_{QM}\psi_{\mu,+},(I+\Lambda_B)^{-1}\Lambda_F^2\hat{S}^*_{QM}\psi_{\mu,+})_{L^2([a,b];\mathcal{K})} =$$
$$= (\Lambda_F\hat{S}^*_{QM}\psi_{\mu,+},(I+\Lambda_B)^{-1}\Lambda_F\hat{S}^*_{QM}\psi_{\mu,+})_{L^2([a,b];\mathcal{K})} \le (\Lambda_F\hat{S}^*_{QM}\psi_{\mu,+},\Lambda_F\hat{S}^*_{QM}\psi_{\mu,+})_{L^2([a,b];\mathcal{K})} =$$
$$= (\hat{S}^*_{QM}\psi_{\mu,+},M_F\hat{S}^*_{QM}\psi_{\mu,+})_{L^2([a,b];\mathcal{K})}.$$

In the derivation of this inequality we have used the fact that $[(I-\Lambda_B),\Lambda_F] = [I-(I-M_F)^{1/2}, M_F^{1/2}] = 0$. To summarize, we have

$$\|(I-\Lambda_B)\hat{S}^*_{QM}\psi_{\mu,+}\|^2_{L^2([a,b];\mathcal{K})} \le (\hat{S}^*_{QM}\psi_{\mu,+},M_F\hat{S}^*_{QM}\psi_{\mu,+})_{L^2([a,b];\mathcal{K})}. \quad (4.2.6)$$

Plugging the definition $M_F = P_{[a,b]}\hat{P}_+P_{[a,b]}\big|_{L^2([a,b];\mathcal{K})}$ into the inequality in Eq. (4.2.6) we obtain

$$\|(I-\Lambda_B)\hat{S}^*_{\mathrm{QM}}\psi_{\mu,+}\|^2_{L^2([a,b];\mathcal{K})} \leq (\hat{S}^*_{\mathrm{QM}}\psi_{\mu,+}, P_{[a,b]}\hat{P}_+P_{[a,b]}\hat{S}^*_{\mathrm{QM}}\psi_{\mu,+})_{L^2([a,b];\mathcal{K})} =$$
$$= (P_{[a,b]}\hat{S}^*_{\mathrm{QM}}\psi_{\mu,+}, \hat{P}_+P_{[a,b]}\hat{S}^*_{\mathrm{QM}}\psi_{\mu,+})_{L^2(\mathbb{R};\mathcal{K})} = \|\hat{P}_+P_{[a,b]}\hat{S}^*_{\mathrm{QM}}\psi_{\mu,+}\|^2_{L^2(\mathbb{R};\mathcal{K})}. \tag{4.2.7}$$

Given the resonance state ψ_μ we define

$$\psi^{app}_\mu := \Lambda_+\psi_\mu, \tag{4.2.8}$$

and refer to ψ^{app}_μ as an *approximate resonance state*. Denoting $\psi^{app}_{\mu,+} := \hat{W}^{\mathrm{QM}}_+\psi^{app}_\mu$ and using Eq. (4.2.7), we have

$$\begin{aligned}\|(I-\Lambda_B)\hat{S}^*_{\mathrm{QM}}\psi_{\mu,+}\|_{L^2([a,b];\mathcal{K})} &\leq \|\hat{P}_+P_{[a,b]}\hat{S}^*_{\mathrm{QM}}\psi_{\mu,+}\|_{L^2(\mathbb{R};\mathcal{K})}\\ &\leq \|\hat{P}_+P_{[a,b]}\hat{S}^*_{\mathrm{QM}}(\psi_{\mu,+}-\psi^{app}_{\mu,+})\|_{L^2(\mathbb{R};\mathcal{K})} + \|\hat{P}_+P_{[a,b]}\hat{S}^*_{\mathrm{QM}}\psi^{app}_{\mu,+}\|_{L^2(\mathbb{R};\mathcal{K})}\\ &\leq \|\psi_{\mu,+}-\psi^{app}_{\mu,+}\|_{L^2([a,b];\mathcal{K})} + \|\hat{P}_+P_{[a,b]}\hat{S}^*_{\mathrm{QM}}\psi^{app}_{\mu,+}\|_{L^2(\mathbb{R};\mathcal{K})}\end{aligned} \tag{4.2.9}$$

In order to estimate the second term on the right hand side of Eq. (4.2.9) we express $\hat{S}_{\mathrm{QM}}$ in the form

$$\hat{S}_{\mathrm{QM}} = \hat{S}^{in}_{\mathrm{QM}}\hat{S}_1$$

where $\hat{S}^{in}_{\mathrm{QM}}$ is the inner part of $\hat{S}_{\mathrm{QM}}$. In this case we have

$$\begin{aligned}\|\hat{P}_+P_{[a,b]}\hat{S}^*_{\mathrm{QM}}\psi^{app}_{\mu,+}\|_{L^2(\mathbb{R};\mathcal{K})} &= \|\hat{P}_+P_{[a,b]}\hat{S}^*_1(\hat{S}^{in}_{\mathrm{QM}})^*\psi^{app}_{\mu,+}\|_{L^2(\mathbb{R};\mathcal{K})} =\\ &= \|\hat{P}_+P_{[a,b]}[\hat{S}^*_1(\hat{S}^{in}_{\mathrm{QM}})^* - (\hat{S}^{in}_{\mathrm{QM}})^* + (\hat{S}^{in}_{\mathrm{QM}})^*]\psi^{app}_{\mu,+}\|_{L^2(\mathbb{R};\mathcal{K})}\\ &\leq \|\hat{P}_+P_{[a,b]}(\hat{S}^*_1 - I)(\hat{S}^{in}_{\mathrm{QM}})^*\psi^{app}_{\mu,+}\|_{L^2(\mathbb{R};\mathcal{K})} + \|\hat{P}_+P_{[a,b]}(\hat{S}^{in}_{\mathrm{QM}})^*\psi^{app}_{\mu,+}\|_{L^2(\mathbb{R};\mathcal{K})}\end{aligned} \tag{4.2.10}$$

We shall first estimate the last term on the right hand side of Eq. (4.2.10). We have,

$$\begin{aligned}\|\hat{P}_+P_{[a,b]}(\hat{S}^{in}_{\mathrm{QM}})^*\psi^{app}_{\mu,+}\|^2_{L^2(\mathbb{R};\mathcal{K})} &= (P_{[a,b]}(\hat{S}^{in}_{\mathrm{QM}})^*\psi^{app}_{\mu,+}, \hat{P}_+P_{[a,b]}(\hat{S}^{in}_{\mathrm{QM}})^*\psi^{app}_{\mu,+})_{L^2(\mathbb{R};\mathcal{K})} =\\ &= \int_{-\infty}^{\infty} ((P_{[a,b]}(\hat{S}^{in}_{\mathrm{QM}})^*\psi^{app}_{\mu,+})(\sigma), (\hat{P}_+P_{[a,b]}(\hat{S}^{in}_{\mathrm{QM}})^*\psi^{app}_{\mu,+})(\sigma))_{\mathcal{K}}\,d\sigma =\\ &= -\frac{1}{2\pi i}\int_{-\infty}^{\infty}\int_{-\infty}^{\infty}\frac{1}{\sigma-\sigma'+i0+}((P_{[a,b]}(\hat{S}^{in}_{\mathrm{QM}})^*\psi^{app}_{\mu,+})(\sigma), (P_{[a,b]}(\hat{S}^{in}_{\mathrm{QM}})^*\psi^{app}_{\mu,+})(\sigma'))_{\mathcal{K}}\,d\sigma' d\sigma =\\ &= -\frac{1}{2\pi i}\int_a^b\int_a^b\frac{1}{E-E'+i0+}((\hat{S}^{in}_{\mathrm{QM}})^*(E)\psi^{app}_{\mu,+}(E), ((\hat{S}^{in}_{\mathrm{QM}})^*(E')\psi^{app}_{\mu,+}(E'))_{\mathcal{K}}\,dE' dE\,.\end{aligned} \tag{4.2.11}$$

By the unitary equivalence of the semigroups $\{Z_{QM}(t)\}_{t\geq 0}$ and $\{\hat{Z}_{LP}(t)\}_{t\geq 0}$, every eigenstate ψ_μ of $Z_{QM}(t)$ is given by $\psi_\mu = R_+^*\psi_\mu^{har}$, where ψ_μ^{har} is an eigenstate of $\hat{Z}_{LP}(t)$. Note also that, since $\{\hat{Z}_{LP}(t)\}_{t\geq 0}$ is a Lax–Phillips type semigroup, then an eigenstate ψ_μ^{har} corresponding to a pole at the point μ has the form

$$\psi_\mu^{har}(\sigma) = \frac{\mathbf{h}}{\sigma-\mu}, \quad \sigma\in\mathbb{R} \tag{4.2.12}$$

where $\mathbf{h}\in\mathcal{K}$ is defined by the inner function $\hat{S}_{QM}^{in}$. Now observe that

$$\psi_\mu^{app} = \Lambda_+\psi_\mu = \Lambda_+ R_+^*\psi_\mu^{har} = \Lambda_+(\hat{W}_+^{QM})^{-1}R^*\psi_\mu^{har} = (\hat{W}_+^{QM})^{-1}\Lambda_F\hat{W}_+^{QM}(\hat{W}_+^{QM})^{-1}R^*\psi_\mu^{har} = \\ = (\hat{W}_+^{QM})^{-1}\Lambda_F R^*\psi_\mu^{har}$$

Furthermore, by Eq. (4.1.10) for every function $f\in\Omega_f L^2([a,b];\mathcal{K})$ we have

$$R^* f = \Lambda_F^{-1}\Omega_f^* f.$$

and hence for every $f\in\Omega_f L^2([a,b];\mathcal{K})$ we obtain

$$\Lambda_F R^* f = \Lambda_F\Lambda_F^{-1}\Omega_f^* f = \Omega_f^* f.$$

Since $\Omega_f L^2([a,b];\mathcal{K})$ is dense in $\mathcal{H}_+^2(\mathbb{R};\mathcal{K})$ the above relation can be extended to the whole of $\mathcal{H}_+^2(\mathbb{R};\mathcal{K})$. Thus, we get that

$$\psi_\mu^{app} = (\hat{W}_+^{QM})^{-1}\Lambda_F R^*\psi_\mu^{har} = (\hat{W}_+^{QM})^{-1}\Omega_f^*\psi_\mu^{har} \Rightarrow \psi_{\mu,+}^{app} = \hat{W}_+^{QM}\psi_\mu^{app} = \Omega_f^*\psi_\mu^{har},$$

and hence by Eqs. (3.1.14) and (4.2.12),

$$\psi_{\mu,+}^{app}(E) = \hat{W}_+^{QM}\psi_\mu^{app} = \frac{\mathbf{h}}{E-\mu}, \quad E\in[a,b]. \tag{4.2.13}$$

Note also that

$$\|\psi_\mu^{app}\| = \|\psi_{\mu,+}^{app}\| = \left(\int_a^b \frac{\|\mathbf{h}\|_{\mathcal{K}}^2}{|E-\mu|^2}dE\right)^{1/2}, \quad \|\psi_\mu\| = \|\psi_{\mu,+}\| = \|\psi_\mu^{har}\| = \left(\int_{-\infty}^{\infty} \frac{\|\mathbf{h}\|_{\mathcal{K}}^2}{|E-\mu|^2}dE\right)^{1/2}. \tag{4.2.14}$$

We shall use these norms in our proof below. Inserting the above expression for $\psi_{\mu,+}^{app}(E)$ into Eq. (4.2.11) we obtain

$$\|\hat{P}_+ P_{[a,b]}(\hat{S}^{in}_{QM})^* \psi^{app}_{\mu,+}\|^2_{L^2(\mathbb{R};\mathcal{K})} =$$
$$= -\frac{1}{2\pi i}\int_a^b\int_a^b \frac{1}{E-E'+i0^+}\left\langle (\hat{S}^{in}_{QM})^*(E)\frac{\mathbf{h}}{E-\mu}, (\hat{S}^{in}_{QM})^*(E')\frac{\mathbf{h}}{E'-\mu}\right\rangle_{\mathcal{K}} dE'dE =$$
$$= -\frac{1}{2\pi i}\int_a^b\int_a^b \frac{1}{E-\overline{\mu}}\cdot\frac{1}{E-E'+i0^+}\cdot\frac{1}{E'-\mu}\left\langle (\hat{S}^{in}_{QM})^*(E)\mathbf{h}, (\hat{S}^{in}_{QM})^*(E')\mathbf{h}\right\rangle_{\mathcal{K}} dE'dE \tag{4.2.15}$$

At this point we make use of the fact that $\psi^{har}_{\mu}(E) = \frac{\mathbf{h}}{E-\mu}$, $E \in \mathbb{R}$ and the fact that $\hat{S}^{in}_{QM}(\cdot)$ is an inner operator such that $(\hat{S}^{in}_{QM})^*\psi^{har}_{\mu} \in \mathcal{H}^2_-(\mathbb{R};\mathcal{K})$ and therefore $\hat{P}_+(\hat{S}^{in}_{QM})^*\psi^{har}_{\mu} = 0$. Thus, if we extend the integration over E' to the whole of $\mathbb{R}$ we obtain

$$\|\hat{P}_+ P_{[a,b]}(\hat{S}^{in}_{QM})^* \psi^{app}_{\mu,+}\|^2_{L^2(\mathbb{R};\mathcal{K})} =$$
$$= -\frac{1}{2\pi i}\int_a^b\int_{-\infty}^{\infty} \frac{1}{E-E'+i0^+}\left\langle (\hat{S}^{in}_{QM})^*(E)\frac{\mathbf{h}}{E-\mu}, (\hat{S}^{in}_{QM})^*(E')\frac{\mathbf{h}}{E'-\mu}\right\rangle_{\mathcal{K}} dE'dE =$$
$$+ \frac{1}{2\pi i}\int_a^b\int_{\mathbb{R}\setminus[a,b]} \frac{1}{E-E'+i0^+}\left\langle (\hat{S}^{in}_{QM})^*(E)\frac{\mathbf{h}}{E-\mu}, (\hat{S}^{in}_{QM})^*(E')\frac{\mathbf{h}}{E'-\mu}\right\rangle_{\mathcal{K}} dE'dE =$$
$$= \frac{1}{2\pi i}\int_a^b\int_{\mathbb{R}\setminus[a,b]} \frac{1}{E-E'+i0^+}\left\langle (\hat{S}^{in}_{QM})^*(E)\frac{\mathbf{h}}{E-\mu}, (\hat{S}^{in}_{QM})^*(E')\frac{\mathbf{h}}{E'-\mu}\right\rangle_{\mathcal{K}} dE'dE \tag{4.2.16}$$

Next, we note that

$$\|\hat{P}_+ P_{[a,b]}(\hat{S}^{in}_{QM})^* \psi^{app}_{\mu,+}\|^2_{L^2(\mathbb{R},\mathcal{K})} =$$
$$= \mathrm{Re}\left[\frac{1}{2\pi i}\int_a^b dE \int_{\mathbb{R}\setminus[a,b]} dE' \frac{1}{E-\overline{\mu}}\cdot\frac{1}{E-E'+i0^+}\cdot\frac{1}{E'-\mu}\langle(\hat{S}^{in}_{QM})^*(E)\mathbf{h}, (\hat{S}^{in}_{QM})^*(E')\mathbf{h}\rangle_{\mathcal{K}}\right] =$$
$$= \lim_{\epsilon\to 0^+}\mathrm{Re}\left[\frac{1}{2\pi i}\int_a^b dE \int_{\mathbb{R}\setminus[a,b]} dE' \frac{1}{E-\overline{\mu}}\cdot\frac{(E-E')-i\epsilon}{(E-E')^2+\epsilon^2}\cdot\frac{1}{E'-\mu}\langle(\hat{S}^{in}_{QM})^*(E)\mathbf{h}, (\hat{S}^{in}_{QM})^*(E')\mathbf{h}\rangle_{\mathcal{K}}\right] =$$
$$= -\frac{1}{2\pi}\lim_{\epsilon\to 0^+}\left[\int_a^b dE \int_{\mathbb{R}\setminus[a,b]} dE' \frac{(E-E')}{(E-E')^2+\epsilon^2}\times\right.$$
$$\left[\mathrm{Re}\left(\frac{1}{E-\overline{\mu}}\frac{1}{E'-\mu}\right)\mathrm{Im}\langle(\hat{S}^{in}_{QM})^*(E)\mathbf{h}, (\hat{S}^{in}_{QM})^*(E')\mathbf{h}\rangle_{\mathcal{K}}\right.$$
$$\left.\left.+\,\mathrm{Im}\left(\frac{1}{E-\overline{\mu}}\frac{1}{E'-\mu}\right)\mathrm{Re}\langle(\hat{S}^{in}_{QM})^*(E)\mathbf{h}, (\hat{S}^{in}_{QM})^*(E')\mathbf{h}\rangle_{\mathcal{K}}\right]\right.$$

Furthermore, we have

$$-\frac{1}{2\pi}\lim_{\epsilon\to 0^+}\int_a^b dE\int_{\mathbb{R}\setminus[a,b]} dE'\frac{(E-E')}{(E-E')^2+\epsilon^2}\mathrm{Re}\left(\frac{1}{E-\overline{\mu}}\frac{1}{E'-\mu}\right)\mathrm{Im}\langle(\hat{S}^{in}_{QM})^*(E)\mathbf{h},(\hat{S}^{in}_{QM})^*(E')\mathbf{h}\rangle_{\mathcal{K}}=$$

$$=-\frac{1}{2\pi}\lim_{\epsilon\to 0^+}\Bigg[\int_a^b dE\int_{\mathbb{R}\setminus[a,b]} dE'\cdot\frac{(E-E')^2}{(E-E')^2+\epsilon^2}\mathrm{Re}\Big(\frac{1}{E-\overline{\mu}}\cdot\frac{1}{E'-\mu}\Big)\times$$
$$\mathrm{Im}\Big\langle\mathbf{h},\frac{(\hat{S}^{in}_{QM})(E)-(\hat{S}^{in}_{QM})(E')}{E-E'}(\hat{S}^{in}_{QM})^*(E')\mathbf{h}\Big\rangle_{\mathcal{K}}\Bigg]=$$

$$=-\frac{1}{2\pi}\int_a^b dE\int_{\mathbb{R}\setminus[a,b]} dE'\,\mathrm{Re}\Big(\frac{1}{E-\overline{\mu}}\cdot\frac{1}{E'-\mu}\Big)\mathrm{Im}\Big\langle\mathbf{h},\frac{(\hat{S}^{in}_{QM})(E)-(\hat{S}^{in}_{QM})(E')}{E-E'}(\hat{S}^{in}_{QM})^*(E')\mathbf{h}\Big\rangle_{\mathcal{K}}=$$

$$=-2\mathrm{Re}\Big[\frac{1}{8\pi i}\int_a^b dE\int_{\mathbb{R}\setminus[a,b]} dE'\Big\langle\frac{\mathbf{h}}{E-\mu},G(E,E')\frac{\mathbf{h}}{E'-\mu}\Big\rangle_{\mathcal{K}}\Big]$$
$$-2\mathrm{Re}\Big[\frac{1}{8\pi i}\int_a^b dE\int_{\mathbb{R}\setminus[a,b]} dE'\Big\langle\frac{\mathbf{h}}{E-\overline{\mu}},G(E,E')\frac{\mathbf{h}}{E'-\overline{\mu}}\Big\rangle_{\mathcal{K}}\Big]$$

where

$$G(E,E'):=\frac{(\hat{S}^{in}_{QM})(E)-(\hat{S}^{in}_{QM})(E')}{E-E'}(\hat{S}^{in}_{QM})^*(E').$$

Now define an operator $\hat{G}\,:\,L^2(\mathbb{R};\mathcal{K})\mapsto L^2(\mathbb{R};\mathcal{K})$ by

$$\langle g,\hat{G}f\rangle_{L^2(\mathbb{R};\mathcal{K})}:=\int_{-\infty}^{\infty} dE\int_{-\infty}^{\infty} dE'\langle g(E),G(E,E')f(E')\rangle_{\mathcal{K}},\quad f,g\in L^2(\mathbb{R};\mathcal{K}).$$

and assume that $\hat{G}$ is bounded, i.e., there exists $C_1>0$ such that $\|\hat{G}\|<C_1$. Under this assumption we obtain

$$\left|\frac{1}{2\pi}\lim_{\epsilon\to 0^+}\int_a^b dE\int_{\mathbb{R}\setminus[a,b]} dE'\frac{(E-E')}{(E-E')^2+\epsilon^2}\mathrm{Re}\left(\frac{1}{E-\overline{\mu}}\frac{1}{E'-\mu}\right)\mathrm{Im}\langle(\hat{S}^{in}_{QM})^*(E)\mathbf{h},(\hat{S}^{in}_{QM})^*(E')\mathbf{h}\rangle_{\mathcal{K}}\right|=$$

$$\le\frac{1}{4\pi}\Big|\int_a^b dE\int_{\mathbb{R}\setminus[a,b]} dE'\Big\langle\frac{\mathbf{h}}{E-\mu},G(E,E')\frac{\mathbf{h}}{E'-\mu}\Big\rangle_{\mathcal{K}}\Big|$$
$$+\frac{1}{4\pi}\Big|\int_a^b dE\int_{\mathbb{R}\setminus[a,b]} dE'\Big\langle\frac{\mathbf{h}}{E-\overline{\mu}},G(E,E')\frac{\mathbf{h}}{E'-\overline{\mu}}\Big\rangle_{\mathcal{K}}\Big|$$
$$\le\frac{C_1}{2\pi}\Big(\int_a^b\frac{\|\mathbf{h}\|^2_{\mathcal{K}}}{|E-\mu|^2}dE\Big)^{1/2}\Big(\int_{\mathbb{R}\setminus[a,b]}\frac{\|\mathbf{h}\|^2_{\mathcal{K}}}{|E'-\mu|^2}dE'\Big)^{1/2}.$$

In addition, we have the following estimate

$$\Bigg|\frac{1}{2\pi}\lim_{\epsilon\to 0^+}\Bigg[\int_a^b dE\int_{\mathbb{R}\setminus[a,b]} dE'\frac{(E-E')}{(E-E')^2+\epsilon^2}\times$$

$$\mathrm{Im}\left(\frac{1}{E-\overline{\mu}}\frac{1}{E'-\mu}\right)\mathrm{Re}\langle(\hat{S}^{in}_{QM})^*(E)\mathbf{h},(\hat{S}^{in}_{QM})^*(E')\mathbf{h}\rangle_{\mathcal{K}}\Bigg|$$

$$\leq\frac{|\mathrm{Im}\,\mu|}{2\pi\|\mathbf{h}\|^2_{\mathcal{K}}}\left(\int_a^b dE\frac{\|\mathbf{h}\|^2_{\mathcal{K}}}{|E-\mu|^2}\right)\left(\int_{\mathbb{R}\setminus[a,b]} dE'\frac{\|\mathbf{h}\|^2_{\mathcal{K}}}{|E'-\mu|^2}\right)$$

Hence we get that

$$\|\hat{P}_+P_{[a,b]}(\hat{S}^{in}_{QM})^*\psi^{app}_{\mu,+}\|^2_{L^2(\mathbb{R},\mathcal{K})}\leq\frac{C_1}{2\pi}\left(\int_a^b dE\frac{\|\mathbf{h}\|^2_{\mathcal{K}}}{|E-\mu|^2}\right)^{1/2}\left(\int_{\mathbb{R}\setminus[a,b]} dE'\frac{\|\mathbf{h}\|^2_{\mathcal{K}}}{|E'-\mu|^2}\right)^{1/2}$$

$$+\frac{|\mathrm{Im}\,\mu|}{2\pi\|\mathbf{h}\|^2_{\mathcal{K}}}\left(\int_a^b dE\frac{\|\mathbf{h}\|^2_{\mathcal{K}}}{|E-\overline{\mu}|^2}\right)\left(\int_{\mathbb{R}\setminus[a,b]} dE'\frac{\|\mathbf{h}\|^2_{\mathcal{K}}}{|E'-\mu|^2}\right) \tag{4.2.17}$$

Going back to Eq. (4.2.10), we want to estimate the first term on the right hand side of the inequality. We have

$$\|\hat{P}_+P_{[a,b]}(\hat{S}^*_1-I)(\hat{S}^{in}_{QM})^*\psi^{app}_{\mu,+}\|_{L^2(\mathbb{R};\mathcal{K})}=\|\hat{P}_+P_{[a,b]}\hat{S}^*_1(I-\hat{S}_1)(\hat{S}^{in}_{QM})^*\psi^{app}_{\mu,+}\|_{L^2(\mathbb{R};\mathcal{K})}$$

$$\leq\|(I-\hat{S}_1)(\hat{S}^{in}_{QM})^*\psi^{app}_{\mu,+}\|_{L^2([a,b];\mathcal{K})}=\left(\int_a^b\|(I-\hat{S}_1(E))(\hat{S}^{in}_{QM})^*(E)\psi^{app}_{\mu,+}(E)\|^2_{\mathcal{K}}\,dE\right)^{1/2}$$

$$\leq\left(\int_a^b\|(I-(\hat{S}^{in}_{QM}(E))^*\hat{S}_{QM}(E)\|^2_{\mathcal{B}(\mathcal{K})}\|\psi^{app}_{\mu,+}(E)\|^2_{\mathcal{K}}\,dE\right)^{1/2} \tag{4.2.18}$$

Equations (4.2.17) and (4.2.18) provide upper bounds on the two terms on the right hand side of Eq. (4.2.10). We are left with the task of estimating the first term on the right hand side of Eq. (4.2.9). First, we have

$$\|\psi_{\mu,+}-\psi^{app}_{\mu,+}\|_{L^2([a,b];\mathcal{K})}=\|\psi_\mu-\psi^{app}_\mu\|\leq\|(I+\Lambda_+)(\psi_\mu-\psi^{app}_\mu)\|=$$
$$=\|\psi_\mu+\Lambda_+\psi_\mu-\psi^{app}_\mu-\Lambda_+\psi^{app}_\mu\|=\|\psi_\mu+\psi^{app}_\mu-\psi^{app}_\mu-\Lambda_+\psi^{app}_\mu\|=\|\psi_\mu-\Lambda_+\psi^{app}_\mu\|.$$

Using the orthogonal projection operator on the subspace spanned by ψ_μ we get that

$$\Lambda_+\psi^{app}_\mu=b(\psi^{app}_\mu)+\frac{1}{\|\psi_\mu\|^2}\langle\psi_\mu,\Lambda_+\psi^{app}_\mu\rangle\psi_\mu=b(\psi^{app}_\mu)+\frac{\|\psi^{app}_\mu\|^2}{\|\psi_\mu\|^2}\psi_\mu$$

where $b(\psi_\mu^{app}) := \Lambda_+\psi_\mu^{app} - \|\psi_\mu\|^{-2}\langle\psi_\mu, \Lambda_+\psi_\mu^{app}\rangle\psi_\mu$. Moreover, by the orthogonality of $b(\psi_\mu^{app})$ and ψ_μ we obtain

$$\|\psi_\mu\|^2 \geq \|\psi_\mu^{app}\|^2 \geq \|\Lambda_+\psi_\mu^{app}\|^2 = \|b(\psi_\mu^{app})\|^2 + \frac{\|\psi_\mu^{app}\|^4}{\|\psi_\mu\|^2}$$
$$\Rightarrow \|b(\psi_\mu^{app})\|^2 \leq \|\psi_\mu\|^2 - \frac{\|\psi_\mu^{app}\|^4}{\|\psi_\mu\|^2} = \left(1 - \frac{\|\psi_\mu^{app}\|^4}{\|\psi_\mu\|^4}\right)\|\psi_\mu\|^2 \leq 2\left(1 - \frac{\|\psi_\mu^{app}\|^2}{\|\psi_\mu\|^2}\right)\|\psi_\mu\|^2$$

where we recall that $\|\psi_\mu^{app}\| \leq \|\psi_\mu\|$. Hence, we get that

$$\|\psi_{\mu,+} - \psi_{\mu,+}^{app}\|_{L^2(\mathbb{R}_+)} \leq \|\psi_\mu - \Lambda_+\psi_\mu^{app}\| = \left\|\psi_\mu - \left(b(\psi_\mu^{app}) + \frac{\|\psi_\mu^{app}\|^2}{\|\psi_\mu\|^2}\psi_\mu\right)\right\|$$
$$\leq \left(1 - \frac{\|\psi_\mu^{app}\|^2}{\|\psi_\mu\|^2}\right)\|\psi_\mu\| + \sqrt{2}\left(1 - \frac{\|\psi_\mu^{app}\|^2}{\|\psi_\mu\|^2}\right)^{1/2}\|\psi_\mu\| \leq (1+\sqrt{2})\left(1 - \frac{\|\psi_\mu^{app}\|^2}{\|\psi_\mu\|^2}\right)^{1/2}\|\psi_\mu\| \tag{4.2.19}$$

Note that the bound in Eq. (4.2.19) applies also to the first term on the right hand side of Eq. (4.2.5). Using the estimates in Eqs. (4.2.5), (4.2.9), (4.2.10), (4.2.17), (4.2.18) and (4.2.19) we obtain

$$\begin{aligned}\|\psi_\mu - \Lambda_+\Lambda_-\psi_\mu\| \leq\ & (1+\sqrt{2})\left(1 - \frac{\|\psi_\mu^{app}\|^2}{\|\psi_\mu\|^2}\right)^{1/2}\|\psi_\mu\| + (1+\sqrt{2})\left(1 - \frac{\|\psi_\mu^{app}\|^2}{\|\psi_\mu\|^2}\right)^{1/2}\|\psi_\mu\| \\ & + \left(\int_a^b \|(I - (\hat{S}_{QM}^{in}(E))^*\hat{S}_{QM}(E)\|^2_{\mathcal{B}(\mathcal{K})}\|\psi_{\mu,+}^{app}(E)\|^2_{\mathcal{K}}\,dE\right)^{1/2} \\ & + \left[\frac{C_1}{2\pi}\left(\int_a^b dE\,\frac{\|\mathbf{h}\|^2_{\mathcal{K}}}{|E-\mu|^2}\right)^{1/2}\left(\int_{\mathbb{R}\setminus[a,b]} dE'\,\frac{\|\mathbf{h}\|^2_{\mathcal{K}}}{|E'-\mu|^2}\right)^{1/2}\right. \\ & \left. + \frac{|\mathrm{Im}\,\mu|}{2\pi\|\mathbf{h}\|^2_{\mathcal{K}}}\left(\int_a^b dE\,\frac{\|\mathbf{h}\|^2_{\mathcal{K}}}{|E-\overline{\mu}|^2}\right)\left(\int_{\mathbb{R}\setminus[a,b]} dE'\,\frac{\|\mathbf{h}\|^2_{\mathcal{K}}}{|E'-\mu|^2}\right)\right]^{1/2}\end{aligned} \tag{4.2.20}$$

Finally, we use the norms in Eq. (4.2.14) to obtain

$$\begin{aligned}\|\psi_\mu - \Lambda_+\Lambda_-\psi_\mu\| \leq\ & 2(1+\sqrt{2})\left(1 - \frac{\|\psi_\mu^{app}\|^2}{\|\psi_\mu\|^2}\right)^{1/2}\|\psi_\mu\| \\ & + \left(\int_a^b \|(I - (\hat{S}_{QM}^{in}(E))^*\hat{S}_{QM}(E)\|^2_{\mathcal{B}(\mathcal{K})}\|\psi_{\mu,+}^{app}(E)\|^2_{\mathcal{K}}\,dE\right)^{1/2} \\ & + \left[\frac{C_1}{2\pi}\|\psi_\mu^{app}\|(\|\psi_\mu\|^2 - \|\psi_\mu^{app}\|^2)^{1/2} + \frac{1}{2}\|\psi_\mu\|^{-2}\|\psi_\mu^{app}\|^2(\|\psi_\mu\|^2 - \|\psi_\mu^{app}\|^2)\right]^{1/2} =\end{aligned}$$

$$\leq 2(1+\sqrt{2})\left(1-\frac{\|\psi_\mu^{app}\|^2}{\|\psi_\mu\|^2}\right)^{1/2}\|\psi_\mu\|$$
$$+\left(\int_a^b \|(I-(\hat{S}^{in}_{QM}(E))^*\hat{S}_{QM}(E)\|^2_{\mathcal{B}(\mathcal{K})}\|\psi^{app}_{\mu,+}(E)\|^2_{\mathcal{K}}\,dE\right)^{1/2}$$
$$+\|\psi_\mu\|\left[\frac{C_1}{2\pi}\left(1-\frac{\|\psi_\mu^{app}\|^2}{\|\psi_\mu\|^2}\right)^{1/2}+\frac{1}{2}\left(1-\frac{\|\psi_\mu^{app}\|^2}{\|\psi_\mu\|^2}\right)\right]^{1/2}$$
$$\leq C\left(1-\frac{\|\psi_\mu^{app}\|^2}{\|\psi_\mu\|^2}\right)^{1/4}\|\psi_\mu\|+\left(\int_a^b \|(I-(\hat{S}^{in}_{QM}(E))^*\hat{S}_{QM}(E)\|_{\mathcal{B}(\mathcal{K})}\|\psi^{app}_{\mu,+}(E)\|^2_{\mathcal{K}}\,dE\right)^{1/2} \tag{4.2.21}$$

Dividing by the norm of ψ_μ we finally obtain

$$\|\tilde{\psi}_\mu-\Lambda_+\Lambda_-\tilde{\psi}_\mu\|\leq C\left(1-\frac{\|\psi_\mu^{app}\|^2}{\|\psi_\mu\|^2}\right)^{1/4}$$
$$+\left(\int_a^b \|(I-(\hat{S}^{in}_{QM}(E))^*\hat{S}_{QM}(E)\|^2_{\mathcal{B}(\mathcal{K})}\frac{\|\psi^{app}_{\mu,+}(E)\|^2_{\mathcal{K}}}{\|\psi_\mu^{app}\|^2}\,dE\right)^{1/2} \tag{4.2.22}$$

■

Let us consider the two terms on the right hand side of Eq. (4.2.4). First, note that since $\psi_\mu^{app}=\Lambda_+\psi_\mu$ and since Λ_+ is contractive we always have $\|\psi_\mu^{app}\|_{\mathcal{H}'_{ac}}\leq\|\psi_\mu\|_{\mathcal{H}'_{ac}}$ and hence the first term on the right hand side of Eq. (4.2.4) is always well defined. Moreover, for a sharp resonance, i.e., a resonance pole close to the real axis in the complex energy plane and such that $\mathrm{Re}\,\mu\in(a,b)$, the first term on the right hand side of Eq. (4.2.4) is small. This is clear from the norms given in Eq. (4.2.14),

$$\|\psi_\mu^{app}\|_{\mathcal{H}'_{ac}}=\left(\int_a^b\frac{\|\mathbf{h}\|^2_{\mathcal{K}}}{|E-\mu|^2}dE\right)^{1/2},\quad \|\psi_\mu\|_{\mathcal{H}'_{ac}}=\left(\int_{-\infty}^{\infty}\frac{\|\mathbf{h}\|^2_{\mathcal{K}}}{|E-\mu|^2}dE\right)^{1/2},$$

and the fact that, for $\mathrm{Re}(\mu)\in(a,b)$ constant, we have $\lim_{\mathrm{Im}\,\mu\to 0}\|\psi_\mu^{app}\|_{\mathcal{H}'_{ac}}=\|\psi_\mu\|_{\mathcal{H}'_{ac}}$ (see Strauss 2005b). As for the second term on the right hand side of Eq. (4.2.4) we recall assumption (iv) above. In the Lax–Phillips theory, the fact that the scattering matrix is an inner function and the fact that the spectrum of the generator of evolution is $\mathbb{R}$ produce ideal, Breit–Wigner shaped, resonance profiles producing ideal phase shifts. In the quantum mechanical case, if the scattering matrix is equal to its inner part, i.e.

$$\hat{S}_{QM}(E)=\hat{S}^{in}_{QM}(E),\quad E\geq 0,$$

then the produced phase shifts, at least for sharp resonances whose energy is distributed mainly in the interval $[a,b]$, are very close to these ideal phase shifts. Note that in this case we would have

$$I - (\hat{S}^{in}_{QM}(E))^* \hat{S}_{QM}(E) = I - (\hat{S}^{in}_{QM}(E))^* \hat{S}^{in}_{QM}(E) = 0$$

and the second term on the right hand side of Eq. (4.2.4) is zero. Thus the second term on the right hand side of Eq. (4.2.4) measures the extent to which the phase shifts produced by the resonance pole at $z = \mu$ deviate from ideal phase shifts. In the integral defining the second term on the right hand side of Eq. (4.2.4) the factor $\|I - (\hat{S}^{in}_{QM}(E))^* \hat{S}_{QM}(E)\|^2_{\mathcal{B}(\mathcal{K})}$ is multiplied by the probability density function $\|\psi^{app}_\mu\|^{-2}_{\mathcal{H}'_{ac}} |\psi^{app}_{\mu,+}(E)|^2$ which has a peak at the resonance energy. Hence the multiplication with the energy probability density of ψ^{app}_μ implies that the deviations from ideal phase shifts are evaluated in the vicinity of the resonance energy. We conclude that the second term on the right hand side of Eq. (4.2.4) provides a good measure of the deviation of the phase shifts of the resonance at $z = \mu$ from the phase shifts of an ideal resonance. Note that even if the scattering matrix $\hat{S}_{QM}$ has the best possible form, i.e., that of an inner function, we still do not get an exact semigroup behavior when we apply the approximate Lax–Phillips semigroup to the resonance state ψ_μ unless the interval $[a, b]$ can be extended to the full real line $\mathbb{R}$.

4.3 Modified Lax–Phillips Theory: Explicit Representations

To demonstrate the usefulness of the Modified Lax–Phillips formalism we consider the important case of a scattering system satisfying assumptions (i)–(iii) with $\sigma_{ac}(H) = \sigma_{ac}(H_0) = \mathbb{R}_+$, where we assume that the spectrum is of uniform multiplicity on $\mathbb{R}_+$ and take $[a, b] = \overline{\mathbb{R}}_+$. In the context of such a scattering system we shall find explicit formula for the central objects of the modified Lax–Phillips structure, such as the operators M_+, M_-, Λ_+, Λ_-, the incoming and outgoing transition representations, the approximate Lax–Phillips semigroup $\{Z_{app}(t)\}_{t\geq 0}$ etc., and apply the explicit expressions we obtain to a simple scattering problem. The starting point for this discussion are the spectral representations of the Lyapunov operators M_F and M_B. Under the restrictions above we have the following result (see Strauss et al. 2011a):

Theorem 4.10 *Let $\sigma_{ac}(H) = \sigma_{ac}(H_0) = \mathbb{R}_+$ with uniform multiplicity and $[a, b] = \overline{\mathbb{R}}_+$. Then the spectrum of M_F satisfies $\sigma(M_F) = \sigma_{ac}(M_F) = [0, 1]$. For $m \in (0, 1)$ and $\xi \in \Xi$, let $\mathbf{g}^f_{m,\xi}$ be the function defined on $\mathbb{R}_+$ by*

$$\mathbf{g}^f_{m,\xi}(E) = \frac{E^{-\frac{i}{2\pi}\ln\left(\frac{1-m}{m}\right)-\frac{1}{2}}}{2\pi\sqrt{m\,(1-m)}}\,|\mathbf{e}_{E,\xi}\rangle, \quad m \in (0, 1), \; E > 0, \tag{4.3.1}$$

(ξ being the degeneracy indices for the energy E) then $\mathbf{g}^f_{m,\xi}$ is a generalized eigenfunction of M_F satisfying $M_F\,\mathbf{g}^f_{m,\xi} = m\,\mathbf{g}^f_{m,\xi}$. These eigenfunctions are normalized in such a way that $\langle \mathbf{g}^f_{m,\xi}, \mathbf{g}^f_{m',\xi'}\rangle_{L^2(\mathbb{R}_+;\mathcal{K})} = \delta(m-m')\delta_{\xi\xi'}$ and (using the Dirac notation) we have the eigenfunction expansion

$$M_F = \sum_{\xi\in\Xi}\int_0^1 dm\, m\, |\mathbf{g}^f_{m,\xi}\rangle\,\langle\mathbf{g}^f_{m,\xi}|, \tag{4.3.2}$$

where summation over ξ stands for discrete summation for discrete indices and appropriate integration over continuous indices. □

Proof Theorem 3.4 states that M_F is positive and contractive, implying that $M_F \subset [0,1]$. Consider the eigenvalue equation $M_F\,\mathbf{g}_{m,\xi} = m\,\mathbf{g}_{m,\xi}$. According to Eq. (3.1.17), in the basis $\{\mathbf{e}_{E,\xi}\}_{E\geq 0;\xi\in\Xi}$ the kernel of M_F is given by

$$\langle \mathbf{e}_{E,\xi}, M_F\mathbf{e}_{E',\xi'}\rangle = -\frac{1}{2\pi i}\frac{1}{E-E'+i0^+}\delta_{\xi\xi'}\,.$$

Hence the eigenvalue equation for M_F assumes in this case the form

$$-\frac{1}{2\pi i}\int_0^\infty \frac{1}{E-E'+i0^+}\,g_{m,\xi}(E') = m\,g_{m,\xi}(E), \quad E\geq 0, \tag{4.3.3}$$

where $\langle \mathbf{e}_{E,\xi'}, \mathbf{g}_{m,\xi}\rangle = g_{m,\xi}(E)\delta_{\xi\xi'}$ and m is an arbitrary real number in the interval $(0,1)$. The left hand side of Eq. (4.3.3) implies that any non trivial solution of this equation is necessarily the boundary value on $\mathbb{R}^+$ from above of an analytic function defined on $\mathbb{C}\backslash\mathbb{R}^+$. Let $\tilde{g}_{m\xi}(z)$ be such an analytic continuation of an arbitrary solution $g_{m,\xi}(E)$ so that $g_{m,\xi}(E) = \tilde{g}_{m,\xi}(E+i0^+)$, $\forall E\in\mathbb{R}_+$. We can now analytically continue Eq. (4.3.3) into the cut plane to obtain

$$-\frac{1}{2\pi i}\int_0^\infty \frac{1}{z-E'}\,g_{m,\xi}(E')dE' = m\,\tilde{g}_{m,\xi}(z), \quad z\in\mathbb{C}\backslash\mathbb{R}^+. \tag{4.3.4}$$

Taking the difference between the limits from above and below $\mathbb{R}^+$ in Eq. (4.3.4) we get

$$g_{m,\xi}(E) = \tilde{g}_{m,\xi}(E+i0^+) = m(\tilde{g}_{m,\xi}(E+i0^+) - \tilde{g}_{m,\xi}(E-i0^+))\,.$$

The function $\tilde{g}_{m\xi}(z)$ can now be continued across the cut along $[0,\infty)$. Denoting the analytically continued two sheeted function again by $\tilde{g}_{m,\xi(x)}$ we obtain the equation

$$\tilde{g}_{m,\xi}(e^{2\pi i}z) = -\frac{1-m}{m}\,\tilde{g}_{m,\xi}(z)\,. \tag{4.3.5}$$

Equation (4.3.5) admits solutions of the form $\tilde{g}_{m,\xi}(z) = N_m z^\beta$ with $\beta = k + \frac{1}{2} - \frac{i}{2\pi} \ln\left(\frac{1-m}{m}\right)$ where $k \in \mathbb{Z}$ and N_m is a normalization factor. Setting $k = -1$ and $N_m = (4\pi^2)m(1-m))^{-1/2}$ the solutions to the eigenvalue equation, Eq. (4.3.3) satisfy $\int_0^\infty \overline{g_{m,\xi}(E)} g_{m',\xi'}(E) dE = \delta(E - E')\delta_{\xi\xi'}$. With this choice of k and M_m the generalized eigenfunctions of M_F are given by Eq. (4.3.1). Note that these generalized eigenfunctions are not valid for the values $m = 0$ and $m = 1$.

Next we shall show that the set of functions in Eq. (4.3.1) forms a complete set of generalized eigenfunction of M_F. We have

$$
\begin{aligned}
&\sum_{\xi \in \Xi} \int_0^1 m \, |\mathbf{g}^f_{m,\xi}\rangle \langle \mathbf{g}^f_{m,\xi}| \, dm = \\
&= \sum_{\xi\Xi} \int_0^1 dm\, m \int_0^\infty dE \, \frac{E^{-\frac{i}{2\pi}\ln\left(\frac{1-m}{m}\right) - \frac{1}{2}}}{2\pi\sqrt{m(1-m)}} |\mathbf{e}_{E,\xi}\rangle \int_0^\infty dE' \, \frac{E'^{\frac{i}{2\pi}\ln\left(\frac{1-m}{m}\right) - \frac{1}{2}}}{2\pi\sqrt{m(1-m)}} \langle \mathbf{e}_{E',\xi}| = \\
&= \sum_{\xi\Xi} \int_0^\infty dE \int_0^\infty dE' \frac{|\mathbf{e}_{E,\xi}\rangle\langle \mathbf{e}_{E',\xi}|}{4\pi^2\sqrt{EE'}} \int_0^1 dm \frac{1}{1-m}\left(\frac{m}{1-m}\right)^{\frac{i}{2\pi}\ln(E/E')} = \\
&= \sum_{\xi\Xi} \int_0^\infty dE \int_0^\infty dE' \frac{|\mathbf{e}_{E,\xi}\rangle\langle \mathbf{e}_{E',\xi}|}{4\pi^2\sqrt{EE'}} \int_0^1 dm \, m^{\frac{i}{2\pi}\ln(E/E')}(1-m)^{-\frac{i}{2\pi}\ln(E/E') - 1}
\end{aligned}
\tag{4.3.6}
$$

Recall the definition of the Euler beta function

$$
B(x, y) = \int_0^1 dm \, m^{x-1}(1-m)^{y-1} = \frac{\Gamma(x)\Gamma(y)}{\Gamma(x+y)}. \tag{4.3.7}
$$

The function $B(x, y)$ is well defined for $\operatorname{Re} x > 0$ and $\operatorname{Re} y > 0$ and can be analytically continued to other parts of the complex x and y planes. However, it is not well defined for $x = 0$ and $y = 0$. This situation occurs in the integral above when $E = E'$. To avoid this difficulty we shift E or E' away from the real axis. Thus, in the integration over m on the right hand side of Eq. (4.3.6) we take the limit

$$
\begin{aligned}
&\frac{1}{4\pi^2\sqrt{EE'}} \lim_{\theta\to 0^+} \int_0^1 dm \, m^{\frac{i}{2\pi}\log(Ee^{i\theta}/E')}(1-m)^{-\frac{i}{2\pi}\log(Ee^{i\theta}/E') - 1} = \\
&= \frac{1}{4\pi^2\sqrt{EE'}} \lim_{\theta\to 0^+} B\left(1 + \frac{i}{2\pi}\log\left(\frac{Ee^{i\theta}}{E'}\right), -\frac{i}{2\pi}\log\left(\frac{Ee^{i\theta}}{E'}\right)\right) = \\
&= \frac{1}{4\pi^2\sqrt{EE'}} \lim_{\epsilon\to 0^+} \frac{\pi}{\sin\left(-\frac{i}{2}\log\left(\frac{Ee^{i\theta}}{E'}\right)\right)} = -\frac{1}{2\pi i}\frac{1}{E - E' + i0^+}.
\end{aligned}
\tag{4.3.8}
$$

where we have used the identity $\Gamma(z)\Gamma(1-z) = \frac{\pi}{\sin(\pi z)}$. From Eqs. (4.3.6) and (4.3.8) we obtain

$$\sum_{\xi\in\Xi}\int_0^1 m\,|\mathbf{g}^f_{m,\xi}\rangle\,\langle\mathbf{g}^f_{m,\xi}|\,dm = -\frac{1}{2\pi i}\sum_{\xi\in\Xi}\int_0^\infty dE\int_0^\infty dE'|\mathbf{e}_{E,\xi}\rangle\frac{1}{E-E'+i0^+}\langle\mathbf{e}_{E',\xi}| = M_F \tag{4.3.9}$$

Thus, we have reconstructed the Lyapunov operator M_F using the generalized eigenfunctions in Eq. (4.3.1). Equation (4.3.9) shows also that M_F has no point spectrum at $m=0$ and $m=1$ and $\sigma(M_F)=\sigma_{ac}(M_F)=[0,1]$. ■

An analogous theorem (proved similarly) holds for the backward Lyapunov operator M_B:

Theorem 4.11 *Let $\sigma_{ac}(H)=\sigma_{ac}(H_0)=\mathbb{R}_+$ with uniform multiplicity and $[a,b]=\overline{\mathbb{R}}_+$. Then the spectrum of M_B satisfies $\sigma(M_B)=\sigma_{ac}(M_B)=[0,1]$. For $m\in(0,1)$ and $\xi\in\Xi$ let $\mathbf{g}^b_{m,\xi}$ be the function defined on $\mathbb{R}_+$ by*

$$\mathbf{g}^b_{m,\xi}(E) = \frac{E^{\frac{i}{2\pi}\ln\left(\frac{1-m}{m}\right)-\frac{1}{2}}}{2\pi\sqrt{m\,(1-m)}}|\mathbf{e}_{E,\xi}\rangle,\quad m\in(0,1),\ E>0, \tag{4.3.10}$$

then $\mathbf{g}^b_{m,\xi}$ is a generalized eigenfunction of M_B satisfying $M_B\,\mathbf{g}^b_{m,\xi}=m\,\mathbf{g}^b_{m,\xi}$. These eigenfunctions are normalized in such a way that $\langle\mathbf{g}^b_{m,\xi},\mathbf{g}^b_{m',\xi'}\rangle_{L^2(\mathbb{R}_+;\mathcal{K})}=\delta(m-m')\delta_{\xi\xi'}$ and (using the Dirac notation) we have the eigenfunction expansion

$$M_B = \sum_{\xi\in\Xi}\int_0^1 dm\,m\,|\mathbf{g}^b_{m,\xi}\rangle\,\langle\mathbf{g}^b_{m,\xi}|, \tag{4.3.11}$$

where summation over ξ stands for discrete summation for discrete indices and appropriate integration over continuous indices. □

With the help of the generalized eigenfunctions of M_F in Eq. (4.3.1) we are able to find explicit formula for functions of M_F. In particular, we are able to find explicit formulas for various powers M_F^α of M_F. Given an element $\psi\in L^2(\mathbb{R}^+;\mathcal{K})$ and the orthogonal basis $\{\mathbf{e}_{E,\xi}\}_{E\in\mathbb{R}^+;\xi\in\Xi}$, we have an expansion $\psi=\sum_{\xi\in\Xi}\int_0^\infty dE\,\psi(E,\xi)|\mathbf{e}_{E,\xi}\rangle$, where we set $\psi(E,\xi)=\langle\mathbf{e}_{E,\xi},\psi\rangle$. Then we can write

$$(M_F^\alpha\psi)(E,\xi) = \sum_{\xi'\in\Xi}\int_0^\infty dE' M_F^\alpha(E,\xi;E',\xi')\psi(E',\xi'),$$

where $M_F^\alpha(E,\xi;E',\xi')=\langle\mathbf{e}_{E,\xi},M_F^\alpha\,\mathbf{e}_{E',\xi'}\rangle$. Using Eqs. (4.3.1), (4.3.2) we obtain

$$
\begin{aligned}
M_F^{\alpha}(E,\xi;E',\xi') &= \langle \mathbf{e}_{E,\xi}, M_F^{\alpha}\,\mathbf{e}_{E',\xi'}\rangle = \\
&= \delta_{\xi\xi'}\frac{1}{4\pi^2}(E\,E')^{-1/2}\lim_{\theta\to 0^+}\int_0^1 dm\, m^{\alpha}(1-m)^{-\frac{i}{2\pi}[\log\frac{E\,e^{i\theta}}{E'}]-1} m^{\frac{i}{2\pi}[\log\frac{E\,e^{i\theta}}{E'}]-1} = \\
&= \delta_{\xi\xi'}\frac{1}{4\pi^2}(E\,E')^{-1/2}\lim_{\theta\to 0^+}\int_0^1 dm\,(1-m)^{-\frac{i}{2\pi}[\log\frac{E\,e^{i\theta}}{E'}]-1} m^{\alpha+\frac{i}{2\pi}[\log\frac{E\,e^{i\theta}}{E'}]-1} = \\
&= \delta_{\xi\xi'}\frac{1}{4\pi^2}(E\,E')^{-1/2}\lim_{\theta\to 0^+} B\left(\alpha+\frac{i}{2\pi}\log\frac{E\,e^{i\theta}}{E'}, -\frac{i}{2\pi}\log\frac{E\,e^{i\theta}}{E'}\right) = \\
&= \delta_{\xi\xi'}\frac{1}{4\pi^2}(E\,E')^{-1/2}\lim_{\theta\to 0^+} B\left(\alpha-\frac{\theta}{2\pi}+\frac{i}{2\pi}\ln\frac{E}{E'}, \frac{\theta}{2\pi}-\frac{i}{2\pi}\ln\frac{E}{E'}\right),
\end{aligned}
\tag{4.3.12}
$$

In particular, if we set $\alpha = 1/2$ we obtain

$$
\begin{aligned}
\Lambda_F(E,\xi;E',\xi') &= \langle \mathbf{e}_{E,\xi}, \Lambda_F\,\mathbf{e}_{E',\xi'}\rangle = \langle \mathbf{e}_{E,\xi}, M_F^{1/2}\,\mathbf{e}_{E',\xi'}\rangle = \\
&= \delta_{\xi\xi'}\frac{1}{4\pi^2}(E\,E')^{-1/2}\lim_{\theta\to 0^+} B\left(1/2-\frac{\theta}{2\pi}+\frac{i}{2\pi}\ln\frac{E}{E'}, \frac{\theta}{2\pi}-\frac{i}{2\pi}\ln\frac{E}{E'}\right).
\end{aligned}
\tag{4.3.13}
$$

As in the case of M_F, with the help of the generalized eigenfunctions of M_B in Eq. (4.3.10) we can find explicit formula for functions of M_B. In particular, we have

$$
\begin{aligned}
M_B^{\alpha}(E,\xi;E',\xi') &= \langle \mathbf{e}_{E,\xi}, M_B^{\alpha}\,\mathbf{e}_{E',\xi'}\rangle = \\
&= \delta_{\xi\xi'}\frac{1}{4\pi^2}(E\,E')^{-1/2}\lim_{\theta\to 0^+}\int_0^1 dm\, m^{\alpha}(1-m)^{\frac{i}{2\pi}[\log\frac{E\,e^{i\theta}}{E'}]-1} m^{-\frac{i}{2\pi}[\log\frac{E\,e^{i\theta}}{E'}]-1} = \\
&= \delta_{\xi\xi'}\frac{1}{4\pi^2}(E\,E')^{-1/2}\lim_{\theta\to 0^+}\int_0^1 dm\,(1-m)^{\frac{i}{2\pi}[\log\frac{E\,e^{i\theta}}{E'}]-1} m^{\alpha-\frac{i}{2\pi}[\log\frac{E\,e^{i\theta}}{E'}]-1} = \\
&= \delta_{\xi\xi'}\frac{1}{4\pi^2}(E\,E')^{-1/2}\lim_{\theta\to 0^+} B\left(\alpha-\frac{i}{2\pi}\log\frac{E\,e^{i\theta}}{E'}, \frac{i}{2\pi}\log\frac{E\,e^{i\theta}}{E'}\right) = \\
&= \delta_{\xi\xi'}\frac{1}{4\pi^2}(E\,E')^{-1/2}\lim_{\theta\to 0^+} B\left(\alpha+\frac{\theta}{2\pi}-\frac{i}{2\pi}\ln\frac{E}{E'}, -\frac{\theta}{2\pi}+\frac{i}{2\pi}\ln\frac{E}{E'}\right),
\end{aligned}
\tag{4.3.14}
$$

and

$$
\begin{aligned}
\Lambda_B(E,\xi;E',\xi') &= \langle \mathbf{e}_{E,\xi}, \Lambda_B\,\mathbf{e}_{E',\xi'}\rangle = \langle \mathbf{e}_{E,\xi}, M_B^{1/2}\,\mathbf{e}_{E',\xi'}\rangle = \\
&= \delta_{\xi\xi'}\frac{1}{4\pi^2}(E\,E')^{-1/2}\lim_{\theta\to 0^+} B\left(1/2+\frac{\theta}{2\pi}-\frac{i}{2\pi}\ln\frac{E}{E'}, -\frac{\theta}{2\pi}+\frac{i}{2\pi}\ln\frac{E}{E'}\right).
\end{aligned}
\tag{4.3.15}
$$

We mention here a useful tool for the manipulation of various powers of M_F, i,e., the following integral formula for the combination of beta functions

$$\int_0^\infty \frac{dE}{E} B\left(\alpha - \frac{i}{2\pi}\ln\frac{E_1}{E}, \beta + \frac{i}{2\pi}\ln\frac{E_1}{E}\right) B\left(\gamma - \frac{i}{2\pi}\ln\frac{E}{E_2}, \delta + \frac{i}{2\pi}\ln\frac{E}{E_2}\right) =$$
$$= 4\pi^2 B\left(\alpha+\gamma - \frac{i}{2\pi}\ln\frac{E_1}{E_2}, \beta+\delta+\frac{i}{2\pi}\ln\frac{E_1}{E_2}\right). \quad (4.3.16)$$

where α, β, γ, δ are complex parameters having values such that the beta functions in Eq. (4.3.16) are well defined for all values of E_1, E_2 and E. Equation (4.3.16) can be verified by direct calculation using the definition of the Euler beta function in Eq. (4.3.7) and by noting that

$$\int_0^\infty \frac{dE}{E} E^{\frac{i}{2\pi}\left[\ln\frac{m}{1-m} - \ln\frac{m'}{1-m'}\right]} = 4\pi^2 m(1-m)\delta(m-m').$$

Equation (4.3.16) provides explicit formulas for the combination of powers of M_F. Thus, this equation leads to the identity

$$\frac{1}{16\pi^4}(E_1 E_2)^{-1/2} \lim_{\theta_1,\theta_2\to 0^+} \int_0^\infty \frac{dE}{E} B\Big(\alpha - \frac{\theta_1}{2\pi} + \frac{i}{2\pi}\ln\frac{E_1}{E}, \frac{\theta_1}{2\pi} - \frac{i}{2\pi}\ln\frac{E_1}{E}\Big)$$
$$\times B\Big(\beta - \frac{\theta_2}{2\pi} + \frac{i}{2\pi}\ln\frac{E}{E_2}, \frac{\theta_2}{2\pi} - \frac{i}{2\pi}\ln\frac{E}{E_2}\Big) =$$
$$= \frac{1}{4\pi^2}(E_1 E_2)^{-1/2} \lim_{\theta\to 0^+} B\Big(\alpha+\beta - \frac{\theta}{2\pi} + \frac{i}{2\pi}\ln\frac{E_1}{E_2}, \frac{\theta}{2\pi} - \frac{i}{2\pi}\ln\frac{E_1}{E_2}\Big),$$

corresponding to

$$\langle \mathbf{e}_{E_1,\xi}, M_F^{\alpha+\beta}\mathbf{e}_{E_2,\xi'}\rangle = \sum_{\xi''}\int_0^\infty dE\, \langle \mathbf{e}_{E_1,\xi}, M_F^{\alpha}\mathbf{e}_{E,\xi''}\rangle\langle \mathbf{e}_{E,\xi''}, M_F^{\beta}\mathbf{e}_{E_2,\xi'}\rangle.$$

Of course, one may derive a similar procedure for the combination of powers of M_B.

Note that the formula above provides expressions for powers of the forward and backward Lyapunov operators M_F^α, M_B^α which are operators on the function space $L^2(\mathbb{R}_+;\mathcal{K})$. For a scattering problem satisfying assumptions (i)–(iii) these functional representations correspond to the spectral representations of the Hamiltonian H of the scattering system, i.e., the incoming and outgoing energy representations. Using the isometric isomorphisms $\hat{W}_+^{QM}$ and $\hat{W}_-^{QM}$ mapping $\mathcal{H}_{ac}$, respectively, onto the outgoing and incoming energy representations and following Eqs. (3.1.20) and (3.1.22) we find that

$$M_+^\alpha = \left(\hat{W}_+^{QM}\right)^{-1} M_F^\alpha \hat{W}_+^{QM} =$$
$$= \lim_{\theta\to 0^+}\frac{1}{4\pi^2}\sum_\xi \int_0^\infty \frac{dE}{\sqrt{E}}\int_0^\infty \frac{dE'}{\sqrt{E'}} |\phi_{E,\xi}^-\rangle B\Big(\alpha - \frac{\theta}{2\pi} + \frac{i}{2\pi}\ln\frac{E}{E'}, \frac{\theta}{2\pi} - \frac{i}{2\pi}\ln\frac{E}{E'}\Big)\langle\phi_{E',\xi}^-|, \quad (4.3.17)$$

and in particular, by setting $\alpha = 1/2$, we obtain

$$\Lambda_+ = \lim_{\theta\to 0^+} \frac{1}{4\pi^2} \sum_\xi \int_0^\infty \frac{dE}{\sqrt{E}} \int_0^\infty \frac{dE'}{\sqrt{E'}} |\phi^-_{E,\xi}\rangle B\left(\frac{1}{2} - \frac{\theta}{2\pi} + \frac{i}{2\pi}\ln\frac{E}{E'}, \frac{\theta}{2\pi} - \frac{i}{2\pi}\ln\frac{E}{E'}\right)\langle\phi^-_{E',\xi}|. \tag{4.3.18}$$

Similarly, we have

$$M^\alpha_- = \left(\hat{W}^{QM}_-\right)^{-1} M^\alpha_B \hat{W}^{QM}_- =$$

$$= \lim_{\theta\to 0^+} \frac{1}{4\pi^2} \sum_\xi \int_0^\infty \frac{dE}{\sqrt{E}} \int_0^\infty \frac{dE'}{\sqrt{E'}} |\phi^+_{E,\xi}\rangle B\left(\alpha + \frac{\theta}{2\pi} - \frac{i}{2\pi}\ln\frac{E}{E'}, -\frac{\theta}{2\pi} + \frac{i}{2\pi}\ln\frac{E}{E'}\right)\langle\phi^+_{E',\xi}|, \tag{4.3.19}$$

and

$$\Lambda_- = \lim_{\theta\to 0^+} \frac{1}{4\pi^2} \sum_\xi \int_0^\infty \frac{dE}{\sqrt{E}} \int_0^\infty \frac{dE'}{\sqrt{E'}} |\phi^+_{E,\xi}\rangle B\left(\frac{1}{2} + \frac{\theta}{2\pi} - \frac{i}{2\pi}\ln\frac{E}{E'}, -\frac{\theta}{2\pi} + \frac{i}{2\pi}\ln\frac{E}{E'}\right)\langle\phi^+_{E',\xi}|. \tag{4.3.20}$$

With the help of Eqs. (4.3.18) and (4.3.20) we can construct explicit representations for the incoming and outgoing transition representations (see Definitions 4.4 and 4.5). Moreover, we may obtain also an explicit representation for the elements of the approximate Lax–Phillips semigroup,

$$Z_{app}(t) = \Lambda_+ U(t) \Lambda_- =$$

$$\left[\lim_{\theta_1\to 0^+} \frac{1}{4\pi^2} \sum_\xi \int_0^\infty \frac{dE_1}{\sqrt{E_1}} \int_0^\infty \frac{dE_1'}{\sqrt{E_1'}} |\phi^-_{E_1,\xi}\rangle B\left(\frac{1}{2} - \frac{i}{2\pi}\log\frac{E_1 e^{i\theta_1}}{E_1'}, -\frac{i}{2\pi}\log\frac{E_1 e^{i\theta_1}}{E_1'}\right)\langle\phi^-_{E_1',\xi}|\right] U(t)$$

$$\times \left[\lim_{\theta_2\to 0^+} \frac{1}{4\pi^2} \sum_{\xi'} \int_0^\infty \frac{dE_2'}{\sqrt{E_2'}} \int_0^\infty \frac{dE_2}{\sqrt{E_2}} |\phi^+_{E_2',\xi'}\rangle B\left(\frac{1}{2} - \frac{i}{2\pi}\ln\frac{E_2' e^{i\theta_2}}{E_2}, \frac{i}{2\pi}\ln\frac{E_2' e^{i\theta_2}}{E_2}\right)\langle\phi^+_{E_2,\xi'}|\right] =$$

$$= \lim_{\theta_1\to 0^+} \lim_{\theta_2\to 0^+} \frac{1}{16\pi^4} \sum_{\xi,\xi'} \int_0^\infty \frac{dE_1}{\sqrt{E_1}} \int_0^\infty \frac{dE_2'}{E_2'} \int_0^\infty \frac{dE_2}{\sqrt{E_2}} \times$$

$$|\phi^-_{E_1,\xi}\rangle B\left(\frac{1}{2} - \frac{\theta_1}{2\pi} + \frac{i}{2\pi}\ln\frac{E_1}{E_2'}, \frac{\theta_1}{2\pi} - \frac{i}{2\pi}\ln\frac{E_1}{E_2'}\right) e^{-iE_2' t} \left(\hat{S}_{QM}(E_2')\right)^\xi_{\xi'}$$

$$\times B\left(\frac{1}{2} + \frac{\theta_2}{2\pi} - \frac{i}{2\pi}\ln\frac{E_2'}{E_2}, -\frac{\theta_2}{2\pi} + \frac{i}{2\pi}\ln\frac{E_2'}{E_2}\right)\langle\phi^+_{E_2,\xi'}|, \quad t \ge 0. \tag{4.3.21}$$

We turn now to the task of finding an explicit expression for a resonance state ψ_μ, corresponding to a resonance pole at $z = \mu$. First, note that by the definition of the approximate resonance ψ^{app}_μ in Eq. (4.2.8) we have

$$\psi_\mu = \Lambda_+^{-1} \psi^{app}_\mu .$$

Next, setting $\alpha = -1/2$ in Eq. (4.3.17) we obtain

$$\Lambda_+^{-1} = M_+^{-1/2} =$$
$$= \lim_{\theta\to 0^+} \frac{1}{4\pi^2} \sum_\xi \int_0^\infty \frac{dE}{\sqrt{E}} \int_0^\infty \frac{dE'}{\sqrt{E'}} |\phi^-_{E,\xi}\rangle B\Big(-\frac{1}{2} - \frac{\theta}{2\pi} + \frac{i}{2\pi}\ln\frac{E}{E'},\ \frac{\theta}{2\pi} - \frac{i}{2\pi}\ln\frac{E}{E'}\Big) \langle \phi^-_{E',\xi}|. \tag{4.3.22}$$

Hence, using the representation of ψ_μ^{app} in Eq. (4.2.13) we get that

$$\psi_\mu = \Lambda_+^{-1}\psi_\mu^{app} =$$
$$= \lim_{\theta\to 0^+} \frac{1}{4\pi^2} \sum_\xi \int_0^\infty \frac{dE}{\sqrt{E}} \int_0^\infty \frac{dE'}{\sqrt{E'}} |\phi^-_{E,\xi}\rangle B\Big(-\frac{1}{2} - \frac{\theta}{2\pi} + \frac{i}{2\pi}\ln\frac{E}{E'},\ \frac{\theta}{2\pi} - \frac{i}{2\pi}\ln\frac{E}{E'}\Big) \langle \phi^-_{E',\xi}|\psi_\mu^{app}\rangle =$$
$$= \lim_{\theta\to 0^+} \frac{1}{4\pi^2} \sum_\xi \int_0^\infty \frac{dE}{\sqrt{E}} \int_0^\infty \frac{dE'}{\sqrt{E'}} |\phi^-_{E,\xi}\rangle B\Big(-\frac{1}{2} - \frac{\theta}{2\pi} + \frac{i}{2\pi}\ln\frac{E}{E'},\ \frac{\theta}{2\pi} - \frac{i}{2\pi}\ln\frac{E}{E'}\Big) \frac{(\mathbf{e}_\xi, \mathbf{h})}{E'-\mu} \tag{4.3.23}$$

In principle, Eq. (4.3.23) provides the explicit expression for the resonance state ψ_μ. However, this expression is not so convenient for practical calculations and a much simpler expression can be obtained. An obvious way to simplify Eq. (4.3.23) is to try to perform the integration over E'. However, this turns out to be somewhat difficult and we shall use an indirect method.

Since $\Lambda_+ = M_+^{1/2}$ we may write the operator Λ_+^{-1} in the form $\Lambda_+^{-1} = \Lambda_+ M_+^{-1}$ so that $\psi_\mu = \Lambda_+^{-1}\psi_\mu^{app} = (\Lambda_+ M_+^{-1})\psi_\mu^{app}$. Starting with this expression for the resonance state, we would like to apply the operators M_+^{-1} and then Λ_+ sequentially to ψ_μ^{app}, a procedure that, as we shall see below, gives us eventually a much more accessible expression for ψ_μ than Eq. (4.3.23) above. However, one has to exercise some care when applying this procedure. although Λ_+^{-1} is an unbounded operator, we know from Eq. (4.2.8) that ψ_μ^{app} is in the domain of this operator and that ψ_μ is a well defined state in $\mathcal{H}_{ac}$. The situation is different with M_+^{-1} which is also an unbounded operator on $\mathcal{H}_{ac}$ but for which ψ_μ^{app} may not be in the domain of definition. To avoid this difficulty we apply spectral projections on the spectrum of M_+. Let E_{M_+} be the spectral projection valued measure corresponding to M_+ and, for $0 \le \alpha < 1$, let $P_m(\alpha)$ be the spectral projection on the spectrum of M_+ corresponding to the interval $[\alpha, 1]$, i.e.,

$$P_m(\alpha) := \int_\alpha^1 E_{\mathrm{M}_+}(dm)\,.$$

Then, for any $0 < \alpha < 1$, and any $\psi \in \mathcal{H}_{ac}$ we have $(P_m(\alpha)\psi) \in \mathcal{D}(M_+^{-1})$ where $\mathcal{D}(M_+^{-1})$ is the domain of M_+^{-1}. Using the spectral projection $P_m(\alpha)$ we have

$$\begin{aligned}\psi_\mu = \Lambda_+^{-1}\psi_\mu^{app} = (\Lambda_+ M_+^{-1})\psi_\mu^{app} &= \lim_{\alpha\to 0^+}[(\Lambda_+ M_+^{-1})P_m(\alpha)\psi_\mu^{app}] = \\ &= \lim_{\alpha\to 0^+}[\Lambda_+(M_+^{-1}P_m(\alpha))\psi_\mu^{app}] = \lim_{\alpha\to 0^+}[\Lambda_+ M_{+,\alpha}^{-1}\psi_\mu^{app}],\end{aligned} \tag{4.3.24}$$

where $M_{+,\alpha}^{-1} := P_m(\alpha)M_+^{-1}P_m(\alpha) = M_+^{-1}P_m(\alpha)$ is a bounded operator. In order to make use of Eq. (4.3.24) we turn again to the explicit representation of powers of the Lyapunov operator M_+ in Eq. (4.3.17). Multiplication by the spectral projection operator $P_m(\alpha)$ gives us

$$\begin{aligned}\langle\phi_{E,\xi}^-, M_{+,\alpha}^\beta\phi_{E',\xi'}^-\rangle &= \langle\phi_{E,\xi}^-, P_m(\alpha)M_+^\beta P_m(\alpha)\phi_{E',\xi'}^-\rangle = \\ &= \delta_{\xi\xi'}\frac{1}{4\pi^2}(E\,E')^{-1/2}\int_\alpha^1 dm\, m^{\beta-\frac{\theta}{2\pi}+\frac{i}{2\pi}[\ln\frac{E}{E'}]-1}(1-m)^{\frac{\theta}{2\pi}-\frac{i}{2\pi}[\ln\frac{E}{E'}]-1}.\end{aligned}$$

If we change in the last integral the variable of integration from m to $\tilde m = 1-m$ we get

$$\begin{aligned}&\langle\phi_{E,\xi}^-, M_{+,\alpha}^\beta\phi_{E',\xi'}^-\rangle = \\ &\quad = \delta_{\xi\xi'}\frac{1}{4\pi^2}(E\,E')^{-1/2}\lim_{\theta\to 0^+}\int_0^{1-\alpha} d\tilde m\, \tilde m^{-\frac{i}{2\pi}\log\frac{Ee^{i\theta}}{E'}-1}(1-\tilde m)^{\beta+\frac{i}{2\pi}\log\frac{Ee^{i\theta}}{E'}-1} = \\ &\quad = \delta_{\xi\xi'}\frac{1}{4\pi^2}(E\,E')^{-1/2}\lim_{\theta\to 0^+} B\left(1-\alpha, -\frac{i}{2\pi}\log\frac{Ee^{i\theta}}{E'}, \beta+\frac{i}{2\pi}\log\frac{Ee^{i\theta}}{E'}\right),\end{aligned} \tag{4.3.25}$$

where $B(z, x, y)$ is the incomplete beta function

$$B(z, x, y) = \int_0^z dm\, m^{x-1}(1-m)^{y-1}.$$

Setting in Eq. (4.3.25) $\beta = -1$ we get

$$\begin{aligned}&\langle\phi_{E,\xi}^-, M_{+,\alpha}^{-1}\phi_{E',\xi'}\rangle = \\ &\quad = \delta_{\xi\xi'}\frac{1}{4\pi^2}(E\,E')^{-1/2}\lim_{\theta\to 0^+} B\left(1-\alpha, -\frac{i}{2\pi}\log\frac{Ee^{i\theta}}{E'}, -1+\frac{i}{2\pi}\log\frac{Ee^{i\theta}}{E'}\right).\end{aligned} \tag{4.3.26}$$

Thus we have obtained an explicit representation of the operator $M_{+,\alpha}^{-1}$ in the outgoing energy representation. In order to be able to apply Eq. (4.3.26) we write the beta function in integral form as in Eq. (4.3.25) and define a new variable of integration $x = \tilde m(1-\tilde m)^{-1}$. Performing the change of integration variable we get

$$
\begin{aligned}
&\langle \phi^-_{E,\xi}, M^{-1}_{+,\alpha}\phi_{E',\xi'}\rangle = \\
&= \delta_{\xi\xi'}\frac{1}{4\pi^2}(E\,E')^{-1/2}\lim_{\theta\to 0^+} B\left(1-\alpha, -\frac{i}{2\pi}\log\frac{Ee^{i\theta}}{E'}, -1+\frac{i}{2\pi}\log\frac{Ee^{i\theta}}{E'}\right) = \\
&= \delta_{\xi\xi'}\frac{1}{4\pi^2}(E\,E')^{-1/2}\lim_{\theta\to 0^+}\int_0^{1-\alpha} d\tilde{m}\,\tilde{m}^{-1}(1-\tilde{m})^{-2}\left(\frac{\tilde{m}}{1-\tilde{m}}\right)^{-\frac{i}{2\pi}\log\frac{Ee^{i\theta}}{E'}} = \\
&= \delta_{\xi\xi'}\frac{1}{4\pi^2}(E\,E')^{-1/2}\lim_{\theta\to 0^+}\int_0^{\frac{1-\alpha}{\alpha}}\frac{dx}{x}(1+x)x^{-\frac{i}{2\pi}\log\frac{Ee^{i\theta}}{E'}}\,.
\end{aligned}
\tag{4.3.27}
$$

Using Eqs. (4.2.13) and (4.3.27) we obtain

$$
\begin{aligned}
\langle \phi^-_{E,\xi}, M^{-1}_{+,\alpha}\psi^{app}_\mu\rangle &= \sum_{\xi'}\int_0^\infty dE'\,\langle \phi^-_{E,\xi}, M^{-1}_{+,\alpha}\phi^-_{E',\xi'}\rangle\langle \phi^-_{E',\xi'}, \psi^{app}_\mu\rangle = \\
&= \frac{1}{4\pi^2}E^{-1/2}\lim_{\theta\to 0^+}\int_0^\infty dE'\,E'^{-1/2}\int_0^{\frac{1-\alpha}{\alpha}}\frac{dx}{x}(1+x)x^{-\frac{i}{2\pi}\log\frac{Ee^{i\theta}}{E'}}\frac{\langle \mathbf{e}_\xi, \mathbf{h}\rangle}{E'-\mu} = \\
&= \frac{1}{4\pi^2}E^{-1/2}\lim_{\theta\to 0^+}\int_0^\infty dE'\,E'^{-1/2}\int_0^{\frac{1-\alpha}{\alpha}}\frac{dx}{x}x^{-\frac{i}{2\pi}\log\frac{Ee^{i\theta}}{E'}}\frac{\langle \mathbf{e}_\xi, \mathbf{h}\rangle}{E'-\mu} \\
&+ \frac{1}{4\pi^2}E^{-1/2}\lim_{\theta\to 0^+}\int_0^\infty dE'\,E'^{-1/2}\int_0^{\frac{1-\alpha}{\alpha}}\frac{dx}{x}x^{-\frac{i}{2\pi}\log\frac{Ee^{i(\theta+2\pi)}}{E'}}\frac{\langle \mathbf{e}_\xi, \mathbf{h}\rangle}{E'-\mu}\,.
\end{aligned}
\tag{4.3.28}
$$

In the last term on the right hand side of Eq. (4.3.28) we use Fubini's theorem to change the order of integration. Furthermore, we evaluate the E' integral by performing a contour integration in the complex E' plane along the key hole contour indicated in Fig. 4.1. The integral along the larger arc vanishes in the limit $|z|\to\infty$ while the integral along the smaller arc vanishes in the limit $|z|\to 0$. Note that when writing $\mu = E_\mu - i\Gamma_\mu/2$, with $E_\mu > 0$ and $\Gamma_\mu > 0$, we consider $\arg\mu$ to be negative so that with respect to the contour of integration in Fig. 4.1 the pole of ψ^{har}_μ appears at the point $E' = \mu\, e^{2\pi i}$. Picking up the residue contribution at this point we obtain the result

$$
\begin{aligned}
&\frac{1}{4\pi^2}E^{-1/2}\lim_{\theta\to 0^+}\int_0^\infty dE'\,E'^{-1/2}\int_0^{\frac{1-\alpha}{\alpha}}\frac{dx}{x}x^{-\frac{i}{2\pi}\log\frac{Ee^{i(\theta+2\pi)}}{E'}}\frac{1}{E'-\mu} = \\
&= \frac{1}{4\pi^2}E^{-1/2}\lim_{\theta\to 0^+}\int_0^\infty dE'\,(E'e^{2\pi i})^{-1/2}\int_0^{\frac{1-\alpha}{\alpha}}\frac{dx}{x}x^{-\frac{i}{2\pi}\log\frac{Ee^{i(\theta+2\pi)}}{E'e^{2\pi i}}}\frac{1}{E'e^{2\pi i}-\mu\,e^{2\pi i}} \\
&+ \frac{i}{2\pi}E^{-1/2}\lim_{\theta\to 0^+} Res\left[E'^{-1/2}\int_0^{\frac{1-\alpha}{\alpha}}\frac{dx}{x}x^{-\frac{i}{2\pi}\log\frac{Ee^{i(\theta+2\pi)}}{E'}}\frac{1}{E'-\mu\,e^{2\pi i}},\quad E'=\mu\,e^{2\pi i}\right] = \\
&= -\frac{1}{4\pi^2}E^{-1/2}\lim_{\theta\to 0^+}\int_0^\infty dE'\,E'^{-1/2}\int_0^{\frac{1-\alpha}{\alpha}}\frac{dx}{x}x^{-\frac{i}{2\pi}\log\frac{Ee^{i\theta}}{E'}}\frac{1}{E'-\mu} \\
&\qquad - \frac{i}{2\pi\sqrt{\mu}}\frac{1}{\sqrt{E}}\int_0^{\frac{1-\alpha}{\alpha}}\frac{dx}{x}x^{-\frac{i}{2\pi}\log\frac{E}{\mu}}\,.
\end{aligned}
\tag{4.3.29}
$$

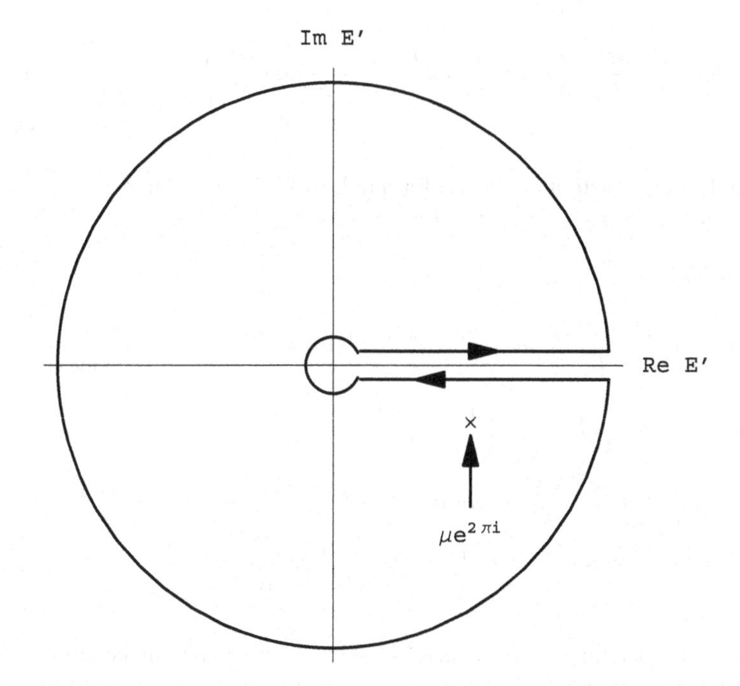

Fig. 4.1 Contour of integration for Eq. (4.3.29). The integral on the large arc vanishes in the limit $|E'| \to \infty$ and the integral on the small arc vanishes in the limit $|E'| \to 0$. The pole at $E' = \mu e^{2\pi i}$ is also indicated

Inserting Eq. (4.3.29) into Eq. (4.3.28) we finally get

$$\langle \phi^-_{E,\xi}, M^{-1}_{+,\alpha}\psi^{app}_\mu \rangle = -\frac{i}{2\pi\sqrt{\mu}}\frac{1}{\sqrt{E}}\int_0^{\frac{1-\alpha}{\alpha}} \frac{dx}{x} x^{-\frac{i}{2\pi}\log\frac{E}{\mu}} \langle \mathbf{e}_\xi, \mathbf{h}\rangle =$$
$$= \frac{1}{\sqrt{\mu}}\frac{1}{\sqrt{E}} \frac{\left(\frac{1-\alpha}{\alpha}\right)^{-\frac{i}{2\pi}\log\frac{E}{\mu}}}{\log\frac{E}{\mu}} \langle \mathbf{e}_\xi, \mathbf{h}\rangle \,. \tag{4.3.30}$$

We are left with the task of applying the operator Λ_+ and taking the limit $\alpha \to 0^+$. Using Eqs. (4.3.18), (4.3.24) and (4.3.30) we obtain

$$\langle \phi^-_{E,\xi}, \psi_\mu \rangle = \lim_{\alpha\to 0^+} \langle \phi^-_{E,\xi}, \Lambda_+ M^{-1}_{+,\alpha}\psi^{app}_\mu \rangle =$$
$$= \lim_{\alpha\to 0^+} \sum_{\xi'} \int_0^\infty dE' \langle \phi^-_{E,\xi}, \Lambda_+ \phi^-_{E',\xi'}\rangle \langle \phi^-_{E',\xi'}, M^{-1}_{+,\alpha}\psi^{app}_\mu \rangle =$$
$$= \frac{1}{4\pi^2\sqrt{\mu}}\frac{1}{\sqrt{E}} \times$$

$$\lim_{\alpha\to 0^+}\lim_{\theta\to 0^+}\int_0^\infty \frac{dE'}{E'} B\left(1/2+\frac{i}{2\pi}\log\frac{E\,e^{i\theta}}{E'},\, -\frac{i}{2\pi}\log\frac{E\,e^{i\theta}}{E'}\right)\frac{\left(\frac{1-\alpha}{\alpha}\right)^{-\frac{i}{2\pi}\log\frac{E'}{\mu}}}{\log\frac{E'}{\mu}}\langle \mathbf{e}_\xi, \mathbf{h}\rangle\,. \tag{4.3.31}$$

Writing the beta function in integral form in Eq. (4.3.31) we obtain

$$\frac{1}{4\pi^2\sqrt{\mu}}\frac{1}{\sqrt{E}}\times$$
$$\lim_{\alpha\to 0^+}\lim_{\theta\to 0^+}\int_0^\infty \frac{dE'}{E'}\int_0^1 dm\, m^{\frac{1}{2}+\frac{i}{2\pi}\log\frac{E\,e^{i\theta}}{E'}-1}(1-m)^{-\frac{i}{2\pi}\log\frac{E\,e^{i\theta}}{E'}-1}\frac{\left(\frac{1-\alpha}{\alpha}\right)^{-\frac{i}{2\pi}\log\frac{E'}{\mu}}}{\log\frac{E'}{\mu}}=$$
$$=\frac{1}{4\pi^2\sqrt{\mu}}\frac{1}{\sqrt{E}}\left(\frac{1-\alpha}{\alpha}\right)^{\frac{i}{2\pi}\log\mu}\times$$
$$\lim_{\alpha\to 0^+}\lim_{\theta\to 0^+}\int_0^\infty \frac{dE'}{E'}\int_0^1 dm\, m^{\frac{1}{2}+\frac{i}{2\pi}\log(E\,e^{i\theta})-1}(1-m)^{-\frac{i}{2\pi}\log(E\,e^{i\theta})-1}\times$$
$$E'^{-\frac{i}{2\pi}[\log(\frac{m}{1-m})-\log(\frac{\alpha}{1-\alpha})]}\frac{1}{\log E'-\log\mu}\,. \tag{4.3.32}$$

Observe that, since $\mathrm{Im}\,\mu<0$, the factor $\left(\frac{1-\alpha}{\alpha}\right)^{\frac{i}{2\pi}\log\mu}$ is singular in the limit $\alpha\to 0^+$. However, the integration on E' cancels this singularity. In Eq. (4.3.32) we use Fubini's theorem to change the order of integration. The integral on E' gives us

$$\int_0^\infty \frac{dE'}{E'}E'^{-\frac{i}{2\pi}[\log(\frac{m}{1-m})-\log(\frac{\alpha}{1-\alpha})]}\frac{1}{\log E'-\log\mu}=$$
$$=\int_{-\infty}^\infty dy\, e^{-\frac{i}{2\pi}y[\log(\frac{m}{1-m})-\log(\frac{\alpha}{1-\alpha})]}\frac{1}{y-\log\mu}=$$
$$=2\pi i\theta(m-\alpha)\left(\frac{1-\alpha}{\alpha}\right)^{-\frac{i}{2\pi}\log\mu}\left(\frac{m}{1-m}\right)^{-\frac{i}{2\pi}\log\mu}, \tag{4.3.33}$$

and we see that Eq. (4.3.33) contains a factor canceling the singular factor in Eq. (4.3.32). Inserting Eq. (4.3.33) into Eq. (4.3.32) we get

$$\frac{i}{2\pi\sqrt{\mu}}\frac{1}{\sqrt{E}}\times$$
$$\lim_{\alpha\to 0^+}\lim_{\theta\to 0^+}\int_\alpha^1 dm\left(\frac{m}{1-m}\right)^{-\frac{i}{2\pi}\log\mu} m^{\frac{1}{2}+\frac{i}{2\pi}\log(E\,e^{i\theta})-1}(1-m)^{-\frac{i}{2\pi}\log(E\,e^{i\theta})-1}=$$
$$=\frac{i}{2\pi\sqrt{\mu}}\frac{1}{\sqrt{E}}\lim_{\alpha\to 0^+}B\left(1-\alpha,\, -\frac{i}{2\pi}\log\frac{E}{\mu},\, \frac{1}{2}+\frac{i}{2\pi}\log\frac{E}{\mu}\right). \tag{4.3.34}$$

The positive additive term $+1/2$ in the third argument of the incomplete beta function in Eq. (4.3.34) renders the $\alpha\to 0^+$ limit in this equation non-singular. Taking this limit we obtain

$$\langle \phi^-_{E,\xi}, \psi_\mu \rangle = \frac{i}{2\pi\sqrt{\mu}} \frac{1}{\sqrt{E}} B\left(\frac{1}{2} + \frac{i}{2\pi}\log\frac{E}{\mu}, -\frac{i}{2\pi}\log\frac{E}{\mu}\right) \langle \mathbf{e}_\xi, \mathbf{h} \rangle\,. \tag{4.3.35}$$

and finally

$$\begin{aligned} \psi_\mu &= \sum_\xi \int_0^\infty dE |\phi^-_{E,\xi}\rangle \langle \phi^-_{E,\xi}, \psi_\mu \rangle = \\ &= \frac{i}{2\pi\sqrt{\mu}} \int_0^\infty dE\, |\phi^-_{E,\xi}\rangle \frac{1}{\sqrt{E}} B\left(\frac{1}{2} + \frac{i}{2\pi}\log\frac{E}{\mu}, -\frac{i}{2\pi}\log\frac{E}{\mu}\right) \langle \mathbf{e}_\xi, \mathbf{h} \rangle\,. \end{aligned} \tag{4.3.36}$$

Equations (4.3.35) and (4.3.36) provide the explicit expressions for the resonance state ψ_μ which we have been searching for.

4.3.1 Example: Scattering from a Square Barrier Potential on a Half-Line

In this section we apply the structures developed in previous sections to a simple one dimensional scattering problem with a square barrier potential (for a nice discussion of this simple model in the context of Rigged Hilbert spaces see de la Madrid and Gadella 2002). Although very simple, this model provides a good illustration of the properties of the Modified Lax–Phillips formalism. The model we consider is a Schrödinger equation in one spatial dimension on the half-line $\mathbb{R}^+$ with square barrier potential. The free Hamiltonian H_0 is $H_0 = -\partial_x^2$ acting on $L^2(\mathbb{R}^+)$ (more precisely H_0 is taken to be the self-adjoint extension in $L^2(\mathbb{R}^+)$ of $-\partial_x^2$ from its original domain of definition $\mathcal{D}(-\partial_x^2) = \{\phi(x) \mid \phi(x) \in W_2^2(\mathbb{R}^+),\ \phi(0) = 0\}$) and the full Hamiltonian is $H = H_0 + V$ where the potential V is a multiplicative operator $(V\psi)(x) = V(x)\psi(x)$ with

$$V(x) = \begin{cases} 0, & 0 < x < a, \\ V_0, & a \le x \le b, \\ 0, & b < x, \end{cases}$$

where $0 < a < b$ and $V_0 > 0$. In this case there are no bound state solutions of the eigenvalue problem for H and we have $\sigma(H) = \sigma_{ac}(H) = \mathbb{R}^+$. In order to find the scattering states and calculate the S-matrix for the problem one solves the eigenvalue problem for the continuous spectrum generalized eigenfunctions $\psi_E(x)$ of H

$$[-\partial_x^2 + V(x)]\psi_E(x) = E\,\psi_E(x), \qquad E \in \mathbb{R}^+ .$$

Imposing appropriate boundary conditions we find that

$$\psi_E(x) = \begin{cases} \alpha_1(k)\sin kx, & 0 < x \le a, \\ \alpha_2(k)e^{ik'x} + \beta_2(k)e^{-ik'x}, & a < x < b, \\ \alpha_3(k)e^{ikx} + \beta_3(k)e^{-ikx}, & b \le x, \end{cases} \tag{4.3.37}$$

where $k = E^{1/2}$ and $k' = \sqrt{E - V_0}$ for $E \ge V_0 > 0$ or $k' = i\sqrt{V_0 - E}$ for $V_0 > E \ge 0$ and. The coefficients in Eq. (4.3.37) are given by[42]

$$\begin{aligned} \alpha_2(k) &= \frac{1}{2}e^{-ik'a}\left[\sin ka + \frac{k}{ik'}\cos ka\right]\alpha_1(k), \\ \beta_2(k) &= \frac{1}{2}e^{ik'a}\left[\sin ka - \frac{k}{ik'}\cos ka\right]\alpha_1(k), \\ \alpha_3(k) = \overline{\beta_3(k)} &= \frac{1}{4}e^{-ikb}\left[(1 + k'/k)e^{ik'(b-a)}\left(\sin ka + \frac{k}{ik'}\cos ka\right)\right. \\ &\quad \left. + (1 - k'/k)e^{-ik'(b-a)}\left(\sin ka - \frac{k}{ik'}\cos ka\right)\right]\alpha_1(k), \end{aligned} \tag{4.3.38}$$

where $\alpha_1(k)$ is determined by normalization conditions.

Given the full set of solutions $\{\psi_E(x)\}_{E\in\mathbb{R}^+}$ for the continuous energy spectrum one can find the sets $\{\psi_E^{\pm}\}_{E\in\mathbb{R}^+}$ of solutions of the Lippmann–Schwinger equation corresponding to incoming and outgoing asymptotic conditions. We have

$$\begin{aligned} \langle x|\phi_E^+\rangle &\equiv \psi_E^+(x) = -\frac{1}{2i}\frac{\psi_E(x)}{\beta_3(k)}, \\ \langle x|\phi_E^-\rangle &\equiv \psi_E^-(x) = \frac{1}{2i}\frac{\psi_E(x)}{\alpha_3(k)}. \end{aligned} \tag{4.3.39}$$

where $\psi_E^+(x)$ and $\psi_E^-(x)$ are, respectively, the incoming and outgoing Lippmann–Schwinger solutions. The normalization conditions for the Lippmann–Schwinger states in Eq. (4.3.39) give us $\alpha_1(k) = (2\pi k)^{-1/2}$. In the energy representation the S-matrix is given by

$$\hat{S}_{QM}(E) = -\frac{\alpha_3(k)}{\beta_3(k)}, \qquad k = E^{1/2} .$$

The above expression for the S-matrix leads to the calculation of the scattering resonances of the problem. Observe that $\alpha_3(k)$ and $\beta_3(k)$ can be extended to analytic functions in the complex k plane and the poles of the analytic continuation of $\hat{S}_{QM}(E)$ across the cut along the positive real axis in the complex energy plane into the lower half-plane are identified with zeros of $\beta_3(k)$ under the condition that $\alpha_3(k)$ does not

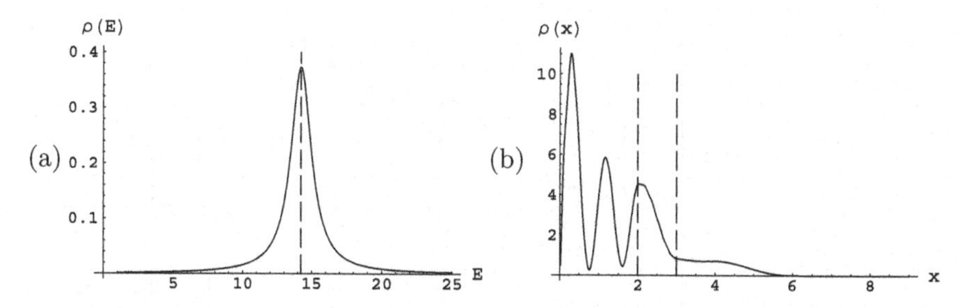

Fig. 4.2 **a** Energy probability distribution $\rho(E) = \|\psi_{\mu_3}\|^{-2}|\langle E^-|\psi_{\mu_3}\rangle|^2$ for the third resonance at $\mu_3 = 14.2336 - i\,0.8923$ of the square barrier potential problem. The vertical dashed line indicates the location of the peak at the resonance energy $E_{\mu_3} = 14.2336$. **b** Spatial probability distribution $\rho(x) = \|\psi_{\mu_3}\|^{-2}|\langle x|\psi_{\mu_3}\rangle|^2$ for the third resonance at $\mu_3 = 14.2336 - i\,0.8923$ of the square barrier potential problem. The vertical dashed lines show the location of the barrier potential

vanish at such points. Note that all of the resonance poles of the system are simple. For a resonance pole $z = \mu_j$ in the lower half-plane below the positive real axis we set $\mu_j = E_{\mu_j} - i\Gamma_{\mu_j}/2$ with $E_{\mu_j} > 0$ the resonance energy and $\Gamma_{\mu_j} > 0$ the resonance width.

For barrier parameters $a = 2$, $b = 3$ and $V_0 = 10$ the three lowest energy resonance poles are located at the points $\mu_1 = 1.8213 - i\,0.0023$, $\mu_2 = 7.0237 - i\,0.0564$ and $\mu_3 = 14.2336 - i\,0.8923$. As an example we consider the resonance pole at μ_3. Figure 4.2a displays the probability density function $\rho(E) = \|\psi_{\mu_3}\|^{-2}|(\phi_E^-, \psi_{\mu_3})|^2$ of the resonance state ψ_{μ_3} in the outgoing energy representation as a function of the energy E. This probability density function is obtained using Eq. (4.3.35). A nice peak can be seen in Fig. 4.2a at the resonance energy $E_{\mu_3} = 14.2336$. We note that in the incoming energy representation the probability density function $|(\phi_E^+, \psi_{\mu_3})|^2$ is the same as in Fig. 4.2a since, by unitarity, for every $E \in \mathbb{R}_+$ the scattering matrix $\hat{S}_{QM}(E)$ is a phase factor. With the help of the outgoing Lippmann–Schwinger eigenfunctions $\psi_E^-(x) = \langle x|\phi_E^-\rangle$, given by Eqs. (4.3.37), (4.3.38) and (4.3.39), the resonance states of the problem can be calculated using Eq. (4.3.36). Thus we obtain

$$\psi_{\mu_3}(x) = \frac{i}{2\pi\sqrt{\mu_3}}\int_0^\infty dE\,\psi_E^-(x)\frac{1}{\sqrt{E}}B\left(\frac{1}{2} + \frac{i}{2\pi}\log\frac{E}{\mu_3}, -\frac{i}{2\pi}\log\frac{E}{\mu_3}\right). \quad (4.3.40)$$

Figure 4.2b displays a numerical calculation, using Eq. (4.3.40), of the spatial probability density $\rho(x) = \|\psi_{\mu_3}\|^{-1}|\psi_{\mu_3}(x)|^2$ of the resonance state ψ_{μ_3}. The two dashed vertical lines in Fig. 4.2b correspond to the location of the barrier potential. Of course, similar calculations can be worked out for all other resonances of the problem.

Figure 4.3 presents the application of the modified Lax–Phillips formalism to the process of scattering of the state ψ_{μ_3} off the square barrier potential. This figure contains a series of snapshots at various points of time in the course of the Schrödinger

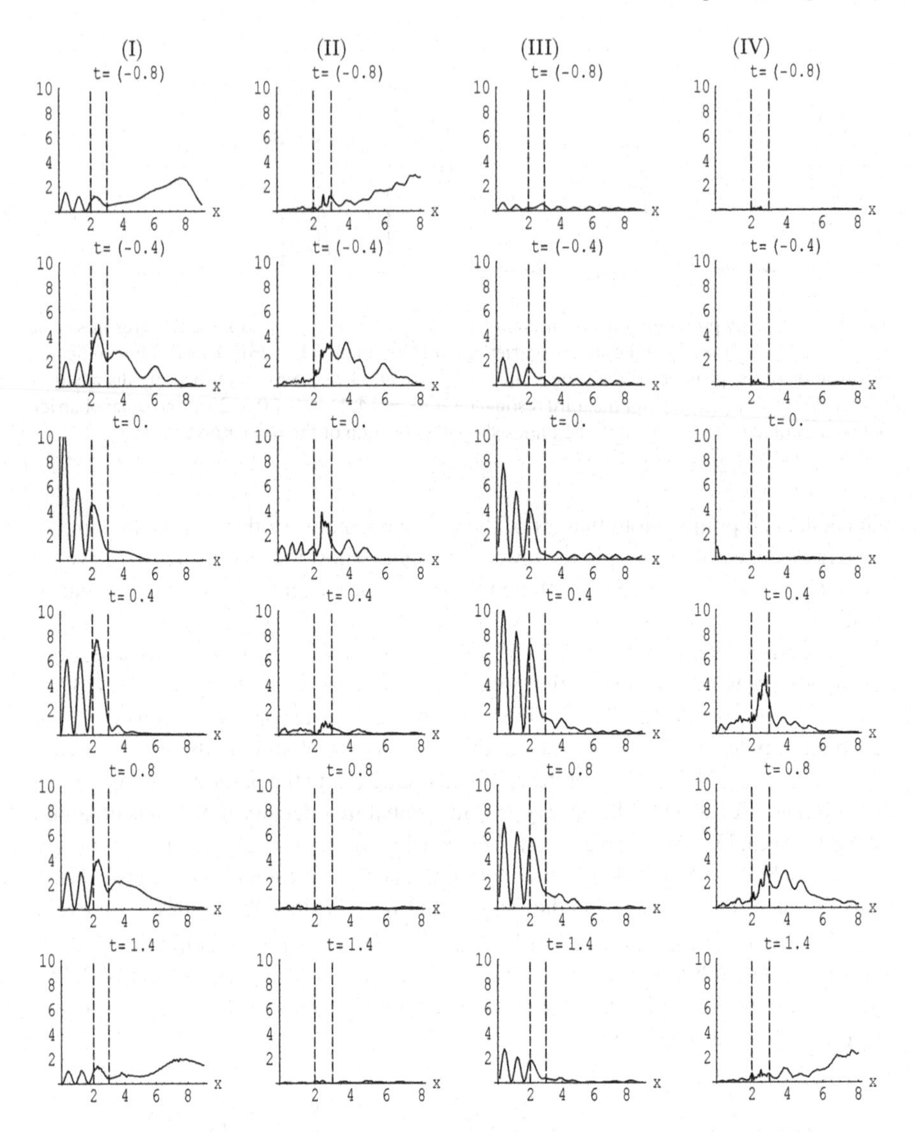

Fig. 4.3 Transition decomposition of $\psi_{\mu_3}(t)$. Each row of graphs is a time frame at time t indicated. In each row: (I)—Spatial probability distribution $\|\psi_{\mu_3}\|^{-2}|\psi_{\mu_3}(x,t)|^2$, (II)—Backward asymptotic component $\|\psi_{\mu_3}\|^{-2}|\psi^b_{\mu_3,-}(x,t)|^2$, (III)—Transient component $\|\psi_{\mu_3}\|^{-2}|\Lambda_+\Lambda_-\psi_{\mu_3}(x;t)|^2$, (IV)—Forward asymptotic component $\|\psi_{\mu_3}\|^{-2}|\psi^f_{\mu_3,+}(x,t)|^2$. The dashed lines show the location of the potential barrier

evolution $\psi_{\mu_3}(t) = e^{-iHt}\psi_{\mu_3}$ of the resonance state ψ_{μ_3}. Each row in Fig. 4.3 is such a snapshot corresponding to a particular time t. Graph number (I) in each row (the left most graph) shows the spatial probability distribution $\|\psi_{\mu_3}\|^{-2}|\psi_{\mu_3}(x,t)|^2$ at time t. Thus, column (I) in Fig. 4.3 shows the Schrödinger evolution of ψ_{μ_3}. Column number (II) in Fig. 4.3 (second graph from the left in each row) shows the time evolution of the spatial probability density $\|\psi_{\mu_3}\|^{-2}|\psi^b_{\mu_3,-}(x,t)|^2$ of $\psi^b_{\mu_3,-}(t) := (I - \Lambda_-)\psi_{\mu_3}(t)$, corresponding to backward asymptotic component of the backward transition representation in the fourth line in Eq. (4.1.26). It can be seen in column (II) in Fig. 4.3 that the time evolution of this component (which in the Lax–Phillips theory corresponds to the projection $P_-^{\perp}\psi(t)$ on the incoming subspace $\mathcal{D}_-$) consists essentially of incoming waves and outgoing waves are not seen in this channel. Hence, this component extracts the incoming part of wave function $\psi_{\mu_3}(t)$ propagating towards the scattering region. Column number (III) (third graph from the left) in Fig. 4.3 shows the evolution of the spatial probability density of the component $\Lambda_+\Lambda_-\psi_{\mu_3}(t)$. This component correspond to an approximate projection of $\psi_{\mu_3}(t)$ on the resonance subspace. Indeed, this component is analogous to the component $P_+P_-\psi(t) = P_{res}\psi(t)$ in the Lax–Phillips case, i.e., the projection of $\psi(t)$ on the subspace of resonances. It can be seen by looking at column (III) in Fig. 4.3 that this channel shows the excitation of the resonance and its subsequent decay but does not show either incoming waves or outgoing waves. Column (IV) in Fig. 4.3 shows the time evolution of (the spatial probability density of) $\psi^f_{\mu_3,+}(t) := (I - \Lambda_+)\psi_{\mu_3}(t)$, corresponding to the forward asymptotic component of the forward transition representation in the third line in Eq. (4.1.26). It can be seen in column (IV) in Fig. 4.3 that the time evolution of this component (which in the Lax–Phillips theory corresponds to the projection $P_+^{\perp}\psi(t)$ on the outgoing subspace $\mathcal{D}_+$) consists essentially of outgoing waves and incoming waves are not seen in this channel. Hence, this component extracts the outgoing part of wave function $\psi_{\mu_3}(t)$ propagating away from the scattering region.

4.4 Summary

The mathematical structure of the Lax–Phillips scattering theory includes certain features which render it an ideal framework for the description of scattering resonances. Notably, such features are the fact that a resonance is associated with a state in Hilbert space and the fact that the evolution of a resonance is given in terms of a continuous semigroup. However, the original Lax–Phillips formalism cannot be directly applied to large classes of quantum mechanical scattering problems, due to the fact that the incoming and outgoing subspaces $\mathcal{D}_\pm$ of the Lax–Phillips theory do not exist in these cases and the construction of essential components of the theory, such as its translation and spectral representations etc. cannot proceed as in the original formalism.

In this chapter we have studied a Modified Lax–Phillips Theory which is applicable to scattering problems for which the original Lax–Phillips theory does not apply. The construction of the modified theory starts with the characterization of the projection operators $P_\pm$ of the original Lax–Phillips theory as Lyapunov operators and rests heavily on the construction of analogous Lyapunov operators $M_\pm$ for a broad class of scattering problems. The Lyapunov operators $M_\pm$ are then used for the definition of mathematical objects, constructions and results analogous to those of the original Lax–Phillips theory. These include the approximate projection operators $\Lambda_\pm$ (the square roots of $M_\pm$), incoming and outgoing translation representations, approximate Lax–Phillips semigroup $\{Z_{app}(t)\}_{t\geq 0}$ and the association of poles of the scattering matrix with resonance states. As its name suggests, the approximate Lax–Phillips semigroup is not an exact semigroup and the resonance states identified within the framework of the modified formalism correspond to semigroup behavior only in an approximate sense. The quality of this approximation, i.e., the extent to which the evolution of a particular resonance state satisfies a semigroup evolution law with respect to the approximate Lax–Phillips semigroup, depends on the quality of the corresponding resonance in terms of its sharpness on the energy axis and the quality of the phase shifts it exhibits (measured by their deviation from those exhibited by an ideal, Breit–Wigner resonance). Recall that the operator $\Lambda_+\Lambda_-$ is the analogue in the modified Lax–Phillips formalism of the projection operator $P_{res} = P_+P_-$ on the resonance subspace in the original Lax–Phillips theory. The basic inequality for a resonance state $\tilde{\psi}_\mu$ in the modified formalism (Eq. (4.2.3))

$$\|Z_{app}(t)\tilde{\psi}_\mu - e^{-i\mu t}\tilde{\psi}_\mu\| \leq \|\tilde{\psi}_\mu - \Lambda_+\Lambda_-\tilde{\psi}_\mu\|,$$

shows that the more the operator $\Lambda_+\Lambda_-$ behaves as a projection on the resonance subspace when applied to $\tilde{\psi}_\mu$ the closer the resonance state $\tilde{\psi}_\mu$ is to be an eigenstate of approximate Lax–Phillips semigroup. Theorem 4.9 shows that the resonance evolution gets closer to the ideal semigroup behavior for resonances close to the real energy axis ("sharper" resonances) and exhibiting better (more Breit–Wigner like) phase shift behavior.

To complement the abstract formalism of the Modified Lax–Phillips Theory presented in this chapter we have included, in Sect. 4.3, a discussion of its implementation in the particularly important case $\sigma_{ac}(H) = \sigma_{ac}(H_0) = \mathbb{R}_+$ and an application, in Sect. 4.3.1, to a specific problem exhibiting the basic features of the formalism. Note that an important property of the modified Lax–Phillips construction, as presented above, is its locality, i.e., it is applicable to an arbitrary subinterval $[a, b] \subseteq \sigma_{ac}(H)$. Thus one may apply an analysis of resonance scattering locally on the spectrum of the Hamiltonian (for intervals not including branching points where the multiplicity of the a.c. spectrum changes).

Chapter 5
Quantization of Geodesic Deviation

5.1 Introduction[1]

The Hamilton equations for Hamiltonian systems of the type discussed, for example, by Gutzwiller (1990) and Curtiss and Miller (1985), with Hamiltonian of the form (which we call *geometrical*)

$$H(x, p) = \frac{1}{2m} g^{ij}(x) p_i p_j \tag{5.1.1}$$

where g^{ij} is invertible, generate geodesic motion. A very sensitive measure of the stability of geodesic motion is the geodesic deviation, the rate of separation of adjacent geodesic curves. As we shall see, the local deviations satisfy an equation of the form of a parametric oscillator, for which its stability provides a sensitive measure of the stability of the motion.

As we discuss in Chap. 6, a Hamiltonian of the standard form

$$H = \frac{p^2}{2m} + V(x) \tag{5.1.2}$$

can be made equivalent to such a geometrical form by introducing an ad hoc structure of the metric

$$g^{ij} = \phi(x)\delta^{ij}, \tag{5.1.3}$$

where (Horwitz and Ben Zion 2007)

$$\phi(x) = \frac{E}{E - V(x)} \tag{5.1.4}$$

[1]This chapter is largely based on our Work reported in Strauss et. al. (2015).

L. Horwitz and Y. Strauss, *Unstable Systems*, Mathematical Physics Studies,
https://doi.org/10.1007/978-3-030-31570-2_5

or rigorously by means of a canonical transformation (Strauss et. al. 2017) (for which g^{ij} may not necessarily have the form (5.1.4)), as we discuss in Chap. 7.

Maassen (1989) has shown that a classical dissipative oscillator can be quantized, and that the quantized excitations may be interpreted as photons emitted from the dissipative system. In this chapter, we apply this procedure to the parametric oscillator arising from the geometrical deviations generated by the Hamiltonian (5.1.1), and interpret these excitations as arising from the interaction of the system with the dynamical medium.

The Hamilton equations associated with the geometrical Hamiltonian (5.1.1) result in the geodesic equations

$$\ddot{x}^i + \Gamma^i{}_{jk}\dot{x}^j\dot{x}^k = 0, \tag{5.1.5}$$

where $\dot{x}^i = \frac{dx^i}{dt}$ (in the geometrical picture t is understood as the arc length parameter), and $\Gamma^i{}_{jk}$ is the compatible connection form associated with g^{ij}. The stability of the geodesic flow is locally determined by the geodesic deviation equation

$$\frac{D^2\xi^i}{dt^2} + R^i{}_{jk\ell}\dot{x}^k\dot{x}^\ell\xi^k = 0, \tag{5.1.6}$$

where $\frac{D}{dt}$ is the covariant derivative constructed with $\Gamma^i{}_{jk}$, $R^i{}_{jk\ell}$ is the corresponding curvature tensor, and ξ^i are the components of the geodesic deviation vector. Equation (5.1.6), discussed further below, is a parametric oscillator equation. The stability of this oscillator, associated with the divergence or convergence of the geodesic curves under small perturbation, corresponds to the local stability of the motion. It has been demonstrated, in a large number of cases (Horwitz et. al. 2007a; Ben Zion and Horwitz 2007, 2008, 2010), that the stability of geodesic motion with a conformal metric of the form (5.1.3) is related to stability of the motion generated by the Hamiltonian (5.1.2) (as seen by means of the Lyapunov exponent and Poincaré plots). Our discussion of the canonical transformation to the geometric form, which we study in Chap. 7, makes this relation clear.

Casseti et al. (1996) have pointed out that this oscillator is essentially parametric due to curvature variations along the motion; they associate dynamical instability with parametric instability of the oscillator. In the next section, we discuss a rigorous representation of the geodesic deviation in terms of such a parametric oscillator. Under an adiabatic assumption, assuming slow changes in the coefficient of ξ^j, the motion generated by this equation can be imbedded into a *unitary evolution* in a Hilbert space through the process of dilation; a more complete discussion of dilation can be found in the book of Nagy and Foias (1970), which provides extra degrees of freedom corresponding to the dynamical environment. One can then understand the corresponding oscillatory, stable (decaying) or unstable behavior of the oscillator, in terms of the effective interaction of the system with a physical environment. The second quantization of the embedding Hilbert space then represents, as in the study of Maassen (1989), the excitation modes of the system in a dynamical environment.

This phenomenon is a remarkable property of complex systems which lends itself to a rigorous description in terms of dilation and second quantization.

We remark that, as will be seen, the natural setting for the application of the dilation and second quantization procedures are function spaces *defined along the geodesic with respect to which the geodesic deviation occurs.*

5.2 Parametric Oscillator Representations of the Geodesic Deviation Equations

In the following we introduce the basic mathematical definitions for the description of the dynamical properties of the geodesic deviation associated with a geodesic flow on a Riemannian manifold. These methods apply directly to the geodesic flow generated by the geometric Hamiltonian (5.1.1).

Consider an n dimensional Riemannian manifold $\mathcal{M}$ and a geodesic curve γ in $\mathcal{M}$. Let $T\mathcal{M}$ be the tangent bundle of $\mathcal{M}$ and let $T_p\mathcal{M}$ be the tangent space to $\mathcal{M}$ at the point $p \in \mathcal{M}$. We denote the part of $T\mathcal{M}$ over the geodesic γ by $T_\gamma\mathcal{M}$, and let $p_0 \in \gamma$ be an arbitrary point on the geodesic γ. We define an arc length parametrization $\gamma(\cdot) : \mathbf{R} \mapsto \mathcal{M}$ of γ,

$$\gamma(s) = \exp_{p_0}(s\mathbf{v}), \tag{5.2.1}$$

where, as indicated by the subscript on the exponent, $\mathbf{v} \in T_{p_0}\mathcal{M}$ is a unit tangent vector to γ at the point p_0. For $s \geq 0$, $\gamma(s)$ is a point an arc length distance s from the point p_0 along the geodesic curve that starts at p_0 and has tangent vector $\mathbf{v}$ at p_0. For $s < 0$, we take $\gamma(s)$ as a point at arc length $|s|$ from p_0 for which $-\mathbf{v}$ is the tangent vector at p_0. We shall assume for our present purposes that $\mathcal{M}$ is geodesically complete, i.e., that all of the geodesics starting at an arbitrary point in $\mathcal{M}$ can be continued indefinitely.

Let us now define a scalar product $\langle \cdot, \cdot \rangle_{T_p\mathcal{M}}$ in the tangent space $T_p\mathcal{M}$ by

$$\langle \mathbf{X}, \mathbf{Y} \rangle_{T_p\mathcal{M}} = g(\mathbf{X}, \mathbf{Y}), \tag{5.2.2}$$

where g is the metric tensor of $\mathcal{M}$ and $\mathbf{X}, \mathbf{Y} \in T_p\mathcal{M}$ are two tangent vectors at the point p.

We furthermore assume that the connection on $\mathcal{M}$ is given by the Christoffel symbols, so that the covariant derivative of the metric tensor vanishes. If we denote the covariant derivative along the smooth curve $\gamma(t)$, parametrized by t, by $\frac{\nabla}{dt}$ (corresponding to $\frac{D}{dt}$ in (5.1.6) in a more geometric notation), for $\mathbf{X}(t)$ and $\mathbf{Y}(t)$ two smooth vector valued functions defined along γ, we have

$$\frac{d}{dt}\langle \mathbf{X}, \mathbf{Y} \rangle_{T_{\gamma(t)}\mathcal{M}} = \left\langle \frac{\nabla \mathbf{X}(t)}{dt}, \mathbf{Y}(t) \right\rangle_{T_{\gamma(t)}\mathcal{M}} + \left\langle \mathbf{X}(t), \frac{\nabla \mathbf{Y}(t)}{dt} \right\rangle_{T_{\gamma(t)}\mathcal{M}}. \tag{5.2.3}$$

In particular, if the two functions $\mathbf{X}(t)$ and $\mathbf{Y}(t)$ are obtained by parallel transport along γ of two vectors $\mathbf{X}, \mathbf{Y} \in T_{p_0}\mathcal{M}$, their covariant derivatives vanish, and therefore the t-derivative of the scalar product $\langle \mathbf{X}, \mathbf{Y} \rangle_{T_{\gamma(t)}\mathcal{M}}$ vanishes.

Let us denote by $\phi(s)\mathbf{X}$ the parallel transport of $\mathbf{X} \in T_{\gamma(t)\mathcal{M}}$ to the tangent space $T_{\gamma(t+s)\mathcal{M}}$ along the curve γ. Then, $\phi(\cdot)$ defines a continuous surjective mapping $\phi(\cdot) : T_\gamma\mathcal{M} \mapsto T_\gamma\mathcal{M}$ and, under the compatibility assumption above, we have

$$\langle \phi(s)\mathbf{X}, \phi(s)\mathbf{Y} \rangle_{T_{\gamma(t+s)\mathcal{M}}} = \langle \mathbf{X}, \mathbf{Y} \rangle_{T_{\gamma(t)\mathcal{M}}} \quad \mathbf{X}, \mathbf{Y} \in T_{\gamma(t)}\mathcal{M}. \tag{5.2.4}$$

Therefore, parallel transport is a *unitary mapping* between the tangent spaces along γ.

Now, let γ be a geodesic curve parametrized by arc length parameter s as in Eq. (5.2.1). Further, let $C^1(\mathbf{R}; T_{p_0}\mathcal{M})$ be the space of all C^1 vector valued functions defined on the real axis $\mathbf{R}$ taking values in the vector space $T_{p_0}\mathcal{M}$. Let $C^1(\mathbf{R}; T_\gamma\mathcal{M})$ be the space of all C^1 vector valued functions defined on the real axis $\mathbf{R}$ taking values in $T_\gamma\mathcal{M}$ and satisfying the condition that

$$\mathbf{X}(\cdot) \in C^1(\mathbf{R}; T_\gamma\mathcal{M}) \Rightarrow \mathbf{X}(\mathbf{s}) \in T_{\gamma(s)}\mathcal{M}, \quad \forall s \in \mathbf{R}. \tag{5.2.5}$$

Let $\mathbf{X}(\cdot), \mathbf{Y}(\cdot) \in C^1(\mathbf{R}; T_\gamma\mathcal{M})$ be arbitrary vector valued functions. Setting $\tilde{\mathbf{X}}(s) = \phi^{-1}(s)\mathbf{X}(s)$ and $\tilde{\mathbf{Y}}(s) = \phi^{-1}(s)\mathbf{Y}(s)$, we have $\tilde{\mathbf{X}}(\cdot), \tilde{\mathbf{Y}}(\cdot) \in T_{p_0}\mathcal{M}$ and so the functions $\mathbf{X}(\cdot), \mathbf{Y}(\cdot)$ belong to $C^1(\mathbf{R}; T_{p_0}\mathcal{M})$.

By the unitarity of the parallel transport, for such arbitrary vector valued functions $\mathbf{X}(\cdot), \mathbf{Y}(\cdot) \in C^1(\mathbf{R}; T_\gamma\mathcal{M})$, we have

$$\langle \mathbf{X}(s), \mathbf{Y}(s) \rangle_{T_{\gamma(s)\mathcal{M}}} = \langle \phi^{-1}(s)\mathbf{X}(s), \phi^{-1}(s)\mathbf{Y}(s) \rangle_{T_{p_0}\mathcal{M}} = \langle \tilde{\mathbf{X}}(s), \tilde{\mathbf{Y}}(s) \rangle_{T_{p_0}\mathcal{M}}. \tag{5.2.6}$$

We may therefore define a mapping $W_\gamma : C^1(\mathbf{R}; T_{p_0}\mathcal{M}) \mapsto C^1(\mathbf{R}; T_\gamma\mathcal{M})$ by

$$[W_\gamma \tilde{\mathbf{X}}(\cdot)](s) = \phi(s)\tilde{\mathbf{X}}(s), \quad \tilde{\mathbf{X}}(\cdot) \in C^1(\mathbf{R}; T_{p_0}\mathcal{M}) \tag{5.2.7}$$

and according to (5.2.6), we then have

$$\langle \tilde{\mathbf{X}}(s), \tilde{\mathbf{Y}}(s) \rangle_{T_{p_0}\mathcal{M}} = \langle [W_\gamma \tilde{\mathbf{X}}(\cdot)](s), [W_\gamma \tilde{\mathbf{Y}}(\cdot)](s) \rangle_{T_{\gamma(s)\mathcal{M}}} \tag{5.2.8}$$

Note that the inverse of W_γ is given by

$$[W_\gamma{}^{-1}\mathbf{X}(\cdot)](s) = \phi^{-1}(s)\mathbf{X}(s), \quad \mathbf{X}(\cdot) \in C^1(\mathbf{R}; T_\gamma\mathcal{M}). \tag{5.2.9}$$

Now, let $\mathbf{Y}(\cdot) \in C^1(\mathbf{R}; T_\gamma\mathcal{M})$ be arbitrary and let $\mathbf{X}(\cdot) \in C^1(\mathbf{R}; T_\gamma\mathcal{M})$ be a vector valued function defined by $\mathbf{X}(s) = \phi(s)\mathbf{X}_0$, where $\mathbf{X}_0 \in= T_{p_0}\mathcal{M} = T_\gamma\mathcal{M}$ i.e., $\mathbf{X}(\cdot)$ is obtained by the parallel transport along γ of a fixed vector $\mathbf{X}_0 \in T_{p_0}\mathcal{M}$. Using (5.2.3), we then have

$$\frac{d}{ds}\langle \mathbf{X}(s), \mathbf{Y}(s)\rangle_{T_{\gamma(s)}\mathcal{M}} = \big\langle \frac{\mathbf{X}}{(} s), \nabla \mathbf{Y}(s) ds\big\rangle_{T_{\gamma(s)}\mathcal{M}} = \big\langle \phi(s)\mathbf{X}_0, \frac{\nabla \mathbf{Y}(s)}{ds}\big\rangle_{T_{\gamma(s)}\mathcal{M}}. \tag{5.2.10}$$

and hence

$$\begin{aligned}\big\langle \phi(s)\mathbf{X}_0, \frac{\nabla \mathbf{Y}(s)}{ds}\big\rangle_{T_{\gamma(s)}\mathcal{M}} &= \frac{d}{ds}\langle \mathbf{X}(s), \mathbf{Y}(s)\rangle_{T_{\gamma(s)}\mathcal{M}} = \frac{d}{ds}\langle \phi^{-1}(s)\mathbf{X}(s), \phi^{-1}\mathbf{Y}(s)\rangle_{T_{p_0}\mathcal{M}} \\ &= \frac{d}{ds}\langle \mathbf{X}_0, \phi^{-1}\mathbf{Y}(s)\rangle_{T_{p_0}\mathcal{M}} = \langle \mathbf{X}_0, \frac{d}{ds}[\phi^{-1}(s)\mathbf{Y}(s)]\rangle_{T_{p_0}\mathcal{M}}\end{aligned} \tag{5.2.11}$$

Furthermore, since

$$\big\langle \phi(s)\mathbf{X}_0, \frac{\nabla \mathbf{Y}(s)}{ds}\big\rangle_{T_{\gamma(s)}\mathcal{M}} = \big\langle \mathbf{X}_0, \phi^{-1}(s)\frac{\nabla \mathbf{Y}(s)}{ds}\big\rangle_{T_{p_0}\mathcal{M}}, \tag{5.2.12}$$

we obtain

$$\big\langle \mathbf{X}_0, \phi^{-1}(s)\frac{\nabla \mathbf{Y}(s)}{ds}\big\rangle_{T_{p_0}\mathcal{M}} = \big\langle \mathbf{X}_0, \frac{d}{ds}[\phi^{-1}(s)\mathbf{Y}(s)]\big\rangle_{T_{p_0}\mathcal{M}}, \tag{5.2.13}$$

and since $\mathbf{X}_0 \in T_{p_0}\mathcal{M}$ is arbitrary, it follows that

$$\phi^{-1}(s)\frac{\nabla \mathbf{Y}(s)}{ds} = \frac{d}{ds}[\phi^{-1}(s)\mathbf{Y}(s)]. \tag{5.2.14}$$

If we set $\tilde{\mathbf{Y}}(s) = \phi^{-1}(s)\mathbf{Y}(s)$ and write in short form $\mathbf{Y} \equiv \mathbf{Y}(\cdot)$, $\tilde{\mathbf{Y}} \equiv \tilde{\mathbf{Y}}(\cdot)$, we can write Eq. (5.2.14) in the form

$$\frac{\nabla}{ds}(W_\gamma \tilde{\mathbf{Y}})(s) = \big(W_\gamma \frac{d\tilde{\mathbf{Y}}}{ds}\big)(s). \tag{5.2.15}$$

The mapping W_γ then *intertwines the covariant derivative with the ordinary derivative.*

In order to discuss geodesic deviation, we now extend the mapping W_γ to more general tensor valued functions. Let us denote by $\Lambda_p(\ell, k)$ the set of all (ℓ, k) type, ℓ times contravariant and k times covariant, tensors, defined at a point $p \in \mathcal{M}$. Then, let $\Lambda_\gamma(\ell, k)$ be the bundle of all (ℓ, k)-type tensors defined at all points of a smooth curve γ, i.e., $\Lambda_\gamma(\ell, k) := \cup_{p\in\gamma}\Lambda_p(\ell, k)$ (the ordering of tensor indices is determined by the context). Now, let $C^1(\mathbf{R}, \Lambda_\gamma(\ell, k))$ be the space of all C^1 tensor valued functions of type (ℓ, k) defined on a smooth curve γ. If $T(\cdot)$ is such a tensor valued function, we denote by $T_{\gamma(t)}$ its value at the point $\gamma(t) \in \gamma$ so that $T_{\gamma(t)} \in \Lambda_\gamma(\ell, k)$. Let γ be a geodesic parametrized by arc length as in (5.2.1) and let $p_0 = \gamma(0)$. Furthermore, let $C^1(\mathbf{R}, \Lambda_{p_0}(\ell, k))$ be the space of all (ℓ, k)-type tensor valued functions defined on the real axis $\mathbf{R}$ and taking values in $\Lambda_{p_0}(\ell, k)$. We define a mapping $W_\gamma : C^1(\mathbf{R}, \Lambda_{p_0}(\ell, k) \mapsto T \in C^1(\mathbf{R}, \Lambda_\gamma(\ell, k)$ such that for each tensor valued function $\tilde{T}(\cdot)) \in C^1(\mathbf{R}, \Lambda_{p_0}(\ell, k)$, we have

$$[W_\gamma \tilde{T}](s) := \phi(s)\tilde{T}_{\gamma(s)}. \tag{5.2.16}$$

If $T \in C^1(\mathbf{R}), \Lambda_\gamma(\ell, k)$ is a tensor valued function defined on the geodesic γ, the action of the inverse of W_γ in T is given by

$$[W_\gamma{}^{-1} T](s) := \phi(-s)T_{\gamma(s)}. \tag{5.2.17}$$

Now, we turn to the geodesic deviation equation. Let γ be a geodesic on $\mathcal{M}$. We vary γ into a family $\{\gamma_\alpha\}_{\alpha \in I}$ of geodesics depending on a parameter $\alpha \in (-\delta, \delta) = I$, with $\gamma_{\alpha=0} \equiv \gamma_0 = \gamma$. We consider all of the geodesics in the family to be parametrized by the arc length parameter s, as in Eq. (5.2.1). Thus, in terms of coordinates $x = (x^1, x^2, \ldots x^n)$ on a coordinate patch in $\mathcal{M}$, the coordinates along γ are $x(s, 0)$; along γ_α they are $x(s, \alpha)$. If we denote symbolically as $\vec{\mathbf{x}}(\alpha, s)$ the points on the surface parametrized by α and s, then the geodesic deviation vector is defined to be

$$\mathbf{J}(s) := \frac{\partial \vec{\mathbf{x}}(s, \alpha)}{\partial \alpha}|_{\alpha=0} \tag{5.2.18}$$

and if the coordinate basis vectors associated with the coordinates x are denoted by $\vec{e}_i \equiv \partial_i, i = 1, 2, \ldots n$, then we have

$$\mathbf{J}(s) = \frac{\partial x^i(s, \alpha)}{\partial \alpha}|_{\alpha=0} \vec{e}_i(s) = \xi^i(s)\vec{e}_i(s), \tag{5.2.19}$$

where $\xi^i(s) = \frac{\partial x^i(\alpha, s)}{\partial \alpha}|_{\alpha=0}$. Since

$$x^i(s, \alpha) - x^i(s, 0) = \alpha \left(\frac{\partial x^i(s, \alpha)}{\partial \alpha}|_{\alpha=0} \right) + O(\alpha^2) \tag{5.2.20}$$

for small α, then $\alpha\mathbf{J}(s)$ is a vector representing the linear approximation to the separation between the geodesic γ and the geodesic γ_α.

Now, let $\mathbf{X}, \mathbf{Y}, \mathbf{Z} \in T_p\mathcal{M}$ be vectors and let $R_p(\mathbf{X}, \mathbf{Y}) : T_p\mathcal{M} \mapsto T_p\mathcal{M}$ be the curvature transformation at the point p, i.e., $\frac{\partial x^i(s,\alpha)}{\partial \alpha}|_{\alpha=0}$ is a linear transformation with matrix elements $[R_p(\mathbf{X}, \mathbf{Y})]^i{}_j = R^i{}_{jk\ell} X^k Y^\ell$ so that

$$R_p(\mathbf{X}, \mathbf{Y})\mathbf{Z} = (R^i{}_{jk\ell} X^k Y^\ell Z^j)\vec{e}_i = (R^i{}_{jk\ell} X^k Y^\ell Z^j)\partial_i. \tag{5.2.21}$$

The quantities $R^i{}_{jk\ell} X^k Y^\ell Z^j$ are the components of the Riemann curvature tensor (evaluated at p). Note also that for $\mathbf{W} \in T_p\mathcal{M}$, we have $< R_p(\mathbf{X}, \mathbf{Y})\mathbf{Z}, \mathbf{W} >_{T_p\mathcal{M}} = R^i{}_{jk\ell} X^k Y^\ell Z^j W_i$, where $W_i = g_{ij} W^j$. Using the above notation for the curvature transformation, the geodesic deviation equation has the form (Strauss 2008)

$$\frac{\nabla^2 \mathbf{J}(s)}{ds^2} + R_{\gamma(s)}(\mathbf{J}(s), (\mathbf{T}(s))(T(s)) = 0 \tag{5.2.22}$$

where $R_{\gamma(s)}$ is the curvature tensor at the point $\gamma(s) \in \gamma$ at the point $\gamma(s)$, $\mathbf{J}(s)$ is the geodesic deviation vector, and $\mathbf{T}(s) \equiv \mathbf{T}_{\gamma(s)}$ is the tangent vector to γ at the point $\gamma(s)$. The component representation of this equation is Eq. (5.1.6) above. Now, take a vector $\mathbf{X}_0 \in T_{p_0}\mathcal{M}$ and parallel transport it along the geodesic γ to obtain a vector valued function $\mathbf{X}(\cdot)$ given by $\mathbf{X}(s) = \phi(s)\mathbf{X}_0$. We then have

$$\langle \mathbf{X}(s), \frac{\nabla^2 \mathbf{J}(s)}{ds^2} + \mathbf{R}_{\gamma(s)}(\mathbf{J}(s), \mathbf{T}(s))(T(s))\rangle_{T_\gamma(s)\mathcal{M}} = 0. \tag{5.2.23}$$

By definition of parallel transport of tensors along γ, we obtain

$$\begin{aligned} 0 &= \langle \mathbf{X}(s), \frac{\nabla^2 \mathbf{J}(s)}{ds^2} + \mathbf{R}_\gamma(\mathbf{J}(s), \mathbf{T}(s))(\mathbf{T}(\mathbf{s})\rangle_{T_\gamma(s)\mathcal{M}} \\ &= \langle W_\gamma{}^{-1}\mathbf{X}(s), W_\gamma{}^{-1}\frac{\nabla^2 \mathbf{J}(s)}{ds^2} + R_{\gamma(s)}(\mathbf{J}(s), \mathbf{T}(s))T(s)\rangle_{T_\gamma(s)\mathcal{M}} \\ &= \langle (\mathbf{X}_0, \frac{d^2(W_\gamma{}^{-1}\mathbf{J})(s)}{ds^2} + (W_\gamma{}^{-1}R_{\gamma(s)})((W_\gamma{}^{-1}\mathbf{J})(s), (W_\gamma{}^{-1}\mathbf{T})(s))(W_\gamma{}^{-1}\mathbf{T}(s)))\rangle_{T_{p_0}\mathcal{M}}. \end{aligned} \tag{5.2.24}$$

Recall that the geometrical form of the geodesic equation for γ (Eq. (5.1.5)) is the component representation for this expression)

$$\frac{\nabla \mathbf{T}(s)}{ds} = \mathbf{0}. \tag{5.2.25}$$

This implies that $(W_\gamma{}^{-1}\mathbf{T})(s)) = \phi(-s)T_{\gamma(s)} = \phi(-s)\mathbf{T}(s) = \mathbf{T}_0$, where $\mathbf{T}_0 \in T_{p_0}\mathcal{M}$ is the tangent vector to γ at the point $p_0 = \gamma(0)$. Hence Eq. (5.2.25) can be written in the form

$$0 = \langle (\mathbf{X}_0, \frac{d^2(W_\gamma{}^{-1}\mathbf{J})(s)}{ds^2} + (W_\gamma{}^{-1}R_{\gamma(s)})(W_\gamma{}^{-1}\mathbf{J})(s), \mathbf{T}_0)(\mathbf{T}_0)\rangle_{T_{p_0}\mathcal{M}}. \tag{5.2.26}$$

Denoting $\tilde{\mathbf{J}}(s) = W_\gamma{}^{-1}\mathbf{J}(s)$ and noting the fact that $\mathbf{X}_0 \in T_{p_0}\mathcal{M}$ is arbitrary, we finally obtain the equation

$$\frac{d^2\tilde{\mathbf{J}}(s)}{ds^2} + ((W_\gamma{}^{-1}R_{\gamma(s)})\tilde{\mathbf{J}}(s), \mathbf{T}_0)\mathbf{T}_0 = \mathbf{0}. \tag{5.2.27}$$

The second term on the left hand side can be regarded as a linear transformation of $\tilde{\mathbf{J}}(s)$. If we set

$$R_s\mathbf{X} := ((W_\gamma{}^{-1}R_{\gamma(s)})\mathbf{X}, \mathbf{T}_0)\mathbf{T}_0 \;\; \forall \mathbf{X} \in T_{p_0}\mathcal{M} \tag{5.2.28}$$

then for each $s \in \mathbf{R}$, $R_s : T_{p_0}\mathcal{M} \mapsto T_{p_0}\mathcal{M}$ is a linear operator on $T_{p_0}\mathcal{M}$.

Therefore, the Jacobi field $\mathbf{J}$ satisfies the geodesic deviation Eq. (5.2.22) if and only if the vector valued function $\dot{\tilde{\mathbf{J}}}(\cdot) \in C^1(\mathbf{R}; T_{p_0}\mathcal{M})$ satisfies the equation

$$\frac{d^2\tilde{\mathbf{J}}(s)}{ds^2} + R_s(\tilde{\mathbf{J}})(s) = \mathbf{0}, \tag{5.2.29}$$

where $R_s : \mathbf{R} \mapsto \mathcal{B}(T_0\mathcal{M})$ is an operator valued function defined on $\mathbf{R}$ and taking values in the space $\mathcal{B}(T_0\mathcal{M})$ of bounded linear operators on $(T_0\mathcal{M})$. We regard (5.2.29) as an operator valued parametric oscillator equation.

The result (5.2.27) shows rigorously that the geodesic deviation equation, containing a second covariant derivative, is exactly representable, via a unitary transformation, by a parametric oscillator equation with ordinary second derivative.

5.3 Dynamical System Representation of the Geodesic Deviation Equation

We have seen that the solutions $\mathbf{J}(s)$ of the geodesic deviation Eq. (5.2.22) are mapped, by the application of $W_\gamma{}^{-1}$, into solutions $\tilde{\mathbf{J}}(s)$ of the (generalized) parametric oscillator equation (5.2.27). Moreover, the mapping W_γ is one-to-one and onto, and hence the behavior of solutions of Eq. (5.2.22) can be studied by an analysis of the corresponding solutions of Eq. (5.2.27). Since this equation is second order, it is the square root of the eigenvalues which are relevant; to avoid the associated ambiguities, we rewrite the system as a pair of first order equations, as is usually done for dynamical systems. Let us first consider the simpler case for which R_s is independent of s, i.e., $R_s = R_0 \ \forall s \in \mathbf{R}$. Then we have, for all $\mathbf{X} \in T_{p_0}\mathcal{M}$,

$$R_s\mathbf{X} = (W_\gamma{}^{-1}R_{\gamma(s)})(\mathbf{X}, \mathbf{T}_0)(\mathbf{T}_0) = R_0\mathbf{X}. \tag{5.3.1}$$

Since $\mathbf{X}$ and $\mathbf{T}_0$ are arbitrary (our previous results apply to arbitrary geodesics starting at the point p_0), this implies that

$$R_{p_0} = R_{\gamma(0)} = W_\gamma{}^{-1}R_{\gamma(s)} \Rightarrow R_{\gamma(s)} = (W_\gamma R_{p_0})(s) = \phi(s)R_{\gamma(0)}, \tag{5.3.2}$$

i.e., R_s is independent of s if $R_{\gamma(s)}$ is the parallel transport of R_{p_0} along γ. In this simple case, (5.2.27) reduces to

$$\frac{d^2\tilde{\mathbf{J}}(s)}{ds^2} + R_0\tilde{\mathbf{J}}(s) = \mathbf{0}. \tag{5.3.3}$$

By the symmetries of the curvature tensor and the Bianchi identity one can prove that R_0 is a self-adjoint operator on the real, finite dimensional Hilbert space $T_{p_0}\mathcal{M}$, i.e., R_0 is symmetric, and therefore its eigenvalues are real. Excluding for now the null eigenvalues (non-trivial kernel), suppose the spectrum $\sigma(R_0)$ consists of positive eigenvalues ${\omega_i}^2 > 0$, $i = 1 \dots q_1$ with corresponding multiplicities k_i, $i = 1, \dots, q_1$ and negative eigenvalues $-{\eta_j}^2 < 0$, $j - 1, \dots q_2$ with corresponding multiplicities ℓ_j, $j = 1, \dots, q_2$ with $\Sigma_{i=1}^{q_1} k_i + \Sigma_{j=1}^{q_2} \ell_j = \dim \mathcal{M} = n$. To fix our conventions we set $\omega_i > 0$ for the case of positive eigenvalues of R_0 and $\eta_j > 0$ in the case of negative eigenvalues. Furthermore, for $i = 1 \dots q_1$, let $\{\hat{\mathbf{v}}_{\omega_i, r_i}\}|_{r_i=1,\dots k_i}$ be an orthonormal basis for the eigenspace $E_{\omega_i} \subseteq T_{p_0}\mathcal{M}$ corresponding to the eigenvalues ${\omega_i}^2 \in \sigma(R_0)$ and, similarly for $j = 1, \dots q_2$, let $\{\hat{\mathbf{w}}_{\eta_j, r_j}\}|_{r_j=1,\dots \ell_j}$ be an orthonormal basis for the eigenspace $E_{\eta_j} \subseteq T_{p_0}\mathcal{M}$ corresponding to the eigenvalues $-{\eta_j}^2 \in \sigma(R_0)$.

Consider a positive eigenvalue ${\omega_i}^2 > 0$ of R_0 and an arbitrary eigenvector $\mathbf{v}_{\omega_i} \in E_{\omega_i}$ corresponding to this eigenvalue. We observe that

$$\tilde{\mathbf{J}}_{\omega_i}(s) = (c_1 e^{i\omega_i s} + c_2 e^{-i\omega_i s})\mathbf{v}_{\omega_i} \tag{5.3.4}$$

is an oscillating solution of (5.3.3). Now, if $-{\eta_j}^2 < 0$ is a negative eigenvalue of R_0 and $\mathbf{w}_{\eta_j} \in E_{\eta_j}$ is an eigenvector corresponding to this eigenvalue, then

$$\tilde{\mathbf{J}}_{\eta_j}(s) = (c_1 e^{\eta_j s} + c_2 e^{-\eta_j s})\mathbf{w}_{\eta_j} \tag{5.3.5}$$

is a solution of Eq.(5.3.3). Thus, $\tilde{\mathbf{J}}_{\eta_j}(s)$ contains in this case both a stable (exponentially decreasing) and an unstable (exponentially increasing) term corresponding to the same eigenvector $\mathbf{w}_{\eta_j}$ uniquely with stable behavior or unstable behavior of the corresponding solution $\tilde{\mathbf{J}}_{\eta_j}(s)$. We treat this problem by constructing a *dynamical system* representation incorporating both Eq. (5.2.27) and the appropriate initial conditions determining a unique solution of the equation. To do this, we rewrite Eq. (5.2.27) in an equivalent form as a non-autonomous linear dynamical system

$$\frac{d}{ds}\begin{pmatrix} \tilde{\mathbf{J}}(s) \\ \frac{d\tilde{\mathbf{J}}(s)}{ds} \end{pmatrix} = \begin{pmatrix} 0 & I \\ -R_s & 0 \end{pmatrix}\begin{pmatrix} \tilde{\mathbf{J}}(s) \\ \frac{d\tilde{\mathbf{J}}(s)}{ds} \end{pmatrix} = \tilde{R}_s \begin{pmatrix} \tilde{\mathbf{J}}(s) \\ \frac{d\tilde{\mathbf{J}}(s)}{ds} \end{pmatrix}, \tag{5.3.6}$$

where

$$\tilde{R}_s = \begin{pmatrix} 0 & I \\ -R_s & 0 \end{pmatrix}. \tag{5.3.7}$$

The dynamical system represented by (5.3.6) is defined on the real vector space $T_{p_0}\mathcal{M} \oplus T_{p_0}\mathcal{M}$. However, unlike R_s, the operator $\tilde{R}_s$ is not self-adjoint, and if we lift it to the complexified Hilbert space $\mathcal{H}_{p_0} = \mathbf{C} \otimes (T_{p_0}\mathcal{M} \oplus T_{p_0}\mathcal{M})$, then its full spectrum is not a subset of $\mathbf{R}$. Therefore, we lift (5.3.6) to be considered as defined on $\mathcal{H}_{p_0}$.

Note that the dynamical system representation of (5.3.6) corresponds to putting Eq. (5.2.22) into the equivalent form

$$\frac{\nabla}{ds}\begin{pmatrix}\tilde{\mathbf{J}}(s)\\ \frac{\nabla\tilde{\mathbf{J}}(s)}{ds}\end{pmatrix}=\begin{pmatrix}0 & I\\ -R_{\gamma(s)}(\cdot,\mathbf{T}(s))(\mathbf{T}(s)) & 0\end{pmatrix}\begin{pmatrix}\tilde{\mathbf{J}}(s)\\ \frac{\nabla\tilde{\mathbf{J}}(s)}{ds}\end{pmatrix} \tag{5.3.8}$$

and applying to this equation the mapping $W_\gamma{}^{-1}$ (more accurately an extension of $W_\gamma{}^{-1}$ to $T_{p_0}\mathcal{M}\oplus T_{p_0}\mathcal{M}$). Equation (5.3.6) is, therefore, a dynamical system representation of the geodesic deviation Eq. (5.2.22).

In the simple case that $R_s = R_0$, $\forall s\in\mathbf{R}$, Eq. (5.3.6) reduces to

$$\frac{d}{ds}\begin{pmatrix}\tilde{\mathbf{J}}(s)\\ \frac{d\tilde{\mathbf{J}}(s)}{ds}\end{pmatrix}=\begin{pmatrix}0 & I\\ -R_s & 0\end{pmatrix}\begin{pmatrix}\tilde{\mathbf{J}}(s)\\ \frac{d\tilde{\mathbf{J}}(s)}{ds}\end{pmatrix}=\tilde{R}_0\begin{pmatrix}\tilde{\mathbf{J}}(s)\\ \frac{d\tilde{\mathbf{J}}(s)}{ds}\end{pmatrix}, \tag{5.3.9}$$

where

$$\tilde{R}_0=\begin{pmatrix}0 & I\\ -R_0 & 0\end{pmatrix}. \tag{5.3.10}$$

Now, if $\hat{\mathbf{v}}_{\omega_i,r_i}\in E_{\omega_i}$ is an (basis) eigenvector ($i=1,\dots q_1$) corresponding to the positive eigenvalue $\omega_i{}^2>0$ of R_0, then the vectors

$$\mathbf{u}_{\omega_i,r_i}=\begin{pmatrix}\hat{\mathbf{v}}_{\omega_i,r_i}\\ i\omega_i\hat{\mathbf{v}}_{\omega_i,r_i}\end{pmatrix},\quad \overline{\mathbf{u}}_{\omega_i,r_i}=\begin{pmatrix}\hat{\mathbf{v}}_{\omega_i,r_i}\\ -i\omega_i\hat{\mathbf{v}}_{\omega_i,r_i}\end{pmatrix},\quad \mathbf{u}_{\omega_i,r_i},\overline{\mathbf{u}}_{\omega_i,r_i}\in\mathcal{H}_{p_0}$$

satisfy

$$\tilde{R}_0\mathbf{u}_{\omega_i,r_i}=i\omega_i\mathbf{u}_{\omega_i,r_i},\quad \tilde{R}_0\overline{\mathbf{u}}_{\omega_i,r_i}=-i\omega_i\overline{\mathbf{u}}_{\omega_i,r_i}.$$

The functions

$$\tilde{\mathbf{J}}_{i\omega_i,r_i}(s)=e^{i\omega_i s}\mathbf{u}_{\omega_i,r_i},\quad \tilde{\mathbf{J}}_{-i\omega_i,r_i}(s)=e^{-i\omega_i s}\overline{\mathbf{u}}_{\omega_i,r_i}\quad \forall s\in\mathbf{r} \tag{5.3.11}$$

are then corresponding oscillating solutions of Eq. (5.3.6).

Next, consider negative eigenvalues of R_0. If $\hat{\mathbf{w}}_{\eta_j,r_j}\in E_{\eta_j}$ is an (basis) eigenvector ($j=1,\dots q_2$) corresponding to the negative eigenvalue $-\eta_j{}^2<0$ of R_0, then the vectors

$$\mathbf{u}^{+}{}_{\eta_j,r_j}=\begin{pmatrix}\hat{\mathbf{w}}_{\eta_j,r_j}\\ \eta_j\hat{\mathbf{w}}_{\eta_j,r_j}\end{pmatrix},\quad \mathbf{u}^{-}{}_{\eta_j,r_j}=\begin{pmatrix}\hat{\mathbf{w}}_{\eta_j,r_j}\\ -\eta_j\hat{\mathbf{w}}_{\eta_j,r_j}\end{pmatrix},\quad \mathbf{u}^{+}{}_{\eta_j,r_j},\,\mathbf{u}^{-}{}_{\eta_j,r_j}\in\mathcal{H}_{p_0}$$

satisfy

$$\tilde{R}_0\mathbf{u}^{+}{}_{\eta_j,r_j}=\eta_j\mathbf{u}^{+}{}_{\eta_j,r_j},\quad \tilde{R}_0\mathbf{u}^{-}{}_{\eta_j,r_j}=-\eta_j\mathbf{u}^{-}{}_{\omega_i,r_i}.$$

The functions

$$\tilde{\mathbf{J}}_{\eta_j,r_j}(s)=e^{\eta_j s}\mathbf{u}^{+}{}_{\eta_j,r_j},\quad \tilde{\mathbf{J}}_{-\eta_j,r_j}(s)=e^{-\eta_j s}\mathbf{u}^{-}{}_{\eta_j,r_j}\quad \forall s\in\mathbf{R} \tag{5.3.12}$$

are then, respectively, an unstable solution and a stable solution of Eq. (5.3.9). We conclude that for each positive eigenvalue ${\omega^2}_i > 0$ of R_0, both $i\omega_i$ and $-i\omega_i$ are eigenvalues of $\tilde{R}_0$ and for each negative eigenvalue $-{\eta^2}_j < 0$ of R_0, both η_j and $-\eta_j$ are eigenvalues of $\tilde{R}_0$. Furthermore, $\tilde{E}_{i\omega_i} = span\{\mathbf{u}_{\omega_i,r_i}\}_{r_i=1,\ldots k_i}$, $\tilde{E}_{-i\omega_i} = span\{\bar{\mathbf{u}}_{\omega_i,r_i}\}_{r_i=1,\ldots k_i}$, $\tilde{E}_{\eta_j} = span\{{\mathbf{u}^+}_{\eta_j,r_j}\}_{r_j=1,\ldots \ell_j}$ and $\tilde{E}_{-\eta_j} = span\{{\mathbf{u}^-}_{\eta_j,r_j}\}_{r_j=1,\ldots \ell_j}$ are, respectively, k_i, k_i, ℓ_j and ℓ_j dimensional subspaces of $\mathcal{H}_{p_0}$ which are eigenspaces, respectively, for the eigenvalues $i\omega_i$, $-i\omega_i$, η_j, and $-\eta_j$ of $\tilde{R}_0$. These eigenspaces satisfy

$$\mathcal{H}_{p_0} = \mathbf{C} \otimes (T\mathcal{M}_{p_0} \oplus T\mathcal{M}_{p_0}) = (\oplus_{i=1}{}^{q_1} \tilde{E}_{i\omega_i}) \oplus (\oplus_{i=1}{}^{q_1} \tilde{E}_{-i\omega_i}) \oplus (\oplus_{j=1}{}^{q_2} \tilde{E}_{\eta_j}) \oplus (\oplus_{j=1}{}^{q_2} \tilde{E}_{-\eta_j}).$$

If we set

$$\mathcal{H}^c{}_{p_0} := (\oplus_{i=1}{}^{q_1} \tilde{E}_{i\omega_i}) \oplus (\oplus_{i=1}{}^{q_1} \tilde{E}_{-i\omega_i}), \quad \mathcal{H}^u{}_{p_0} := \oplus_{j=1}{}^{q_2} \tilde{E}_{\eta_j}, \quad \mathcal{H}^s{}_{p_0} := \oplus_{j=1}{}^{q_2} \tilde{E}_{-\eta_j},$$

then according to Eqs. (5.3.11) and (5.3.12), $\mathcal{H}^s{}_{p_0}$ is a stable manifold, $\mathcal{H}^u{}_{p_0}$ is an unstable manifold, and $\mathcal{H}^c{}_{p_0}$ is an oscillating (or central) manifold for the dynamical system in Eq. (5.3.9) and we have a decomposition of the form

$$\mathcal{H}_{p_0} = \mathcal{H}^c{}_{p_0} \oplus \mathcal{H}^s{}_{p_0} \oplus \mathcal{H}^u{}_{p_0} \tag{5.3.13}$$

We conclude the present section by restating its main result, i.e., the fact that the geodesic deviation Eq. (5.2.22) can be represented in terms of the non-autonomous linear dynamical system in Eq. (5.3.6) defined on the complex Hilbert space $\mathcal{H}_{p_0} = \mathbf{C} \otimes (T\mathcal{M}_{p_0} \oplus T\mathcal{M}_{p_0})$. We shall see that the restriction of the evolution of the dynamical system to $\mathcal{H}^s{}_{p_0}$ and $\mathcal{H}^u{}_{p_0}$ corresponds to semigroups to which the dilation procedure of Nagy and Foias (1970), followed by second quantization, may be applied.

5.4 Isometric Dilation of the Geodesic Deviation

In this section, we discuss in the first part isometric dilation of the stable and unstable solutions for the geodesic deviation, and in the second, dilations of the geodesic evolution. In the next section, we discuss second quantization.

A. Isometric Dilation of Stable and Unstable Evolutions of the Geometric Deviation
We have seen that the geodesic deviation (5.2.22) corresponds to the geodesic flow generated by a geometric Hamiltonian of the form given in Eq.(5.1.1) on a Riemannian manifold $\mathcal{M}$ can be transformed into an operator valued parametric oscillator equation, Eq. (5.2.26), and subsequently into a linear non-autonomous dynamical system, Eq.(5.3.6). Under the assumption that the operator $\tilde{R}_s$ in Eqs.(5.3.6) and (5.3.7) satisfies $\tilde{R}_s = \tilde{R}_0 \;\; \forall s \in \mathbf{R}$, which amounts, as we have seen, to the simple

case where the curvature tensor $R_{\gamma(s)}$ at the point $\gamma(s) \in \gamma$ is the parallel transport along γ of the curvature tensor R_{p_0} at the point $p_0 \in \gamma$, we obtain the simple linear autonomous dynamical system in Eq. (5.3.9). The generator of evolution of this dynamical system is the operator $\tilde{R}_0$ given in Eq. (5.3.10), i.e., if $\tilde{\phi}_0(s)|_{s\in\mathbf{R}}$ denotes the evolution of the dynamical system in Eq. (5.3.9) then, given an initial condition $(\tilde{\mathbf{J}}(0), \frac{d\tilde{\mathbf{J}}}{ds}(0))^T \in \mathcal{H}_{p_0}$, we have

$$\begin{pmatrix} \tilde{\mathbf{J}}(s) \\ \frac{d\tilde{\mathbf{J}}(s)}{ds} \end{pmatrix} = \tilde{\phi}_0(s) \begin{pmatrix} \tilde{\mathbf{J}}(0) \\ \frac{d\tilde{\mathbf{J}}}{ds}(0) \end{pmatrix} = e^{\tilde{R}_0 s} \begin{pmatrix} \tilde{\mathbf{J}}(0) \\ \frac{d\tilde{\mathbf{J}}}{ds}(0) \end{pmatrix}, \tag{5.4.1}$$

An analysis of the spectrum of $\tilde{R}_0$ leads to the decomposition in Eq. (5.3.13) of the Hilbert space $\mathcal{H}_{p_0}$ into stable, unstable, and oscillating (central) subspaces with respect to forward evolution (i.e., for positive values of s). We wish to apply here the procedure of dilation, followed by second quantization, to the stable and unstable parts of the evolution $\tilde{\phi}_0(s)$. As we have mentioned above, the dilation introduces degrees of freedom corresponding to a dynamical environment inducing the stability, in the case of the stable part, and instability, in the case of the unstable part, of the evolution of the dynamical system in Eq. (5.3.9). The procedure of second quantization exhibits the quantum field associated with this dynamical environment in the transition to a quantum mechanical model. The simple case where $\tilde{R}_s = \tilde{R}_0. \forall s \in \mathbf{R}$, convenient for the identification of the additional degrees of freedom corresponding to the dynamical environment affecting the stability of the system, is not the general case and can be considered a first order approximation which is good only in the case for which $\tilde{R}_s$ depends slowly on the parameter s.

We now apply the procedure of dilation to the stable and unstable parts of the evolution of the dynamical system in Eq. (5.3.9). We start with the stable part of this evolution. First, note that for $\tau \geq 0$, the stable subspace $\mathcal{H}^s{}_{p_0} \subset \mathcal{H}_{p_0}$ is invariant under the evolution $\tilde{\phi}_0(\tau)$. For every $\tau \geq 0$, define the operator $\tilde{Z}_f(\tau) : \mathcal{H}^s{}_{p_0} \mapsto \mathcal{H}^s{}_{p_0}$ by

$$\tilde{Z}_f(\tau) := \tilde{\phi}_0(\tau)|_{\mathcal{H}^s{}_{p_0}} = e^{\tilde{R}_0\tau}|_{\mathcal{H}^s{}_{p_0}}, \quad \tau \geq 0, \tag{5.4.2}$$

i.e., for $\tau \geq 0$, $\tilde{Z}_f(\tau)$ is the restriction of the evolution $\tilde{\phi}_0(\tau)$ to the stable subspace $\mathcal{H}^s{}_{p_0}$. By the definition of $\tilde{Z}_f(\tau)$ and $\mathcal{H}^s{}_{p_0}$ we have

$$\tilde{Z}_f(0) = I_{\mathcal{H}^s{}_{p_0}} \quad \tilde{Z}_f(\tau_1)\tilde{Z}_f(\tau_2) = \tilde{Z}_f(\tau_1 + \tau_2)), \quad \tau_1, \tau_2 \geq 0, \tag{5.4.3}$$

where $I_{\mathcal{H}^s{}_{p_0}}$ is the identity operator on $\mathcal{H}^s{}_{p_0}$. In addition, $\forall \mathbf{v} \in \mathcal{H}^s{}_{p_0}$,

$$\|\tilde{Z}_f(\tau)\mathbf{v}\|_{\mathcal{H}_{p_0}} \leq \|\mathbf{v}\|_{\mathcal{H}_{p_0}}, \quad \tau \geq 0, \quad \lim_{\tau\to\infty} \|\tilde{Z}_f(\tau)\mathbf{v}\|_{\mathcal{H}_{p_0}} = 0. \tag{5.4.4}$$

Thus, $\{\tilde{Z}_f(\tau)\}_{\tau\geq 0}$ is a continuous, contractive, semigoup on $\mathcal{H}^s{}_{p_0}$ tending to zero in the limit $\tau \to \infty$ (in fact, by the finite dimensionality of the Hilbert pace, the limit may be taken in the strong, weak or operator norm sense, which are all equivalent in this case).

According to the Sz.-Nagy–Foias theory of contraction operators and contractive semigroups on Hilbert space (Sz.-Nagy and Foias 1970) for such a semigroup, *there exists a minimal isometric dilation, i.e.,* there exists on a Hilbert space $\mathcal{R}^f{}_+$ an isometric semigroup $\{U_+(\tau)\}_{\tau\geq 0}$ defined on $\mathcal{R}^f{}_+$, a subspace $\mathcal{H}^s{}_{p_0} \subset \mathcal{R}^f{}_+$, and an isometric isomorphism $V_+ : \mathcal{H}^s{}_{p_0} \mapsto \mathcal{H}^s{}_{p_0}$ such that

$$\tilde{Z}_f(\tau) = V^*{}_+ Z_f(\tau) V_+ \quad \tau \geq 0, \tag{5.4.5}$$

where

$$Z_f(\tau) := P_{\mathcal{H}^s{}_{p_0},+} U_+(\tau) P_{\mathcal{H}^s{}_{p_0},+} \quad \tau \geq 0, \tag{5.4.6}$$

and $P_{\mathcal{H}^s{}_{p_0},+}$ is the orthogonal projection in $\mathcal{R}^f{}_+$ on the subspace $P_{\mathcal{H}^s{}_{p_0},+}$. Therefore, for $\tau \geq 0$, $\tilde{Z}_f(\tau)$ is unitarily equivalent to the projection of $U_+(\tau)$ onto the subspace $\mathcal{H}^s{}_{p_0+} \subset \mathcal{R}^f{}_+$, representing $\mathcal{H}^s{}_{p_0}$. The minimality of this isometric dilation means that $\mathcal{R}^f{}_+$ equals $\overline{\vee_{\tau\geq 0} U_+(\tau)\mathcal{H}^s{}_{p_0+}}$. We refer to $\mathcal{R}^f{}_+$ as the dilation Hilbert space for the (minimal) isometric dilation of $\tilde{Z}_f(\tau)_{\tau\geq 0}$ and to $U_+(\tau)_{\tau\geq 0}$ as a (minimal) isometric dilation of the semigroup $\tilde{Z}_f(\tau)|_{\tau\geq 0}$. The dilation Hilbert space $\mathcal{R}^f{}_+$ is naturally decomposed into two orthogonal subspaces,

$$\mathcal{R}^f{}_+ = \mathcal{H}^s{}_{p_0+} \oplus \mathcal{D}^f{}_+. \tag{5.4.7}$$

The subspace $\mathcal{D}^f{}_+$, generated in the process of construction of the isometric dilation, corresponds to the dynamical environment discussed at the beginning of this section, inducing the instability of the evolution corresponding to any initial condition in $\mathcal{H}^s{}_{p_0}$.

We now proceed with a construction of an explicit representation of the isometric dilation of the semigroup $\tilde{Z}_f(\tau)_{\tau\geq 0}$ following the procedure of Nagy and Foias (Nagy 1976). For this, we consider the Hilbert space $L^2(\mathbf{R}; \mathcal{H}^s{}_{p_0}) = L^2(\mathbf{R}; C \otimes (T_{p_0}\mathcal{M} \oplus T_{p_0}\mathcal{M}))$ of Lebesgue square integrable functions defined on the real axis and taking values in $\mathcal{H}^s{}_{p_0}$. We define the inner product in this space to be

$$< \tilde{\mathbf{F}}, \tilde{\mathbf{G}} > |_{L^2(\mathbf{R};\mathcal{H}^s{}_{p_0})} = \int_{-\infty}^{\infty} < \tilde{\mathbf{F}}(s), \tilde{\mathbf{G}}(s) >_{\mathcal{H}^s{}_{p_0}} ds, \quad \tilde{\mathbf{F}}, \tilde{\mathbf{G}} \in L^2(\mathbf{R}; \mathcal{H}^s{}_{p_0}) \tag{5.4.8}$$

for which the corresponding norm is

$$\|\tilde{\mathbf{F}}\||_{L^2(\mathbf{R};\mathcal{H}^s{}_{p_0})} = \left(\int_{-\infty}^{\infty} \|\tilde{\mathbf{F}}(s)\|^2{}_{\mathcal{H}^s{}_{p_0}} ds\right)^{\frac{1}{2}}, \quad \tilde{\mathbf{F}} \in L^2(\mathbf{R}; \mathcal{H}^s{}_{p_0}). \tag{5.4.9}$$

Following the decomposition of $\mathcal{H}^s{}_{p_0}$ in (5.3.13) above, we shall be particularly concerned with the two subspaces of $L^2(\mathbf{R}; \mathcal{H}_{p_0})$ corresponding to the stable subspace

$\mathcal{H}^s{}_{p_0}$ and the unstable subspace $\mathcal{H}^u{}_{p_0}$ of $\mathcal{H}_{p_0}$, i.e., the function spaces $L^2(\mathbf{R}; \mathcal{H}^s{}_{p_0})$ and $L^2(\mathbf{R}; \mathcal{H}^u{}_{p_0})$.

Let $\tilde{B}_f = i\tilde{R}_0|_{\mathcal{H}^s{}_{p_0}}$ denote the generator of $\{\tilde{Z}_f(\tau)\}_{\tau\geq 0}$, i.e., $\tilde{Z}_f(\tau) = e^{\tilde{R}_0\tau}|_{\mathcal{H}^s{}_{p_0}} = e^{-i\tilde{B}_f\tau}$, $\tau \geq 0$; then there exists a representation of the isometric dilation $\{\tilde{Z}_f(\tau)\}_{\tau\geq 0}$ known as an outgoing representation, in which the dilation Hilbert space is $\mathcal{R}^{f,out}{}_+ = \mathcal{H}^{s,out}{}_{p_0} \oplus \mathcal{D}^{f,out}{}_+ \subset L^2(\mathbf{R}; \mathcal{H}^s{}_{p_0})$ where $\mathcal{D}^{f,out}{}_+ = \hat{V}_+\mathcal{H}^s{}_{p_0} \subset L^2(\mathbf{R}_-; \mathcal{H}^s{}_{p_0})$ is a unitary embedding of $\mathcal{H}^s{}_{p_0}$ into $\mathcal{R}^{f,out}{}_+$ given by

$$[\hat{V}_+\psi](s) = \begin{cases} (-2\tilde{B}_f)^{\frac{1}{2}}\tilde{Z}_f(-s)\psi, & s \leq 0 \\ 0, & s > 0 \end{cases}, \quad \psi \in \mathcal{H}^s{}_{p_0} \tag{5.4.10}$$

If $\hat{U}(\tau) : L^2(\mathbf{R}; \mathcal{H}^s{}_{p_0}) \mapsto L^2(\mathbf{R}; \mathcal{H}^s{}_{p_0})$ is translation to the right by τ units, i.e.,

$$[\hat{U}(\tau)\hat{\mathbf{F}}](t) = \hat{\mathbf{F}}(t-\tau), \quad \hat{\mathbf{F}} \in \mathcal{H}^s{}_{p_0}, \tag{5.4.11}$$

then we have

$$< \hat{V}_+\phi, \hat{U}(\tau)\hat{V}_+\psi >_{L^2(\mathbf{R};\mathcal{H}^s{}_{p_0})} = < \phi, \tilde{Z}_f(\tau)\psi >_{\mathcal{H}^s{}_{p_0}}, \quad \forall\phi, \ \psi \in \mathcal{H}^s{}_{p_0}, \tau \geq 0. \tag{5.4.12}$$

so that $\hat{U}(\tau)$ is an isometric dilation of $\{\tilde{Z}_f(\tau)\}_{\tau\geq 0}$ on $\mathcal{R}^{f,out}{}_+$.

The dilation of the unstable part of the evolution of the dynamical system in Eq. (5.3.9) proceeds along lines similar to the dilation of the stable part of that evolution. Noting that the unstable subspace $\mathcal{H}^u{}_{p_0} \subset \mathcal{H}_{p_0}$ is invariant under the evolution $\tilde{\phi}_0(\tau)$ for $\tau \leq 0$, we define the operator $\tilde{Z}_b(\tau) : \mathcal{H}^u{}_{p_0} \mapsto \mathcal{H}^u{}_{p_0}$ by

$$\tilde{Z}_b(\tau) := \tilde{\phi}_0(\tau)|_{\mathcal{H}^u{}_{p_0}} = e^{\tilde{R}_0\tau}|_{\mathcal{H}^u{}_{p_0}}, \quad \tau \leq 0. \tag{5.4.13}$$

By the definitions of $\tilde{Z}_b(\tau)$ and $\mathcal{H}^u{}_{p_0}$, we have

$$\tilde{Z}_b(0) = I_{\mathcal{H}^u{}_{p_0}}, \quad \tilde{Z}_b(\tau_1)\tilde{Z}_b(\tau_2) = \tilde{Z}_b(\tau_1+\tau_2), \quad \tau_1, \tau_2 \leq 0, \tag{5.4.14}$$

where $I_{\mathcal{H}^u{}_{p_0}}$ is the identity operator on $\mathcal{H}^u{}_{p_0}$, and, furthermore, $\forall \mathbf{v} \in \mathcal{H}^u{}_{p_0}$,

$$\|\tilde{Z}_b(\tau)\mathbf{v}\|_{\mathcal{H}_{p_0}} \leq \|\mathbf{v}\|_{\mathcal{H}_{p_0}}, \quad \tau \leq 0, \quad \lim_{\tau\to-\infty} \|\tilde{Z}_b(\tau)\mathbf{v}\|_{\mathcal{H}_{p_0}} = 0. \tag{5.4.15}$$

Therefore $\{\tilde{Z}_b(\tau)\}_{\tau\leq 0}$ is a continuous, contractive semigroup on $\mathcal{H}^u{}_{p_0}$ tending to zero in the limit $\tau \to -\infty$. The minimal isometric dilation of this semigroup consists of a dilation Hilbert space $\mathcal{R}^b{}_+$, an isometric evolution semigroup $\{U_-(\tau)\}_{\tau\leq 0}$ defined on $\mathcal{R}^b{}_+$, a subspace of $\mathcal{H}^u{}_{p_0,+} \subset \mathcal{R}^b{}_+$, and an isometric isomorphism $V_- : \mathcal{H}^u{}_{p_0} \mapsto \mathcal{H}^u{}_{p_0,-}$ such that

$$\tilde{Z}_b(\tau) = V^*{}_-Z_b(\tau)V_-, \quad \tau \leq 0, \tag{5.4.16}$$

where

$$Z_b(\tau) = P_{\mathcal{H}^u{}_{p_0,+}} U_-(\tau) P_{\mathcal{H}^u{}_{p_0,+}}, \quad \tau \leq 0, \tag{5.4.17}$$

and where $P_{\mathcal{H}^u{}_{p_0,+}}$ is the orthogonal projection in $\mathcal{R}^b{}_+$ on the subspace $\mathcal{H}^u{}_{p_0,+}$ representing $\mathcal{H}^u{}_{p_0}$. The minimality of the isometric dilation means that $\mathcal{R}^b{}_+ = \overline{\vee_{\tau \leq 0} U_-(\tau) \mathcal{H}^u{}_{p_0,-}}$. The dilation Hilbert space $\mathcal{R}^b{}_+$ decomposes into two orthogonal pieces

$$\mathcal{R}^b{}_+ = \mathcal{D}^b{}_+ \oplus \mathcal{H}^u{}_{p_0,+}. \tag{5.4.18}$$

The subspace $\mathcal{D}^b{}_+$, generated by the procedure of dilation, corresponds to a dynamical environment inducing the instability of the evolution (equivalently, the stability of evolution in the backward direction) corresponding to any initial condition in $\mathcal{H}^u{}_{p_0}$.

Our next step is the construction of an explicit representation of the isometric dilation of the $\{\tilde{Z}_b(\tau)\}_{\tau \leq 0}$ following the procedure described in (Nagy 1976). If $\tilde{B}_b = i\tilde{R}_0|_{\mathcal{H}^u{}_{p_0}}$ denotes the generator of $\{\tilde{Z}_b(\tau)\}_{\tau \leq 0}$, i.e., $\tilde{Z}_b(\tau) = e^{\tilde{R}_0 \tau}|_{\mathcal{H}^u{}_{p_0}} = e^{-i\tilde{B}_b \tau}$, $\tau \leq 0$, then there exists an outgoing representation of the isometric dilation of $\{\tilde{Z}_b(\tau)\}_{\tau \leq 0}$ where the dilation Hilbert space is $\mathcal{R}_+{}^{b,out} = \mathcal{H}^{u,out}{}_{p_0} \oplus \mathcal{D}^{b,out}{}_+ \subset L^2(\mathbf{R}, \mathcal{H}^u{}_{p_0})$, $\mathcal{D}^{b,out}{}_+ = L^2(\mathbf{R}_-, \mathcal{H}^u{}_{p_0})$ and $\mathcal{H}^{u,out}{}_{p_0} = \hat{V}\mathcal{H}^u{}_{p_0} \subset L^2(\mathbf{R}, \mathcal{H}^u{}_{p_0})$ is a unitary embedding of $\mathcal{H}^u{}_{p_0}$ into $\mathcal{R}_+{}^{b,out}$ given by

$$[\hat{V}_-\psi](s) = \begin{cases} (2\tilde{B}_b)^{\frac{1}{2}} \tilde{Z}_b(-s)\psi, & s \geq 0 \\ 0, & s < 0. \end{cases} \tag{5.4.19}$$

If $\hat{U}(\tau) : L^2(\mathbf{R}, \mathcal{H}^u{}_{p_0}) \mapsto L^2(\mathbf{R}, \mathcal{H}^u{}_{p_0})$ is translation to the right by τ units, i.e., if

$$[\hat{U}(\tau)\hat{\mathbf{F}}(t) = \hat{\mathbf{F}}(t - \tau), \quad \hat{\mathbf{F}} \in L^2(\mathbf{R}, \mathcal{H}^u{}_{p_0}), \tag{5.4.20}$$

we have

$$< \hat{V}_-\phi, \hat{U}(\tau)\hat{V}_-\psi >_{L^2(\mathbf{R}, \mathcal{H}^u{}_{p_0})} = < \phi, \tilde{Z}_b(\tau)\psi >_{\mathcal{H}^u{}_{p_0}} \quad \forall \phi, \psi \in \mathcal{H}^u{}_{p_0}, \ \tau \leq 0, \tag{5.4.21}$$

so that $\hat{U}(\tau)$ is an isometric dilation of $\tilde{Z}_b(\tau)|\tau \leq 0$ on $\mathcal{R}_+{}^{b,out}$. We therefore obtain functional representations of the isometric dilations of the stable semigroup $\{\tilde{Z}_f(\tau)\}_{\tau \geq 0}$ and the unstable semigroup $\{\tilde{Z}_b(\tau)\}\}^{\tau \leq 0}$ on subspaces of the function spaces $\mathcal{H}^s{}_{p_0}$ and $\mathcal{H}^u{}_{p_0}$ respectively, which, in turn, are orthogonal subspaces of $L^2(\mathbf{R}, \mathcal{H}_{p_0})$.

B. Isometric Dilations on Geodesics

In this subsection, we consider Hilbert spaces of vector valued functions defined along a geodesic γ which may carry representations of the isometric dilation of the stable semigroup $\{\tilde{Z}_f(\tau)\}_{\tau \geq 0}$ and the unstable semigroup $\{\tilde{Z}_b(\tau)\}_{\tau \leq 0}$.

Let γ be a geodesic parametrized by an arc length parametrization as in Eq.(5.2.1) so that $T_{\gamma(s)}\mathcal{M}$ is the tangent space to $\mathcal{M}$ at the point $\gamma(s) \in \gamma$. Let $T\mathcal{M} \oplus T\mathcal{M}$ denote the direct sum of the tangent bundle of $\mathcal{M}$ with itself. Denote by $T_\gamma\mathcal{M} \oplus T_\gamma\mathcal{M}$

the restriction of $T\mathcal{M} \oplus T\mathcal{M}$ to the geodesic γ. Denote by $\mathbf{C} \otimes (T\mathcal{M} \oplus T\mathcal{M})$ the complexification of $T\mathcal{M} \oplus T\mathcal{M}$. The leaf of this complexified bundle at $p \in \mathcal{M}$ is the complex Hilbert space $\mathbf{C} \otimes (T_p\mathcal{M} \oplus T_p\mathcal{M})$. Let $\mathbf{C} \otimes (T_\gamma\mathcal{M} \oplus T_\gamma\mathcal{M})$ be the restriction of $\mathbf{C} \otimes (T\mathcal{M} \oplus T\mathcal{M})$ to the geodesic γ. A section of $\mathbf{C} \otimes (T_\gamma\mathcal{M} \oplus T_\gamma\mathcal{M})$ is a function $\mathbf{F}$ assigning to each point $p \in \gamma$ a vector $\mathbf{F} \in \mathbf{C} \otimes (T_p\mathcal{M} \oplus T_p\mathcal{M})$. We fix the parametrization of γ to be the arc length parametrization and consider such a section $\mathbf{F}$ to be a function of the arc length parameter s. Hence we may use the short notation $\mathbf{F}(s) \equiv \mathbf{F}(\gamma(s))$ and consider the section $\mathbf{F}$ to be a function defined on $\mathbf{R}$. Finally, denote by $L^2(\mathbf{R}; \mathbf{C} \otimes (T_\gamma\mathcal{M} \oplus T_\gamma\mathcal{M}))$ the Hilbert space of all sections of $\mathbf{C} \otimes (T_\gamma\mathcal{M} \oplus T_\gamma\mathcal{M})$ which are Lebesgue square integrable with respect to the arc length parameter. If $\mathbf{G}, \mathbf{F} \in \mathbf{C} \otimes (T_\gamma\mathcal{M} \oplus T_\gamma\mathcal{M})$ are two such sections then their inner product is

$$< \mathbf{F}, \mathbf{G} > |_{L^2(\mathbf{R};\mathbf{C}\otimes(T_\gamma\mathcal{M}\oplus T_\gamma\mathcal{M}))} = \int_{-\infty}^{\infty} < \mathbf{F}(s), \mathbf{G}(s) > |_{L^2(\mathbf{R};\mathbf{C}\otimes(T_{\gamma(s)}\mathcal{M}\oplus T_{\gamma(s)}\mathcal{M}))} ds, \tag{5.4.22}$$

and if $\mathbf{F} \in \mathbf{C} \otimes (T_\gamma\mathcal{M} \oplus T_\gamma\mathcal{M})$, then its norm is given by

$$\|\mathbf{F}\|_{L^2(\mathbf{R};\mathbf{C}\otimes(T_{\gamma(s)}\mathcal{M}\oplus T_{\gamma(s)}\mathcal{M}))} = \Big(\int_{-\infty}^{\infty} \|\mathbf{F}(s)\|^2{}_{L^2(\mathbf{R};\mathbf{C}\otimes(T_{\gamma(s)}\mathcal{M}\oplus T_{\gamma(s)}\mathcal{M}))} ds\Big)^{\frac{1}{2}}. \tag{5.4.23}$$

We know from (5.2.8) that the mapping W_γ is locally a unitary mapping of $T_{p_0}\mathcal{M}$ onto $T_\gamma\mathcal{M}$. We now use this property to extend this mapping to a unitary map $\hat{W}_\gamma = W_\gamma \oplus W_\gamma : L^2(\mathbf{R}; \mathcal{H}_{p_0}) \mapsto L^2(\mathbf{R}; \mathbf{C} \otimes (T_\gamma\mathcal{M} \oplus T_\gamma\mathcal{M}))$, i.e. for each function $\tilde{\mathbf{F}} = (\tilde{\mathbf{F}}_1, \tilde{\mathbf{F}}_2)^T \in L^2(\mathbf{R}; \mathcal{H}_{p_0})$, we define

$$[\hat{W}_\gamma \tilde{\mathbf{F}}](s) = \begin{pmatrix} [W_\gamma \tilde{\mathbf{F}}_1](s) \\ [W_\gamma \tilde{\mathbf{F}}_2](s) \end{pmatrix} = \begin{pmatrix} \phi(s)\tilde{\mathbf{F}}_1(s) \\ \phi(s)\tilde{\mathbf{F}}_2(s) \end{pmatrix}, \tag{5.4.24}$$

where $\phi(s)$ is parallel transport along γ. For $\tilde{\mathbf{F}} = (\tilde{\mathbf{F}}_1, \tilde{\mathbf{F}}_2)^T \in L^2(\mathbf{R}; \mathcal{H}_{p_0})$, $\tilde{\mathbf{G}} = (\tilde{\mathbf{G}}_1, \tilde{\mathbf{G}}_2)^T \in L^2(\mathbf{R}; \mathcal{H}_{p_0})$, we then have, using Eq. (5.2.8),

$$\begin{aligned} &< \tilde{\mathbf{F}}, \tilde{\mathbf{G}} >_{L^2(\mathbf{R};\mathcal{H}_{p_0})} = \\ &= \int_{-\infty}^{\infty} < \tilde{\mathbf{F}}(s), \tilde{\mathbf{G}}(s) >_{\mathcal{H}_{p_0}} = \int_{-\infty}^{\infty} (< \bar{\tilde{\mathbf{F}}}_1(s), \tilde{\mathbf{G}}_1(s) >_{T_{p_0}\mathcal{M}} + < \bar{\tilde{\mathbf{F}}}_2(s), \tilde{\mathbf{G}}_2(s) >_{T_{p_0}\mathcal{M}}) ds \\ &= \int_{-\infty}^{\infty} (< \overline{[W_\gamma \tilde{\mathbf{F}}_1]}(s), [W_\gamma \tilde{\mathbf{G}}_1](s) >_{T_{\gamma(s)}\mathcal{M}} + < \overline{[W_\gamma \tilde{\mathbf{F}}_2]}(s), [W_\gamma \tilde{\mathbf{G}}_2](s) >_{T_{\gamma(s)}\mathcal{M}}) ds \\ &\qquad = < W_\gamma \tilde{\mathbf{F}}, W_\gamma \tilde{\mathbf{G}} >_{L^2(\mathbf{R};\mathbf{C}\otimes(T_\gamma\mathcal{M}\oplus T_\gamma\mathcal{M}))} . \end{aligned} \tag{5.4.25}$$

It is easy to check that $\hat{W}_\gamma$ is surjective, and hence it is unitary.

We note here an important observation associated with the unitary mapping $\hat{W}_\gamma$. Let $\hat{U}(\tau) : L^2(\mathbf{R}; \mathcal{H}_{p_0}) \mapsto L^2(\mathbf{R}; \mathcal{H}_{p_0})$ be the operator of translation to the right by τ units, i.e.,

$$[\hat{U}(\tau)\hat{\mathbf{F}}](s) = \hat{\mathbf{F}}(s-\tau), \quad s \in \mathbf{R}, \quad \hat{\mathbf{F}} \in L^2(\mathbf{R}; \mathcal{H}_{p_0}). \tag{5.4.26}$$

This operator is unitary on $L^2(\mathbf{R}; \mathcal{H}_{p_0})$. To see how this operator transforms under the unitary mapping $\hat{W}_\gamma$, we denote $U_\gamma(\tau) := \hat{W}_\gamma \hat{U}(\tau)\hat{W}_\gamma^{-1}$; we then have

$$\begin{aligned}
[\hat{W}_\gamma \hat{U}(\tau)\hat{\mathbf{F}}](s) &= \begin{pmatrix} [W_\gamma(\hat{U}(\tau)\tilde{\mathbf{F}}_1)](s) \\ [W_\gamma(\hat{U}(\tau)\tilde{\mathbf{F}}_2)](s) \end{pmatrix} = \begin{pmatrix} \phi(s)(\hat{U}(\tau)\tilde{\mathbf{F}}_1)(s) \\ \phi(s)(\hat{U}(\tau)\tilde{\mathbf{F}}_2)(s) \end{pmatrix} = \begin{pmatrix} \phi(s)\tilde{\mathbf{F}}_1(s-\tau) \\ \phi(s)\tilde{\mathbf{F}}_2(s-\tau) \end{pmatrix} = \\
&= \begin{pmatrix} \phi(\tau)\phi(s-\tau)\tilde{\mathbf{F}}_1(s-\tau) \\ \phi(\tau)\phi(s-\tau)\tilde{\mathbf{F}}_2(s-\tau) \end{pmatrix} = \phi(\tau) \begin{pmatrix} \phi(s-\tau)\tilde{\mathbf{F}}_1(s-\tau) \\ \phi(s-\tau)\tilde{\mathbf{F}}_2(s-\tau) \end{pmatrix} = \\
&= \phi(\tau)[\hat{W}_\gamma \tilde{\mathbf{F}}](s-\tau) = [U_\gamma(\tau)\hat{W}_\gamma \tilde{\mathbf{F}}](s).
\end{aligned} \tag{5.4.27}$$

Thus, we have obtained the result that translation to the right by τ units on $L^2(\mathbf{R}; \mathcal{H}_{p_0})$ is transformed into parallel transport by τ units on $L^2(\mathbf{R}; \mathbf{C} \otimes (T_\gamma \mathcal{M} \oplus T_\gamma \mathcal{M}))$.

Now, consider the representation of the isometric dilation of $\{\tilde{Z}_f(\tau)\}_{\tau \geq 0}$ on the function space $\mathcal{R}^{f,out}{}_+ = \mathcal{H}^{s,out}{}_{p_0+} \oplus \mathcal{D}^{f,out}{}_+ \subset L^2(\mathcal{H}^s{}_{p_0})$ introduced above. We apply the mapping $\hat{W}_\gamma$ to $\mathcal{R}^{f,out}{}_+$ and set

$$\mathcal{R}^{s,out}{}_{+,\gamma} := \hat{W}_\gamma \mathcal{R}^{f,out}{}_+, \quad \mathcal{H}^{s,out}{}_{p_0,\gamma} := \hat{W}_\gamma \mathcal{H}^{f,out}{}_{p_0+}, \quad \mathcal{D}^{s,out}{}_\gamma := \hat{W}_\gamma \mathcal{D}^{f,out}{}_{+,\gamma}. \tag{5.4.28}$$

Then by the unitarity of $\hat{W}_\gamma$, we have

$$\mathcal{R}^{s,out}{}_{+,\gamma} = \mathcal{H}^{s,out}{}_{p_0,\gamma} \oplus \mathcal{D}^{f,out}{}_{+,\gamma} \subset L^2(\mathbf{R}; \mathbf{C} \otimes T_\gamma \mathcal{M} \oplus T_\gamma \mathcal{M}) \tag{5.4.29}$$

with

$$\mathcal{D}^{f,out}{}_{+,\gamma} = \hat{W}_\gamma \mathcal{D}^{f,out}{}_{+,\gamma} \subset L^2(\mathbf{R}; \mathbf{C} \otimes T_\gamma \mathcal{M} \oplus T_\gamma \mathcal{M}) \tag{5.4.30}$$

and where $\mathcal{H}^{s,out}{}_{p_0,\gamma}$ is a unitary embedding of $\mathcal{H}^s{}_{p_0}$ into $L^2(\mathbf{R}; \mathbf{C} \otimes T_\gamma \mathcal{M} \oplus T_\gamma \mathcal{M})$ given explicitly by

$$[\hat{V}_{+,\gamma}\psi](s) = [\hat{W}_\gamma \hat{V}_+ \psi](s) = \begin{cases} \phi(s)(-2\tilde{B}_f)^{1/2}\tilde{Z}_f(-s)\psi, & s \leq 0 \\ 0, & s < 0, \end{cases} \tag{5.4.31}$$

where $\hat{V}_{+,\gamma} : \mathcal{H}^s{}_{p_0} \mapsto \mathcal{H}^{s,out}{}_{p_0,\gamma}$ is defined by $\hat{V}_{+,\gamma} := \hat{W}_\gamma \hat{V}_+$. We then have

$$\begin{aligned}
&< \hat{V}_{+,\gamma}\phi, U_\gamma(\tau)\hat{V}_{+,\gamma}\psi >_{L^2(\mathbf{R}; \mathbf{C}\otimes T_\gamma \mathcal{M} \oplus T_\gamma \mathcal{M})} = < \hat{W}_\gamma \hat{V}_+\phi, U_\gamma(\tau)\hat{W}_\gamma \hat{V}_+\psi >_{L^2(\mathbf{R}, \mathcal{H}^s{}_{p_0})} = \\
&= < \hat{V}_+\phi, \hat{W}^*_\gamma U_\gamma(\tau)\hat{W}_\gamma \hat{V}_{+,\gamma}\psi >_{L^2(\mathbf{R}, \mathcal{H}^s{}_{p_0})} = < \hat{V}_+\phi, \hat{W}_\gamma^{-1} U_\gamma(\tau)\hat{W}_\gamma \hat{V}_{+,\gamma}\psi >_{L^2(\mathbf{R}, \mathcal{H}^s{}_{p_0})} = \\
&= < \hat{V}_+\phi, \hat{W}_\gamma^{-1} U_\gamma(\tau)\hat{W}_\gamma \hat{V}_+\psi >_{L^2(\mathbf{R}, \mathcal{H}^s{}_{p_0})} = \\
&= < \hat{V}_+\phi, \hat{U}(\tau)\hat{V}_+\psi >_{L^2(\mathbf{R}, \mathcal{H}^s{}_{p_0})} = < \phi, \tilde{Z}_f(\tau)\psi >_{\mathcal{H}^s{}_{p_0}}, \quad \forall \phi, \psi \in \mathcal{H}^s{}_{p_0}, \tau \geq 0,
\end{aligned} \tag{5.4.32}$$

so that $U_\gamma(\tau)$ is an isometric dilation of the stable semigroup $\{\tilde{Z}_f(\tau)\}_{\tau\geq 0}$ on $\mathbf{R}^{f,out}{}_{+,\gamma} \subset L^2(\mathbf{R}; \mathbf{C} \otimes T_\gamma\mathcal{M} \oplus T_\gamma\mathcal{M})$. Hence, we obtain an isometric dilation of $\{\tilde{Z}_f(\tau)\}_{\tau\geq 0}$ on a Hilbert space of functions defined on the geodesic γ. By essentially repeating the same procedure, we may obtain an isometric dilation of the unstable semigroup $\{\tilde{Z}_b(\tau)\}_{\tau\leq 0}$ on a function space $\mathcal{R}^{b,out}{}_{+,\gamma} \subset L^2(\mathbf{R}; \mathbf{C} \otimes (T_\gamma\mathcal{M} \oplus T_\gamma\mathcal{M}))$ defined by $\mathcal{R}^{b,out}{}_{+,\gamma} := \hat{W}_\gamma \mathcal{R}^{b,out}{}_+$.

As the geodesic deviation equation is an equation of motion along a given geodesic γ, it seems natural to make use of isometric dilations in function spaces defined over γ since they also utilize motion along the geodesic γ. However, we emphasize that, much in the same way that the geodesic deviation Eq. (5.2.22) and its dynamical system representation in Eq. (5.3.6) are unitarily equivalent isometric dilations of $\{\tilde{Z}_f(\tau)\}_{\tau\geq 0}$ and of $\{\tilde{Z}_b(\tau)\}_{\tau\leq 0}$ embedded in the function space $L^2(\mathbf{R}, \mathcal{H}^s{}_{p_0})$ of $\mathcal{H}^s{}_{p_0}$ valued functions are completely equivalent by the unitarity of the mapping $\hat{W}_\gamma$ and there is no fundamental reason to favor one of these representations over the other. Although the isometric dilation in the function space defined over the geodesic γ is more natural and conceptually important, for the sake of simplicity, the procedure of second quantization of the isometric dilation is applied in the next section within the $L^2(\mathbf{R}, \mathcal{H}^s{}_{p_0})$ setting.

5.5 Second Quantization

Consider the isometric dilation of the stable semigroup $\{\tilde{Z}_f(\tau)\}_{\tau\geq 0}$ constructed in the function space $\{\mathbf{R}^{f,out}\}_{+,\gamma} \subset L^2(\mathbf{R}, \{\mathcal{H}^s\}_{p_0})$ in Sect. 5.4 above.

The subspace $\mathcal{H}^{s,out}{}_{p_0} \subset \mathcal{R}^{f,out}{}_+$ is unitarily equivalent to $\mathcal{H}^s{}_{p_0}$ via the mapping $\hat{V}_+$ and the restriction $\{\tilde{Z}_f(\tau)\}_{\tau\geq 0}, \tilde{Z}_f(\tau) = P_{\mathcal{H}^{s,out}{}_{p_0}} \hat{U}(\tau) P_{\mathcal{H}^{s,out}{}_{p_0}}$ of the isometric evolution $\{\hat{U}(\tau)\}_{\tau\geq 0}$ in $\mathcal{R}^{f,out}{}_+$ to $\mathcal{H}^{s,out}{}_{p_0}$ is unitarily equivalent to $\{\tilde{Z}_f(\tau)\}_{\tau\geq 0}$, which in turn, as we recall, is a first order approximation to the stable part of the forward evolution of the geodesic devation Eq. (5.2.22). As opposed to $\mathcal{H}^{s,out}{}_{p_0}$, the subspace $\mathcal{D}^{f,out}$ is generated by the dilation procedure and represents new degrees of freedom which are not part of the original system. In order to understand the meaning of these new degrees of freedom and the way they influence the evolution generated by the geodesic deviation equation, we apply a procedure of second quantization, identify a quantum field associated with $\mathcal{D}^{f,out}$, and observe how the interaction of this field with the system induces the evolution of the stable semigroup $\{\tilde{Z}_f(\tau)\}_{\tau\geq 0}$. More specifically, we apply coherent state second quantization to the isometric dilation Hilbert space $\mathcal{R}^{f,out}{}_+ = \mathcal{H}^{s,out}{}_{p_0} \oplus \mathcal{D}^{f,out}$. For this, we define a positive kernel $K(\cdot,\cdot) : \mathcal{R}^{f,out}{}_+ \times \mathcal{R}^{f,out}{}_+ \mapsto \mathbf{C}$ by

$$K(u, v) = e^{<u,v>_{\mathcal{R}^{f,out}{}_+}} = e^{<u,v>_{L^2(\mathbf{R},\mathcal{H}^{s,out}{}_{p_0})}}, \quad u, v \in \mathcal{R}^{f,out}{}_+ \tag{5.5.1}$$

and perform a Kolmogorov dilation (see, for example, Nagy 1976) of $\mathcal{R}^{f,out}{}_+$ with respect to $K(\cdot,\cdot)$. The procedure of Kolmogorov dilation introduces a symmetric

Fock space $\Gamma_s(\mathcal{R}^{f,out}{}_+) = \Sigma_{n=0}^{\infty} \oplus \left[\left(\mathcal{R}^{f,out}{}_+ \right)^{\otimes^n} \right]_{sym}$ such that to every state $u \in \mathcal{R}^{f,out}{}_+$, there is assigned an exponential vector (a coherent state up to a normalization factor) $e(u) \in \Gamma_s(\mathcal{R}^{f,out}{}_+)$

$$e(u) = \Sigma_{n=0}^{\infty} \frac{1}{\sqrt{n!}} u^{\otimes^n}, \tag{5.5.2}$$

and we have

$$< e(u), e(v) >_{\Gamma_s(\mathcal{R}^{f,out}{}_+)} = \Sigma_{n=0}^{\infty} \frac{1}{n!} < u, v >^n_{\mathcal{R}^{f,out}{}_+} = \\ = e^{<u,v>_{\mathcal{R}^{f,out}{}_+}} = K(u, v), \quad u, v \in \mathcal{R}^{f,out}{}_+. \tag{5.5.3}$$

We define a representation of the canonical commutation relations (CCR) on $\Gamma_s(\mathcal{R}^{f,out}{}_+)$ through the algebra of Weyl operators (see, for example, Parhasarathy 1992) $W(u, U) : \Gamma_s(\mathcal{R}^{f,out}{}_+) \mapsto \Gamma_s(\mathcal{R}^{f,out}{}_+)$, with $u \in \mathcal{R}^{f,out}{}_+$ and $U \in \mathcal{U}(\mathcal{R}^{f,out}{}_+)$, where $\mathcal{U}(\mathcal{R}^{f,out}{}_+)$ is the group of unitary operators on $\mathcal{R}^{f,out}{}_+$. The action of a Weyl operator on exponential vectors is given by

$$W(u, U)e(v) = e^{-\frac{1}{2}\|u\|^2_{\mathcal{R}^{f,out}{}_+} + <u,Uv>_{\mathcal{R}^{f,out}{}_+}} e(Uv + u) \tag{5.5.4}$$

The Weyl operators are unitary operators on $\Gamma_s(\mathcal{R}^{f,out}{}_+)$. They have the composition rule (see, for example, Frankel 1997)

$$W(u_2, U_2)W(u_1, U_1) = e^{-Im<u_2,U_2u_1>} W(U_2u_1 + u_2, U_2U_1) \tag{5.5.5}$$

We consider in the following some special cases. First, suppose that we define

$$W(u) := W(u, I_{\mathcal{R}^{f,out}{}_+}), \quad \Gamma(U) := W(0, U). \tag{5.5.6}$$

Using the composition rule (5.5.5) we obtain

$$\begin{aligned} W(u)W(v) &= e^{-i\mathrm{Im}<u,v>} W(u + v), \\ W(u)W(v) &= e^{-2i\mathrm{Im}<u,v>} W(v)W(u), \\ \Gamma(U_2)\Gamma(u_1) &= \Gamma(U_2U_1), \\ \Gamma(U)W(u)\Gamma^{-1}(U) &= W(Uu), \\ W(su)W(tu) &= W((s + t)u), \quad s, t \in \mathbf{R}. \end{aligned} \tag{5.5.7}$$

Basic observables on $\Gamma_s(\mathcal{R}^{f,out}{}_+)$ are defined through one parameter subgroups in the algebra of Weyl operators. Thus, if u is an element of $\mathcal{R}^{f,out}{}_+$, the last of Eq. (5.5.7) implies that $\{W(tu)\}_{t\in\mathbf{R}}$ is a continuous, one parameter unitary semigroup on $\Gamma_s(\mathcal{R}^{f,out}{}_+)$. This group has a self-adjoint generator $p(u)$ so that

$$W(tu) = e^{-ip(u)t} \tag{5.5.8}$$

and $p(u)$ is a basic observable on the Fock space $\Gamma_s(\mathcal{R}^{f,out}{}_+)$. Setting

$$q(u) := -p(iu), \tag{5.5.9}$$

we may define the operators

$$a(u) = \frac{1}{2}(q(u) + ip(u)), \quad a^\dagger(u) = \frac{1}{2}(q(u) - ip(u)), \tag{5.5.10}$$

which act, respectively, as annihilation and creation operators on $\Gamma_s(\mathcal{R}^{f,out}{}_+)$.

Another type of basic observable on $\Gamma_s(\mathcal{R}^{f,out}{}_+)$ follows from the fact that if $\{U(t)\}_{t\in\mathbf{R}}, U(t) = e^{-iHt}$ is a continuous one-parameter unitary group on $\Gamma_s(\mathcal{R}^{f,out}{}_+)$ with a self-adjoint generator H, then Eq. (5.5.7) implies that

$$\{\Gamma(U(t))\}_{t\in\mathbf{R}} = \{\Gamma(e^{-iHt})\}_{t\in\mathbf{R}}$$

is a continuous one-parameter unitary group defined on $\Gamma_s(\mathcal{R}^{f,out}{}_+)$. We denote the self-adjoint generator of this group by $\lambda(H)$ so that

$$\Gamma(e^{-iHt}) = e^{-i\lambda(H)t} \tag{5.5.11}$$

and $\lambda(H)$ is a basic observable on $\Gamma_s(\mathcal{R}^{f,out}{}_+)$. Among the variables of this type, we consider, in particular, those associated with orthogonal projections in $\mathcal{R}^{f,out}{}_+$. If $P : \mathcal{R}^{f,out}{}_+ \mapsto \mathcal{R}^{f,out}{}_+$ is an orthogonal projection then $\lambda(P)$ is the corresponding observable on $\Gamma_s(\mathcal{R}^{f,out}{}_+)$. In fact, $\lambda(P)$ has a natural interpretation as the observable that counts the number of quanta in a state $f \in \mathcal{R}^{f,out}{}_+$ for which the question defined by the projection P (i.e., the question of whether they belong to the range of P) is answered in the affirmative. In other words, if $P^\perp = I_{\mathcal{R}^{f,out}{}_+}$ so that $\mathcal{R}^{f,out}{}_+ = \left(P\mathcal{R}^{f,out}{}_+\right) \oplus \left(P^\perp\mathcal{R}^{f,out}{}_+\right)$ and hence

$$\Gamma_s(\mathcal{R}^{f,out}{}_+) = \Gamma_s(P\mathcal{R}^{f,out}{}_+) \otimes \Gamma_s(P^\perp\mathcal{R}^{f,out}{}_+), \tag{5.5.12}$$

then $\lambda(P)$ is the observable counting the number of quanta in $\Gamma_s(P\mathcal{R}^{f,out}{}_+)$ and $\lambda(P^\perp)$ is the observable counting the number of quanta in $\Gamma_s(P^\perp\mathcal{R}^{f,out}{}_+)$.

Now, note that, by the orthogonal direct sum decomposition $\mathcal{R}^{f,out}{}_+ = \mathcal{H}^{s,out}{}_{p_0} \oplus \mathcal{D}^{f,out}_+$, we have a decomposition of the Fock space $\Gamma_s(\mathcal{R}^{f,out}{}_+)$ into a tensor product

$$\Gamma_s(\mathcal{R}^{f,out}{}_+) = \Gamma_s(\mathcal{H}^{s,out}{}_{p_0}) \otimes \Gamma_s(\mathcal{D}^{f,out}_+). \tag{5.5.13}$$

Let P_+ be the orthogonal projection in $L^2(\mathbf{R}, \mathcal{H}_{p_0})$ on the closed subspace $L^2(\mathbf{R}_+, \mathcal{H}_{p_0})$ and let $P_+{}^\perp := I_{L^2(\mathbf{R},\mathcal{H}_{p_0})} - P_+$, so that $P_+{}^\perp$ is the orthogonal projection in $L^2(\mathbf{R}, \mathcal{H}_{p_0})$ on the closed subspace $L^2(\mathbf{R}_-, \mathcal{H}_{p_0})$. Since $\mathcal{R}^{f,out}{}_+ = \mathcal{H}^{s,out}{}_{p_0} \oplus$

$\mathcal{D}_+^{f,out}$ with $\mathcal{D}_+^{f,out} = L^2(\mathbf{R}_+, \mathcal{H}_{p_0})$ and $\mathcal{H}_{p_0}{}^{s,out} \subset L^2(\mathbf{R}_-, \mathcal{H}_{p_0}{}^s)$, we find that $\tilde{P}_+ := P_+|_{\mathcal{R}^{f,out}{}_+}$ is the orthogonal projection in $\mathcal{R}^{f,out}{}_+$ on $\mathcal{D}_+^{f,out}$ and $\tilde{P}_+^\perp := P_+{}^\perp|_{\mathcal{R}^{f,out}{}_+}$ is the orthogonal projection in $\mathcal{R}^{f,out}{}_+$ on $\mathcal{H}^{s,out}{}_{p_0}$ so that, in particular, we have $\tilde{P}_+ + \tilde{P}_+^\perp = I_{\mathcal{R}^{f,out}}$ and

$$\Gamma_s(\mathcal{R}^{f,out}{}_+) = \Gamma_s(\mathcal{H}_{p_0}{}^{s,out}) \otimes \Gamma_s(\mathcal{D}_+^{f,out}) = \Gamma_s(\tilde{P}_+\mathcal{R}_+{}^{f,out}) \otimes \Gamma_s(\tilde{P}_+^\perp\mathcal{R}_+{}^{f,out}). \tag{5.5.14}$$

The variables $\lambda(\tilde{P}_+)$ and $\lambda(\tilde{P}_+^\perp)$ are then observables on $\Gamma(\mathcal{R}_+{}^{f,out})$ counting, respectively, the number of quanta in $\Gamma_s(\mathcal{D}_+^{f,out})$ and $\Gamma(\mathcal{H}_{p_0}{}^{s,out})$. Note also that we have

$$\begin{aligned} e^{-i\lambda(\tilde{P}_+)t}e^{-i\lambda(\tilde{P}_+^\perp)t} &= \Gamma(e^{-i\tilde{P}_+t})\Gamma(e^{-i\tilde{P}_+^\perp t}) = \\ &= \Gamma(e^{-i(\tilde{P}_+ + \tilde{P}_+^\perp)t}) = \Gamma(e^{-iI_{\mathcal{R}^{f,out}}t}) = e^{-i\lambda(I_{\mathcal{R}^{f,out}})t} \end{aligned} \tag{5.5.15}$$

from which we obtain

$$\lambda(\tilde{P}_+) + \lambda(\tilde{P}_+^\perp) = \lambda(I_{\mathcal{R}^{f,out}}), \tag{5.5.16}$$

where $\lambda(I_{\mathcal{R}^{f,out}})$ is the observable counting the total number of quanta in $\Gamma_s(\mathcal{R}^{f,out}{}_+)$.

Our next step following the identification of the observables counting the number of quanta in $\Gamma_s(\mathcal{D}^{f,out}{}_+)$ and $\Gamma_s(\mathcal{H}_{p_0}{}^{s,out})$ is the lifting of the isometric evolution $\hat{U}(\tau)|_{\tau\geq 0}$ from $\mathcal{R}^{f,out}{}_+$ to the Fock space $\Gamma_s(\mathcal{R}^{f,out}{}_+)$ and the analysis of this evolution in $\Gamma_s(\mathcal{D}_+^{f,out})$ and $\Gamma_s(\mathcal{H}_{p_0}{}^{s,out})$. For a general contraction operator $C : \mathcal{R}^{f,out}{}_+ \mapsto \mathcal{R}^{f,out}{}_+$, the lifting of the action of C into the Weyl algebra, is denoted by $\Gamma_0(C)$ and defined by

$$\Gamma_0(C)W(u) = e^{\frac{1}{2}\left(\|Cu\|^2_{\mathcal{R}^{f,out}{}_+} - \|u\|^2_{\mathcal{R}^{f,out}{}_+}\right)} W(Cu). \tag{5.5.17}$$

Note that if C_1 and C_2 are contractions, we have

$$\begin{aligned} \Gamma_0(C_2)\Gamma_0(C_1)W(u) &= e^{\frac{1}{2}\left(\|C_1u\|^2_{\mathcal{R}^{f,out}{}_+} - \|u\|^2_{\mathcal{R}^{f,out}{}_+}\right)}\Gamma_0(C_2)W(C_1u) = \\ &= e^{\frac{1}{2}\left(\|C_1u\|^2_{\mathcal{R}^{f,out}{}_+} - \|u\|^2_{\mathcal{R}^{f,out}{}_+}\right)} \cdot e^{\frac{1}{2}\left(\|C_1C_2u\|^2_{\mathcal{R}^{f,out}{}_+} - \|C_1u\|^2_{\mathcal{R}^{f,out}{}_+}\right)} W(C_2C_1u) = \\ &= \Gamma_0(C_2C_1)W(u), \end{aligned} \tag{5.5.18}$$

from which we obtain the composition rule $\Gamma_0(C_2)\Gamma_0(C_1) = \Gamma_0(C_2C_1)$ extending the composition rule in Eq. (5.5.7). Moreover, since an exponential vector $e(v)$ is given by the action of a Weyl operator on the vacuum state $e(0)$,

$$e(v) = e^{\frac{1}{2}\|v\|^2_{\mathcal{R}^{f,out}{}_+}} W(v)e(0), \tag{5.5.19}$$

we can extend the action of $\Gamma_0(C)$ to exponential vectors by

$$\Gamma_0(C)e(v) = e^{\frac{1}{2}\|v\|^2_{\mathcal{R}^{f,out}_+}}[\Gamma_0(C)W(v)]e(0) =$$
$$= e^{\frac{1}{2}\|v\|^2_{\mathcal{R}^{f,out}_+}} e^{\frac{1}{2}\left(\|Cv\|^2_{\mathcal{R}^{f,out}_+} - \|v\|^2_{\mathcal{R}^{f,out}_+}\right)} e(Cv). \quad (5.5.20)$$

If we now take for the contraction operators the elements $\hat{U}(\tau)$ of the isometric evolution $\{\hat{U}(\tau)\}_{\tau\geq 0}$, we obtain

$$\Gamma_0(\hat{U}(\tau_2))\Gamma_0(\hat{U}(\tau_1)) = \Gamma_0(\hat{U}(\tau_1)\hat{U}(\tau_2)) = \Gamma_0(\hat{U}(\tau_1+\tau_2)), \quad \tau_2, \tau_1 \geq 0, \quad (5.5.21)$$

and

$$\Gamma_0(\hat{U}(\tau))e(v) = e(\hat{U}(\tau)v) \quad \tau \geq 0, v \in \mathcal{R}^{f,out}{}_+. \quad (5.5.22)$$

This defines a lifting of the evolution $\{\hat{U}(\tau)\}_{\tau\geq 0}$ into the Weyl algebra and hence into $\Gamma_s(\mathcal{R}^{f,out}{}_+)$.

Given an eigenvector $\mathbf{u}^-{}_{\eta_j,r_j} \in \mathcal{H}^s_{p_0}$ of the stable semigroup $\{\tilde{Z}_f(\tau)\}_{\tau\geq 0}$ satisfying

$$\tilde{Z}_f(\tau)\mathbf{u}^-{}_{\eta_j,r_j} = e^{-\eta_j\tau}\mathbf{u}^-{}_{\eta_j,r_j}, \quad \tau \geq 0, \quad (5.5.23)$$

we apply the unitary mapping $\hat{V}_+ : \mathcal{H}^s_{p_0} \mapsto \mathcal{H}^{s,out}_{p_0}$ embedding $\mathcal{H}_{p_0}{}^s$ and the semigroup $\{\tilde{Z}_f(\tau)\}_{\tau\geq 0}$ into $\mathcal{R}^{f,out}{}_+$. Thus, if

$$Z_f(\tau) = \hat{V}^*_+\tilde{Z}_f(\tau)\hat{V}_+ = P_{\mathcal{H}_{p_0}{}^{s,out}}\hat{U}(\tau)P_{\mathcal{H}_{p_0}{}^{s,out}} = \tilde{P}^\perp_+\hat{U}(\tau)\tilde{P}^\perp_+, \quad \tau \geq 0, \quad (5.5.24)$$

then

$$Z_f(\tau)\hat{V}_+\mathbf{u}^-{}_{\eta_j,r_j} = e^{-\eta_j\tau}\hat{V}_+\mathbf{u}^-{}_{\eta_j,r_j} \quad \tau \geq 0. \quad (5.5.25)$$

If we now apply second quantization, then $\hat{V}_+\mathbf{u}^-{}_{\eta_j,r_j}$ is mapped into an exponential vector $e(\hat{V}_+\mathbf{u}^-{}_{\eta_j,r_j}) \in \Gamma_s(\mathcal{R}^{f,out}{}_+)$ and the isometric dilation evolution $\{\hat{U}(\tau)\}_{\tau\geq 0}$ is lifted into the evolution $\{\Gamma_0(\hat{U}(\tau))\}_{\tau\geq 0} \in \Gamma_s(\mathcal{R}^{f,out}{}_+)$. For any observable A on $\Gamma_s(\mathcal{R}^{f,out}{}_+)$, the expectation value of A in the evolved state $\Gamma_0(\hat{U}(\tau))e(\mathbf{u}^-{}_{\eta_j,r_j})$ is, of course, given by

$$< \Gamma_0(\hat{U}(\tau))e(\hat{V}_+\mathbf{u}^-{}_{\eta_j,r_j}), A\Gamma_0(\hat{U}(\tau))e(\hat{V}_+\mathbf{u}^-{}_{\eta_j,r_j}) >_{\Gamma_s(\mathcal{R}^{f,out}{}_+)} =$$
$$=< e(\hat{U}(\tau)\hat{V}_+\mathbf{u}^-{}_{\eta_j,r_j}), Ae(\hat{U}(\tau)\hat{V}_+\mathbf{u}^-{}_{\eta_j,r_j}) >_{\Gamma_s(\mathcal{R}^{f,out}{}_+)} . \quad (5.5.26)$$

For $0 < s < 1$ define on $\Gamma_s(\mathcal{H}^s_{p_0})$ the operator $s^{\lambda(\tilde{P}^\perp_+)}$ and recall that $\lambda(\tilde{P}^\perp_+)$ is the operator counting the number of quanta in $\Gamma_s(\mathcal{H}_{p_0}{}^{s,out})$. Calculating the above expectation value with $A = s^{\lambda(\tilde{P}^\perp_+)}$, one obtains

$$< \Gamma_0(\hat{U}(\tau))e(\hat{V}_+\mathbf{u}^-_{\eta_j,r_j}),\, s^{\lambda(\tilde{P}_+^{\perp})}\Gamma_0(\hat{U}(\tau))e(\mathbf{V}_+\mathbf{u}^-_{\eta_j,r_j}) >_{\Gamma_s(\mathcal{R}_+^{f,out})}=$$

$$=< e(\hat{U}(\tau)\hat{V}_+\mathbf{u}^-_{\eta_j,r_j}),\, e(s^{\tilde{P}_+^{\perp}}\hat{U}(\tau)\hat{V}_+\mathbf{u}^-_{\eta_j,r_j}) >_{\Gamma_s(\mathcal{R}_+^{f,out})}=$$

$$=< e(\hat{U}(\tau)\hat{V}_+\mathbf{u}^-_{\eta_j,r_j}),\, e(e^{(\ln s)\tilde{P}_+^{\perp}}\hat{U}(\tau)\hat{V}_+\mathbf{u}^-_{\eta_j,r_j}) >_{\Gamma_s(\mathcal{R}_+^{f,out})}$$

$$=< e(\hat{U}(\tau)\hat{V}_+\mathbf{u}^-_{\eta_j,r_j}),\, e([e^{(\ln s)}\tilde{P}_+^{\perp} + \tilde{P}_+]\hat{U}(\tau)\hat{V}_+\mathbf{u}^-_{\eta_j,r_j}) >_{\Gamma_s(\mathcal{R}_+^{f,out})}$$

$$=< e(\hat{U}(\tau))\hat{V}_+\mathbf{u}^-_{\eta_j,r_j}),\, e([(s-1)\tilde{P}_+^{\perp} + I_{\mathcal{R}_+^{f,out}}]\hat{U}(\tau)\hat{V}_+\mathbf{u}^-_{\eta_j,r_j}) >_{\Gamma_s(\mathcal{R}_+^{f,out})}$$

$$= e^{<\hat{U}(\tau)\hat{V}_+\mathbf{u}^-_{\eta_j,r_j}[(s-1)\tilde{P}_+^{\perp}+I_{\mathcal{R}_+^{f,out}}]\hat{U}(\tau)\hat{V}_+\mathbf{u}^-_{\eta_j,r_j}>_{\mathcal{R}_+^{f,out}}}$$

$$= e^{<\hat{U}(\tau)\hat{V}_+\mathbf{u}^-_{\eta_j,r_j}(s-1)\tilde{P}_+^{\perp}\hat{U}(\tau)\hat{V}_+\mathbf{u}^-_{\eta_j,r_j}>_{\mathcal{R}_+^{f,out}}+<\hat{V}_+\mathbf{u}^-_{\eta_j,r_j},\hat{V}_+\mathbf{u}^-_{\eta_j,r_j}>_{\mathcal{R}_+^{f,out}}}$$

$$= e^{(s-1)\|\tilde{P}_+^{\perp}\hat{U}(\tau)\hat{V}_+\mathbf{u}^-_{\eta_j,r_j}\|^2_{\mathcal{R}_+^{f,out}}+\|\hat{V}_+\mathbf{u}^-_{\eta_j,r_j}\|^2_{\mathcal{R}_+^{f,out}}}$$

$$= e^{(s-1)\|Z_f(\tau)\hat{V}_+\mathbf{u}^-_{\eta_j,r_j}\|^2_{\mathcal{R}_+^{f,out}}+\|\hat{V}_+\mathbf{u}^-_{\eta_j,r_j}\|^2_{\mathcal{R}_+^{f,out}}} = e^{(s-1)e^{-2\eta_j\tau}\|\hat{V}_+\mathbf{u}^-_{\eta_j,r_j}\|^2_{\mathcal{R}_+^{f,out}}+\|\hat{V}_+\mathbf{u}^-_{\eta_j,r_j}\|^2_{\mathcal{R}_+^{f,out}}}$$

$$= e^{\|\hat{V}_+\mathbf{u}^-_{\eta_j,r_j}\|^2_{\mathcal{R}_+^{f,out}}(1-e^{-2\eta_j\tau}(1-s))} = e^{\|\mathbf{u}^-_{\eta_j,r_j}\|^2_{\mathcal{H}^s_{p_0}}(1-e^{-2\eta_j\tau}(1-s))}\,. \tag{5.5.27}$$

The right hand side of this equation is a generating function for a pure death process in which the number of quanta in $\Gamma_s(\mathcal{H}_{p_0}{}^{s,out})$, counted by $\lambda(\tilde{P}_+^{\perp})$ decays monotonically over time. Since $\lambda(\tilde{P}_+) + \lambda(\tilde{P}_+^{\perp}) = \lambda(I_{\mathcal{R}^{f,out}{}_+})$, and $\lambda(I_{\mathcal{R}^{f,out}{}_+})$ counts the total number of quanta in $\Gamma(\mathcal{R}_+{}^{f,out})$, which is conserved under the evolution by $\{\Gamma_0(\hat{U}(\tau))\}$, we conclude that the number of quanta counted by $\lambda(\tilde{P}_+^{\perp})$, i.e., the number of quanta in $\Gamma_s(\mathcal{D}_+{}^{f,out})$ monotonically increases over time, that is the quanta emitted by the system are absorbed in the Fock space $\Gamma_s(\mathcal{D}_+{}^{f,out})$. This shows that the stable motion of $\{Z_f(\tau)\}_{\tau\geq 0}$ is induced by the emission of quanta into the dynamical environment described by the Fock space $\Gamma_s(\mathcal{D}_+{}^{f,out})$. Similar results are obtained with respect to the isometric dilation of the unstable semigroup $\{\tilde{Z}_b(\tau)\}_{\tau\leq 0}$ in the space $\mathcal{R}_+{}^{b,out} = \mathcal{H}^{u,out}{}_{p_0} \oplus \mathcal{D}_+{}^{b,out}$ and the corresponding Fock space obtained in the second quantization,

$$\Gamma_s(\mathcal{R}_+{}^{b,out}) = \Gamma_s(\mathcal{H}^{u,out}{}_{p_0}) \otimes \Gamma_s(\mathcal{D}_+{}^{b,out}). \tag{5.5.28}$$

In this case the quanta emitted from $\Gamma_s(\mathcal{H}^{u,out}{}_{p_0})$, when applying the (lifting of the) backward isometric evolution, are absorbed in the environment $\Gamma_s(\mathcal{D}_+{}^{b,out})$. If we consider this latter emission process in the reversed direction of time, i.e., for forward propagation, we obtain a process of absorption of quanta from the environment $\Gamma_s(\mathcal{D}_+{}^{b,out})$ into the system $\Gamma_s(\mathcal{H}^{u,out}{}_{p_0})$, inducing the instability of the motion associated with the unstable semigroup $\{\tilde{Z}_b(\tau)\}_{\tau\leq 0}$.

5.6 Conclusions

We have studied the stability of the trajectories generated by a Hamiltonian of the geometric form (5.5.1). The local stability of such a system can be characterized in terms of the geodesic deviation associated with these trajectories. The analysis of such a system is based on the second covariant derivative of the geodesic deviation vector, providing the structure of a harmonic oscillator. This result is based on an implicit assumption that in the limit of a locally flat coordinate system in the neighborhood of the point on the trajectory under study, the geodesic deviation equation is well approximated by an oscillator at this point, with structure that does not change in time. However, during the oscillations, the point under study moves along the geodesic curve. We have seen, rigorously, that this motion along the geodesic curve can be transformed by a unitary map of the geodesic equation into a parametric oscillator equation. By transcribing this second order parametric oscillator equation into the corresponding form of a first order dynamical system, one finds that stable and unstable behaviors are clearly separated. This construction, furthermore, supplies a symplectic form for the system of dynamical variables, which then lends itself to a second quantization which admits the identification of the excitation modes with the dynamical behavior of the system.

Assuming that the curvature tensor entering into the geodesic deviation equation is slowly changing, we then locally approximate the behavior of the solution of the dynamical system corresponding to the geodesic deviation equation in terms of a forward contractive semigroup for the stable part of the evolution, and a backward contractive semigroup for the unstable part of the evolution. We then apply a Sz. Nagy–Foias dilation procedure to obtain an isometric dilation of both the forward and backward semigroups. The dilation of these semigroups leads to an understanding of the dynamical behavior of the system with a field representing the environment.

The dilation procedure introduces degrees of freedom associated with the stability of the system. Second quantization of the dilated system provides an interpretation of the dynamical behavior of the original system. The stability of the stable part of the motion of the original system is associated with the emission of quanta into an environment corresponding to the additional degrees of freedom introduced in the dilation process. Similarly, the instability of the unstable part of the evolution is associated with the absorption of quanta.

The structure we have described constitutes an embedding of a (ideally) conservative physical system[2] into a larger system with quantized degrees of freedom, providing an interpretation of the instability of the original system. As in the case of the damped harmonic oscillator, where the quantized degrees of freedom associated with the dilation may be put into correspondence with radiation due to the friction in the oscillator, and of more general dilations of Markov processes, one could

[2]Such a system may be thought of as an idealized model for a real unstable physical system coupled, for example, to the vacuum fluctuations associated with a real physical system. Geodesic instability is associated with fluctuations of orbit due to effects not included in such a model. The theory presented here takes such fluctuations into account in a systematic way.

imagine that the quantized degrees of freedom of the instabilities of a dynamical system have observable consequences which might be seen in intrinsic thermodynamic properties of the system. The treatment carried out here could, moreover, provide a rigorous framework for the considerations of Kandrup et al. (2001) based on the work of Caiani et al. (1997) on the association of the behavior of chaotic systems with thermodynamics properties. thermodynamics

Chapter 6
Classical Hamiltonian Instability

6.1 Introduction

In this chapter we show that the characterization of unstable Hamiltonian systems in terms of the curvature associated with a Riemannian metric tensor in terms of the structure of a geometric type Hamiltonian can be applied to a wide class of potential models of standard form through definition of a conformal metric.[1] The geodesic equations reproduce the Hamilton equations of the original potential model when a transition is made to an associated system of coordinates for which the geodesics coincide with the orbits of the Hamiltonian potential model. We find in this way a geometrical description of the time development of a Hamiltonian potential model. The second covariant derivative of the geodesic deviation (described in detail in Chap. 5) in these associated coordinates generates a dynamical curvature, resulting in (energy dependent) criteria for unstable behavior different from the usual Lyapunov criteria.

The relation between the coordinates describing the motion induced by the geometric Hamilonian and the coordinates on which the standard Hamiltonian acts is defined in a non-integrable relation on the tangent space. The correspondence between the two sets of coordinates is therefore not uniquely determined. However, we show that functions on one set of coordinates are determined by Taylor series expansion in terms of corresponding functions on the other set on common domains of analyticity. As we shall see in Chap. 7, the construction of general canonical transformations from standard to geometrical form is also carried out on domains of analyticity for the potential function and the generating function of the transformation.

We discuss some examples of unstable Hamiltonian systems in two dimensions giving, as a particular illustration, detailed results for a potential obtained from a fifth order expansion of a Toda lattice Hamiltonian.

---

[1] This chapter is based on the work presented in Horwitz et al. (2007a, 2017).

L. Horwitz and Y. Strauss, *Unstable Systems*, Mathematical Physics Studies,
https://doi.org/10.1007/978-3-030-31570-2_6

6.2 The Embedding

A Hamiltonian system of the form (we use the summation convention)

$$H(x, p) = \frac{1}{2M} g_{ij}(x) p^i p^j, \tag{6.2.1}$$

where g_{ij} is a function of the coordinates alone, generates (at least locally) instability if there is goedesic instability along the orbits it generates. One can easily see, as done explicitly in Sect. 6.3, that the orbits described by the Hamilton equations for (6.2.1) correspond to the geodesics on a Riemannian space associated with the metric $g_{ij}(x)$, i.e. it follows directly from the Hamilton equations associated with (6.2.1) that (using (6.2.12) and the time derivative of (6.2.10) below)

$$\ddot{x}_\ell = -\Gamma_\ell^{mn} \dot{x}_m \dot{x}_n, \tag{6.2.2}$$

where the connection form Γ_ℓ^{mn} is given by

$$\Gamma_\ell^{mn} = \frac{1}{2} g_{\ell k} \left(\frac{\partial g^{km}}{\partial x_n} + \frac{\partial g^{kn}}{\partial x_m} - \frac{\partial g^{nm}}{\partial x_k} \right), \tag{6.2.3}$$

and g^{ij} is the inverse of g_{ij}.

The second covariant derivative of the geodesic deviation[2] depends on the curvature (Gutzwiller 1990; Curtiss and Miller 1985)

$$R_i^{jk\ell} = \frac{\partial \Gamma_i^{jk}}{\partial x_\ell} - \frac{\partial \Gamma_i^{j\ell}}{\partial x_k} + \Gamma_m^{jk} \Gamma_i^{\ell m} - \Gamma_m^{j\ell} \Gamma_i^{km}; \tag{6.2.4}$$

i.e., for $\xi_i = x_i' - x_i$ on closely neighboring trajectories at t,

$$\frac{D^2 \xi_i}{Dt^2} = R_i^{j\ell k} \dot{x}_j \dot{x}_k \xi_\ell, \tag{6.2.5}$$

where D/Dt is the covariant derivative along the curve $x_j(t)$. The contraction $R_i^{j\ell k} \dot{x}_j$ absorbs one index of the curvature tensor providing, due to the structure of $R_i^{j\ell k}$, a projection into a direction orthogonal to the velocity; projection the two sides of (6.2.5) into another direction orthogonal to the velocity then provides a second rank matrix on the right hand side as a coefficient of the velocity field orthogonal to the motion, i.e., the rate of deviation of the orbit. The sign of the eigenvalues of this matrix then gives information on the stability of the orbits (Gutzwiller 1990; Ar'nold 1978).

---

[2] See Chap. 5 for a rigorous discussion of geodesic deviation and its quantization based on the theory of dilations of Nagy and Foias (1970).

In this chapter, we point out that this formulation of dynamical stability has application to a wide range of Hamiltonian models; in fact, every potential model Hamiltonian of the form

$$H = \frac{{p^i}^2}{2M} + V(q), \tag{6.2.6}$$

where V is a function of space variables alone, can be put into the form (6.2.1), where the metric tensor is of conformal form (Horwitz 2007). We show, in this construction, that the motion induced by the Hamiltonian (6.2.6) has an embedding into a geometrically defined space, and obtain in this way a geometrical description of the time development for a Hamiltonian potential model.[3]

Casetti et al. (1997) have studied the application of both the Jacobi and Eisenhardt metrics in their analyses of the geometry of Hamiltonian chaos. The Jacobi metric (1884) (of the form $(E - V)\delta_{ij}$) leads to geodesic equations parametrized by the invariant distance associated with this metric on the manifold, in this case, the kinetic energy, thus corresponding to the Hamilton action. Transformation to parametrization by the time t leads to the second order Newton law in the form (6.2.14) below, for which the geometrical structure is no longer evident.

The Eisenhardt metric, leading to geodesic motion in t, involves the addition of an extra dimension. As noted by Caiani et al. (1997), this metric leads to the method commonly used for the computation of Lyapunov exponents in standard Hamiltonian systems. The method that we use here appears to be a more sensitive diagnostic than the computation of Lyapunov exponents of a locally linearized system.

The transformation of Hamiltonian dynamics of the type of Eq. (6.2.6) into the form (6.2.1) may be carried out by requiring that (6.2.6) be dynamically equivalent to (6.2.1). For a metric of conformal form[4]

$$g_{ij} = \varphi\delta_{ij}, \tag{6.2.7}$$

on the hypersurface defined by $H = E = constant$, the requirement of equivalence implies that

$$\varphi = \frac{E}{E - V(q)}. \tag{6.2.8}$$

To see that the Hamilton equations obtained from (6.2.1) can be put into correspondence with those obtained from the Hamiltonian of the potential model (6.2.6), we first note, from the Hamilton equations for (6.2.1), that

$$\dot{x}_i = \frac{\partial H}{\partial p^i} = \frac{1}{M} g_{ij} p^j. \tag{6.2.9}$$

[3]Here we assume a priori that the momenta are equal at all times in both pictures. In the next section, we impose this relation as a condition on dynamical equivalence.

[4]In Chap. 7 we discuss a method for obtaining a more general class of geometric Hamiltonions by canonical transformation.

We then use the geometrical property that $\dot{x}_i$ is a first rank contravariant tensor under local diffeomorphisms that preserve the constraint that H be constant, to define the *velocity field*

$$\dot{q}^j \equiv g^{ji}\dot{x}_i = \frac{1}{M}p^j, \tag{6.2.10}$$

coinciding formally with one of the Hamilton equations implied by (6.2.6), and which we shall therefore identify as associated with the variables of the Hamiltonian (6.2.6). From this definition, we recognize that we are dealing which might be thought of as two sets of coordinates, each characterized, as we shall see, by a different connection form, but related along any orbit, for small increments by

$$dq^j = g^{ji}dx_i \tag{6.2.11}$$

at each point (for which g^{ij} is nonsingular). Note that, while providing a basis for a change of variables along the orbits locally, (6.2.11) cannot be considered as a relation between one-forms since it would not, in general, be integrable. For the conformal form (6.2.7), one may alternatively consider the factor ϕ as inducing a local scale transformation on t along the orbit; we will not pursue this method here, but refer the reader to the discussion in Calderon et al. (2013).

To complete our correspondence with the dynamics induced by (6.2.6), consider the Hamilton equation for $\dot{p}^i$,

$$\dot{p}^\ell = -\frac{\partial H}{\partial x_\ell} = -\frac{1}{2M}\frac{\partial g_{ij}}{\partial x_\ell}p^i p^j. \tag{6.2.12}$$

Under local diffeomorphisms, (6.2.7) changes form. In geometric form, our formulation is covariant, so that, in principle, there is a wide range of coordinate systems representing the same dynamics. The formula (6.2.7) therefore corresponds to a particular choice of coordinates. In the particular coordinate system in which g_{ij} has the form (6.2.7) with φ of the form (6.2.8).

Since g^{ji}, in this choice of coordinates, is $\phi^{-1}\delta^{ji}$, (6.2.11) implies that in (6.2.12), $\frac{\partial}{\partial x_\ell}$ may be replaced by $\phi^{-1}\frac{\partial}{\partial q^\ell}$. It then follows that

$$\frac{\partial g_{ij}}{\partial x_\ell} = \phi^{-1}\frac{\partial \phi}{\partial q^\ell} = \frac{1}{E-V}\frac{\partial V}{\partial q^\ell}. \tag{6.2.13}$$

The coefficient $(E-V)^{-1}$ is cancelled by the remaining factor of $\mathbf{p}^2/2M$ in (6.2.13), which then becomes

$$\dot{p}^\ell = -\frac{\partial V}{\partial q^\ell}, \tag{6.2.14}$$

the second Hamilton equation in the usual form.

As we have pointed out, the $\{q^\ell\}$, which we shall call the *Hamilton coodinates*, are not uniquely defined in terms of the original coordinates $\{x_\ell\}$, which we shall call the *Gutzwiller coodinates*, since (6.2.11) is not an exact differential.

The geodesic Eq. (6.2.2) can be transformed directly from an equation for $\ddot{x}_j$ to an equation for $\ddot{x}^j$, the motion defined in the Hamilton coordinates. From (6.2.10) it follows that

$$\ddot{x}_\ell = g_{\ell j}\ddot{x}^j + \frac{\partial g_{\ell j}}{\partial x_n}\dot{x}_n\dot{q}^j = -\frac{1}{2}g_{\ell k}\left(\frac{\partial g^{km}}{\partial x_n} + \frac{\partial g^{kn}}{\partial x_m} - \frac{\partial g^{nm}}{\partial x_k}\right)\dot{x}_m\dot{x}_n. \tag{6.2.15}$$

Now, using the identity

$$\frac{\partial g_{\ell j}}{\partial x_n} = -g_{\ell k}\frac{\partial g^{km}}{\partial x_n}g_{mj}, \tag{6.2.16}$$

it follows that, with the symmetry of $\dot{x}_n\dot{x}_m$,

$$\frac{\partial g_{\ell j}}{\partial x_n}\dot{x}_n\dot{x}^j = -\frac{1}{2}g_{\ell k}\left(\frac{\partial g^{km}}{\partial x_n} + \frac{\partial g^{kn}}{\partial x_m}\right)\dot{x}_n\dot{x}_m. \tag{6.2.17}$$

Thus, the term on the left side of (6.2.15) containing the derivative of $g_{\ell j}$ cancels the first two terms of the connection form; multiplying the result by the inverse of $g_{\ell j}$, and applying the identity (6.2.16) to lower the indices of g^{nm} in the remaining term on the right side of (6.2.15), one obtains

$$\ddot{q}^\ell = -M^\ell_{mn}\dot{q}^m\dot{q}^n, \tag{6.2.18}$$

where

$$M^\ell_{mn} \equiv \frac{1}{2}g^{\ell k}\frac{\partial g_{nm}}{\partial q^k}. \tag{6.2.19}$$

Equation (6.2.18) has the form of a geodesic equation, with a truncated connection form. In fact, it is straightforward to show that the form (6.2.19) is indeed a connection form, transforming as

$$M'^\ell_{mn} = \frac{\partial q'^\ell}{\partial q^r}\frac{\partial q^p}{\partial q'^m}\frac{\partial q^q}{\partial q'^n}M^r_{pq} + \frac{\partial q'^\ell}{\partial q^r}\frac{\partial^2 q^r}{\partial q'^m\partial q'^n},$$

consistent with the covariance of (6.2.18) under local diffeomorphisms of the Hamilton manifold. This result is worked out explicitly in Sect. 6.3.

Equation (6.2.18) is therefore a covariant form of the Hamilton–Newton law, exhibiting what can be considered as an underlying geometry of standard Hamiltonian motion.

The geometrical structure of the Hamilton coordinates can be understood as follows. Let us write the covariant derivative for a (rank one) covariant tensor on the Gutzwiller manifold (defined as transforming in the same way as $\partial/\partial x_m$), using the full connection form (6.2.3),

$$A^{m;q} = \frac{\partial A^m}{\partial x_q} - \Gamma_k^{mq} A^k. \tag{6.2.20}$$

Lowering the index q with $g_{\ell q}$, we obtain the covariant derivative in the Hamilton coordinates, with connection form (with the help of (6.2.16))

$$\Gamma_{\ell k}^{m\ H} \equiv g_{\ell q} \Gamma_k^{mq} = \frac{1}{2} g^{mq} \left(\frac{\partial g_{\ell q}}{\partial q^k} - \frac{\partial g_{kq}}{\partial q^\ell} - \frac{\partial g_{k\ell}}{\partial q^q} \right). \tag{6.2.21}$$

This induced connection form, in the formula for curvature, would give a curvature corresponding to the Hamilton coordinates. However, it contains terms antisymmetric in the lower indices (ℓ, k) (indicating the possible existence of non-trivial torsion). Evaluated along a line parametrized by t, corresponding to geodesic motion, however, the antisymmetric terms cancel, leaving precisely the symmetric connection form (6.2.19).[5]

Since the coefficients M_{mn}^ℓ constitute a connection form, they can be used to construct a covariant derivative. It is this covariant derivative which must be used to compute the rate of transport of the geodesic deviation $\xi^\ell = q'^\ell - q^\ell$ along the (approximately common) motion of neighboring orbits in the Hamilton coordinates.

The second order geodesic deviation equations[6]

$$\ddot{\xi}^\ell = -2M_{mn}^\ell \dot{q}^m \dot{\xi}^n - \frac{\partial M_{mn}^\ell}{\partial q^q} \dot{q}^m \dot{q}^n \xi^q, \tag{6.2.22}$$

obtained from (6.2.18), can be factorized in terms of this covariant derivative,

$$\xi_{;n}^\ell = \frac{\partial \xi^\ell}{\partial q^n} + M_{nm}^\ell \xi^m. \tag{6.2.23}$$

One obtains

$$\frac{D_M{}^2}{D_M t^2} \xi^\ell = R_M{}^\ell{}_{qmn} \dot{q}^q \dot{q}^n \xi^m, \tag{6.2.24}$$

where the index M refers to the connection (6.2.19), and what we shall call the *dynamical curvature* is given by

[5] Note that since (6.2.19) and (6.2.21) are not directly derived from g_{ij}, they are not metric compatible connections. However, performing parallel transport on the local flat tangent space of the Gutzwiller coordinates, the resulting connection, after raising the tensor index to reach the Hamilton structure, results in exactly the "truncated" connection (6.2.19).

[6] Substituting the conformal metric (6.2.7) into (6.2.22), and taking into account the constraint that both trajectories q'^ℓ and q^ℓ have the same energy E, one sees that (6.2.22) becomes the orbit deviation equation based on (6.2.14).

$$R_M{}^{\ell}{}_{qmn} = \frac{\partial M^{\ell}_{qm}}{\partial q^n} - \frac{\partial M^{\ell}_{qn}}{\partial q^m} + M^k_{qm} M^{\ell}_{nk} - M^k_{qn} M^{\ell}_{mk}. \tag{6.2.25}$$

This expression, as remarked above, is not the curvature of the Hamilton manifold (given by this formula with $\Gamma^{\ell}_{qm}{}^{H}$ in place of M^{ℓ}_{qm}), but a dynamical curvature which is appropriate for geodesic deviation.

We give in the following a general formula for the geodesic deviation in the Hamilton manifold in two dimensions, and then show results of computer simulation for Poincaré plots showing a correspondence with the prediction of instability from the geodesic deviation.

With the conformal metric in noncovariant form (6.2.7), (6.2.8), the dynamical curvature (6.2.25) can be written in terms of derivatives of the potential V, and the geodesic deviation Eq. (6.2.24) becomes

$$\frac{D^2_M \xi}{D_M t^2} = -\mathcal{V} P \xi, \tag{6.2.26}$$

where the matrix $\mathcal{V}$ is given by

$$\mathcal{V}_{\ell i} = \left\{ \frac{3}{M^2 v^2} \frac{\partial V}{\partial q^{\ell}} \frac{\partial V}{\partial q^i} + \frac{1}{M} \frac{\partial^2 V}{\partial q^{\ell} \partial q^i} \right\}. \tag{6.2.27}$$

and

$$P^{ij} = \delta^{ij} - \frac{v^i v^j}{v^2}, \tag{6.2.28}$$

with $v^i \equiv \dot{x}^i$, defining a projection into a direction orthogonal to v^i.

We then find for the component orthogonal to the motion

$$\frac{D^2_M (\mathbf{v}_{\perp} \cdot \xi)}{D_M t^2} = -\left[\lambda_1 \cos^2 \phi + \lambda_2 \sin^2 \phi\right](\mathbf{v}_{\perp} \cdot \xi^{\ell}) \tag{6.2.29}$$

where λ_1 and λ_2 are eigenvalues of the matrix $\mathcal{V}$, and ϕ is the angle between $\mathbf{v}_{\perp}$ and the eigenvector for λ_1.

Instability should occur if at least one of the eigenvalues of $\mathcal{V}$ is negative, in terms of the second covariant derivatives of the transverse component of the geodesic deviation.

One may easily verify that the oscillator potential is predicted to be stable. Our criteria imply that the Duffing oscillator (without perturbation, not a chaotic system) clearly indicates instability in a neighborhood of the unstable fixed point. The potentials discussed by Oloumi and Teychenne (1999) also demonstrate the effectiveness of our procedure; our results in these cases are in agreement with theirs. The relation (6.2.29) provides a clear indication of the local regions of instability giving rise to chaotic motion in the Hénon–Heiles model (1969).

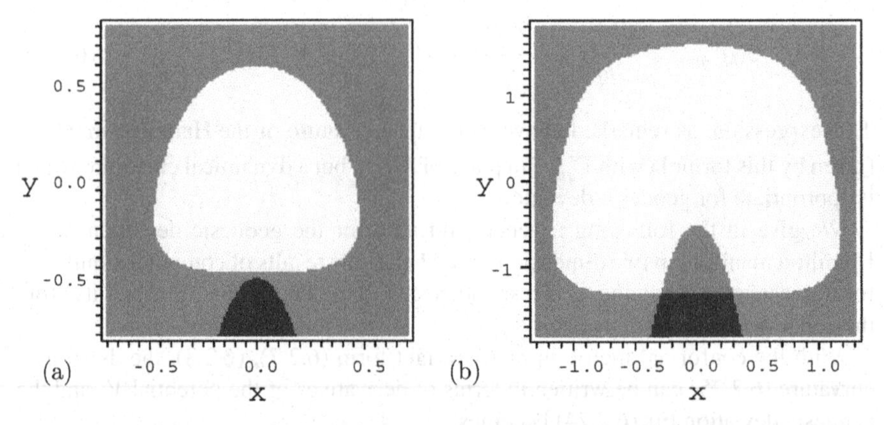

Fig. 6.1 **a** The dark area shows the regions of negative eigenvalues for the matrix $\mathcal{V}$ for a Hamiltonian with the potential (6.2.30). The light area corresponds to physically allowable motion for $E = 1/6$. The region of negative eigenvalues does not penetrate the physically accessible region in this case. **b** The dark area of negative eigenvalues for the matrix $\mathcal{V}$ is seen to penetrate deeply into the light region of physically allowable motion for $E = 3$. Reprinted with permission from Horwitz et. al. (2007a). Copyright 2007 by American Physical Society

We take for a simple illustration here a slight modification of the fifth order expansion of a two body Toda lattice Hamiltonian (for which the fourth order expansion coincides with the Hénon–Heiles model)

$$V(x, y) = \frac{1}{2}(x^2 + y^2) + x^2 y - \frac{1}{3}y^3 + \frac{3}{2}x^4 + \frac{1}{2}y^4. \tag{6.2.30}$$

This provides a new Hamiltonian chaotic system for which the above criterion gives a clear local signal for the presence of instability (see also Ben Zion and Horwitz 2007, 2008, 2010). Figure 6.1a shows that the region of negative eigenvalues does not penetrate the physically accessible region for $E = 1/6$; Fig. 6.2a shows a Poincaré plot in the (y, p_y) plane for this case, indicating completely regular orbits. In Fig. 6.1b, the distribution of negative eigenvalues for $E = 3$ is shown to penetrate deeply into the physical region, and Fig. 6.2b shows the corresponding Poincaré plot displaying a high degree of chaotic behavior. The criterion for instability we have given depends sensitively on the energy of the system. The critical energy for which the negative eigenvalues begin to penetrate the physically accessible region, in this example, is $E \cong 1/5$.

The condition implied by the geodesic deviation Eq. (6.2.26), in terms of covariant derivatives, in which the orbits are viewed geometrically as geodesic motion, is an effective condition for instability, based on the underlying geometry, for a Hamiltonian system of the form (6.2.6). This geometrical picture of Hamiltonian dynamics provides, moreover, new insight into the structure of the unstable and chaotic behavior of Hamiltonian dynamical systems.

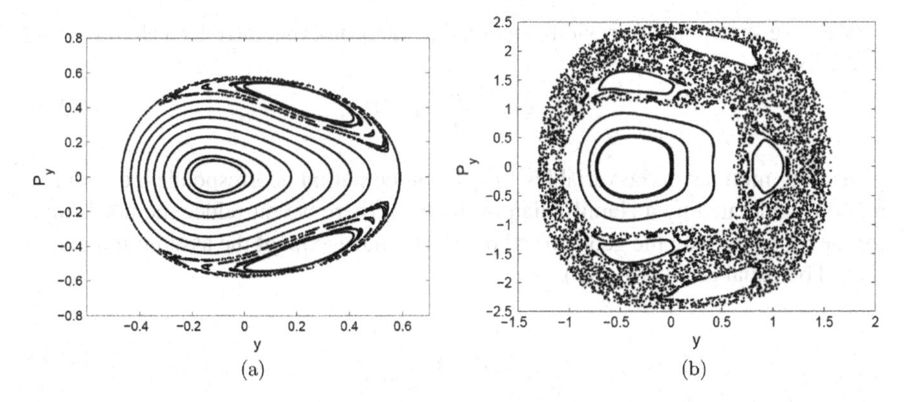

Fig. 6.2 **a** A Poincaré plot in the (y, p_y) plane for $E = 1/6$, indicating regular motion. **b** A Poincaré plot in the (y, p_y) plane for $E = 3$, indicating strongly chaotic behavior. Reprinted with permission from Horwitz et al. (2007a). Copyright 2007 by American Physical Society

6.3 An Underlying Geometrical Manifold for Hamiltonian Mechanics

6.3.1 *Introduction*

Recall that by (6.2.11), the relation between $\{x_i\}$ and $\{q^i\}$ is not uniquely defined globally, since this relation is not integrable. This section[7] investigates the relation between functions defined on these coordinates. We show that there exists, however, an underlying set of coordinates with a conformal metric and compatible connection form, and a metric type Hamiltonian (which we call the *geometrical picture*), that can be put into correspondence with the usual Hamilton–Lagrange mechanics as we have asserted in the previous section. The requirement of dynamical equivalence of the two types of Hamiltonians, that the momenta generated by the two pictures be equal for all times, is sufficient to determine a Taylor series expansion of the conformal factor. This factor is defined on the geometrical coordinate representation, in its domain of analyticity, with coefficients to all orders determined by functions of the potential of the Hamiltonian-Lagrange picture, defined on the Hamilton–Lagrange coordinate representation and its derivatives. Conversely, if the conformal function is known, the potential of a Hamilton–Lagrange picture can be determined in a similar way. We show that arbitrary local variations of the orbits in the Hamilton–Lagrange picture can be generated by variations along geodesics in the geometrical picture and establish a correspondence which provides a basis for understanding how the instability in the geometrical picture is manifested in the instability of the original Hamiltonian motion.

[7]This section is based on the paper of Horwitz et al. (2016).

We have shown in the previous section that the orbits generated by a Hamiltonian of the form

$$H = \delta_{ij}\frac{p^i p^j}{2m} + V(q), \tag{6.3.1}$$

generating motions in the variables $\{q, p\}$, can be put into correspondence with the motions generated by a Hamiltonian of the form (here we introduce the notation π for the momentum in the geometrical picture to make explicit the assumption that π should be equal at every t to p)

$$H_G = g_{ij}(x)\frac{\pi^i \pi^j}{2m}, \tag{6.3.2}$$

where we denote here, as in the previous section, by $\{x\}$ the coordinates of the geometrical manifold (which we have called Gutzwiller coordinates; see, e.g. Horwitz and Ben Zion 2007, Gutzwiller 1990), on which (6.3.2) generates motions through the Hamilton equations for the canonical variables $\{x, \pi\}$.

Equation (6.3.2) implies a geodesic motion associated with the metric g_{ij}. As pointed out in the previous section, the stability analysis of this representation of the dynamics, by means of geodesic deviation, has been shown to be a very sensitive indication of the local stability of the original Hamiltonian (6.3.1) (Horwitz 2011), and in many cases well-correlated with the chaotic behavior of the system (Horwitz et. al. 2011d). In application to the restricted three-body problem (Yahalom et al. 2011; Levitan et. al 2014), the method was shown to be highly effective. In this case, small regions of instability did not affect the overall stability since their effects were suppressed by the "uncertainty relation" (Yahalom et. al. 2015).

A relation between the Hamiltonians (6.3.1) and (6.3.2) may be established, for $p(t) = \pi(t)$ for all t, which shall consider here to correspond to dynamical equivalence, by assuming (Horwitz and Ben Zion 2007) a conformal metric

$$g_{ij}(x) = \phi(x)\delta_{ij}. \tag{6.3.3}$$

In the next Chapter, where we discuss general canonical transformations, we will not make this requirement a priori, but such a picture remains accessible. In this case, it follows that *there is a correspondence (Horwitz 2011) between the variables x and q and the functions $\phi(x)$ and $V(q)$* such that[8]

$$\phi(x) = \frac{E}{E - V(q)} \equiv F(q), \tag{6.3.4}$$

defining the function $F(q)$, which we shall use for brevity in the following, on the energy surfaces $H = H_G = E$.

[8]Both H and H_G are conserved, and can be assigned values E and E'; the results are not affected by this freedom. It only involves a constant shift in the potential V, and we take for convenience $H = H_G = E$.

We can show that the functions $\phi(x)$ and $F(q)$ on each of the manifolds can be put into correspondence through series expansion, and furthermore that the Poisson brackets and Lagrangians in both pictures are essentially equivalent, with variations in the geometrical picture driving variations in the Lagrangian of the Lagrange–Hamilton dynamics. We may therefore think of the geometrical picture in terms of an underlying geometry for the standard Lagrange–Hamilton mechanics.

For our present purposes, we give here a more detailed derivation of the geodesic Eq. (6.2.2). The Hamilton equations

$$\dot{x}_i = \frac{\partial H_G}{\partial \pi^i} \quad \dot{\pi}^i = -\frac{\partial H_G}{\partial x_i}, \tag{6.3.5}$$

applied to the Gutzwiller Hamiltonian (6.3.2), provide the relations

$$\dot{x}_i = \frac{1}{m} g_{ij} \pi^j \tag{6.3.6}$$

and

$$\dot{\pi}^j = -\frac{1}{2m}\frac{\partial g_{k\ell}}{\partial x_j}\pi^k \pi^\ell = -\frac{1}{2m}\frac{\partial g_{k\ell}}{\partial x_j} m^2 g^{km} g^{\ell n} \dot{x}_m \dot{x}_n = \frac{m}{2}\frac{\partial g^{mn}}{\partial x_j}\dot{x}_m \dot{x}_n, \tag{6.3.7}$$

where we have used the identity

$$\frac{\partial g^{ij}}{\partial x_m} = -g^{ik}\frac{\partial g_{k\ell}}{\partial x_m} g^{\ell j}. \tag{6.3.8}$$

Then since, from (6.3.6), we have

$$\ddot{x}_i = \frac{1}{m}\frac{\partial g_{ij}}{\partial x_m}\dot{x}_m \pi^j + \frac{1}{m} g_{ij}\dot{\pi}^j. \tag{6.3.9}$$

Using (6.3.7) for $\dot{\pi}^j$ and taking onto account the symmetry of $\dot{x}_m \dot{x}_n$, one obtains (with some rearrangement of indices) the geodesic equation

$$\ddot{x}_\ell = -\Gamma_\ell{}^{mn} \dot{x}_m \dot{x}_n, \tag{6.3.10}$$

where the coefficients have the structure of the usual connection form (g^{ij} is the inverse of g_{ij})

$$\Gamma_\ell^{mn} = \frac{1}{2} g_{\ell k}\left(\frac{\partial g^{km}}{\partial x_n} + \frac{\partial g^{kn}}{\partial x_m} - \frac{\partial g^{nm}}{\partial x_k}\right). \tag{6.3.11}$$

This connection form is *compatible* with the metric $g_{ij}(x)$ by construction, i.e., the covariant derivative constructed with the Γ_ℓ^{mn} of (6.3.10) of g_{ij} vanishes, and we recognize that the dynamics generated on the manifold $\{x\}$ is a geodesic flow. It can carry, moreover, a tensor structure which may be inferred from the invariance of the form (6.3.2) under local coordinate transformations (to be discussed below). In this framework, the definitions (6.3.3) and (6.3.4) correspond to a special choice of coordinates.

However, this geodesic flow does not precisely follow the form of the orbits generated by (6.3.1) when the special choice of coordinates is used for which (6.3.3) and (6.3.4) are valid. It has been shown that in the conformal case which we are considering, a change in scale along the orbits suffices to bring the motions into correspondence (Calderon et.al. 2013) and achieve the stability criterion of Horwitz and Ben Zion (2007) within a single coordinatization. For many physical applications, however, it is important to have both sets of variables available (see, for example, Gershon and Horwitz 2009; Horwitz, Gershon and Schiffer 2010 in application to the theory of Milgrom 1983; Bekenstein 2004; Bekenstein and Sanders 2006).

As we shall show, the requirement that the momenta $p^i(t)$ and $\pi^i(t)$ are equal for all times is sufficient to determine the function $\phi(x)$ in an open analytic set in terms of nonlinear functions of $F(q)$ and its derivatives, and conversely, if $\phi(x)$ is known, to determine $F(q)$ (and therefore $V(q)$), in an open analytic set, by the same method. The existence of these analytic expansions demonstrates the coexistence and correspondence of the coordinates on the two manifolds.

The relation

$$\dot{x}_i = g_{ij}\dot{q}^j, \tag{6.3.12}$$

as we shall show below, follows immediately from the requirement of dynamical equivalence of the two pictures.[9] The relation

$$dx_i = g_{ij}dq^j, \tag{6.3.13}$$

as discussed in the previous section, is clearly not integrable, so that one may not directly construct a global one to one correspondence between the two manifolds using this relation. One may, however, integrate it along a smooth geodesic curve, a procedure which will be adequate for our purposes. Substituting (6.3.12) into the geodesic formula (6.3.10), one obtains the geodesic type relation

$$\ddot{q}^\ell = -M^\ell_{mn}\dot{q}^m\dot{q}^n, \tag{6.3.14}$$

where

$$M^\ell_{mn} \equiv \frac{1}{2}\frac{\partial g_{nm}}{\partial x_\ell}. \tag{6.3.15}$$

As we have remarked above, the coordinates $\{x\}$ may carry local diffeomorphisms; $\dot{x}_j$ transforms under these diffeomorphisms as a (contravariant) vector. In the special set of coordinates for which (6.3.4) holds, as shown in the previous section, the relation (6.3.14) reduces to precisely the Hamilton equation

[9] This relation was *assumed* in the previous section (Horwitz and Ben Zion 2007) in order to establish a correspondence between the Hamilton and geometric pictures.

$$\ddot{q}^i = -\frac{1}{m}\frac{\partial V(q)}{\partial q^i}, \tag{6.3.16}$$

Therefore, the result (6.3.14) constitutes a *geometric embedding* of the original Hamiltonian motion into a set of coordinates which we call the *Hamiltonian coordinates* $\{q\}$, to which we have assigned, for simplicity in notation, the same symbol as the coordinates for the original Hamiltonian. As was shown in Horwitz and Ben Zion 2007 and subsequent studies (Ben Zion and Horwitz 2007), the computation of the geodesic deviation of the orbits described by (6.3.14) are remarkably predictive of instability in the original Hamiltonian motion, and have provided an effective criterion for the occurrence of such instabilities. Before demonstrating the relation of the functions $\phi(x)$ and $F(q)$, we prove here explicitly that the quantity M_{ij}^{ℓ} is a valid connection form (as asserted in the previous section), i.e., under local diffeomorphisms, it provides the proper compensation terms to insure that (6.3.14) is a covariant relation.

It follows from the Hamilton equations for the geometrical Hamiltonian (6.3.2) that

$$\dot{x}_i = g_{ij}\frac{\pi^j}{m}, \tag{6.3.17}$$

or, using the assumption that g_{ij} is invertible with inverse g^{ij},

$$\pi^j = mg^{ji}\dot{x}_i. \tag{6.3.18}$$

Substituting this back into the Hamiltonian, one obtains

$$H_G = \frac{m}{2}g^{ij}(x)\dot{x}_i\dot{x}_j \tag{6.3.19}$$

This form is strongly suggestive of the existence of a local diffeomorphism invariance for the coordinates $\{x\}$ of the (contravariant) form

$$dx'_i = \frac{\partial x'_i}{\partial x_j}dx_j \tag{6.3.20}$$

Requiring H_G to be invariant under such diffeomorphisms, we see that the metric must transform as

$$g'^{k\ell} = \frac{\partial x_i}{\partial x'_k}\frac{\partial x_j}{\partial x'_\ell}g^{ij}. \tag{6.3.21}$$

We remark that from the result (6.3.18) and the transformation property (6.3.20), one obtains (with inverse Jacobian)

$$\pi'^j = \frac{\partial x_\ell}{\partial x'_j}\pi^\ell, \tag{6.3.22}$$

from which the inverse metric can be easily seen to transform as

$$g'_{kj} = \frac{\partial x'_k}{\partial x_m} \frac{\partial x'_j}{\partial x_\ell} g_{\ell m}. \tag{6.3.23}$$

It then follows from the transformation law (6.3.20) that there is a diffeomorphism induced on the coordinates $\{q\}$ of the form

$$dq'^j = \frac{\partial x_m}{\partial x'_j} dq^m, \tag{6.3.24}$$

which appears to be covariant from the point of view of the $\{x\}$ manifold, but is contravariant on $\{q\}$.

We now study the covariance of Eq. (6.3.14), writing it as

$$\ddot{q}'^j = -M'^j_{k\ell} \dot{q}'^k \dot{q}'^\ell, \tag{6.3.25}$$

where

$$M'^j_{k\ell} \equiv \frac{1}{2} \frac{\partial g'_{k\ell}}{\partial x'_j}. \tag{6.3.26}$$

Formally dividing (6.3.24) by dt on both sides and differentiating with respect to t, one obtains

$$\ddot{q}'^j = \frac{\partial x'_p}{\partial x_k} \frac{\partial^2 x_m}{\partial x'_j \partial x'_p} \dot{x}_k \dot{q}^m + \frac{\partial x_m}{\partial x'_j} \ddot{q}^m. \tag{6.3.27}$$

Clearly, $\ddot{q}^m$ does not transform as a tensor; such a transformation law would involve only the last term of (6.3.27). We now show that the connection form $M'^j_{k\ell}$ develops a compensating term which returns the geodesic equation to its form before the transformation, up to a factor $\frac{\partial x_m}{\partial x'_j}$, i.e., the diffeomorphism induced by the transformation. Replacing $\dot{x}_k$ by $g_{k\ell}\dot{q}^\ell$ in the first term on the right, we have

$$\ddot{q}'^j = g_{k\ell} \frac{\partial x'_p}{\partial x_k} \frac{\partial^2 x_m}{\partial x'_j \partial x'_p} \dot{q}^\ell \dot{q}^m + \frac{\partial x_m}{\partial x'_j} \ddot{q}^m \tag{6.3.28}$$

We now note that

$$\begin{aligned} M'^j_{k\ell} &= \frac{1}{2} \frac{\partial x_p}{\partial x'_j} \frac{\partial}{\partial x_p} \left(\frac{\partial x'_k}{\partial x_m} g_{mn} \frac{\partial x'_\ell}{\partial x_n} \right) = \\ &= \frac{1}{2} \frac{\partial x_p}{\partial x'_j} \left[\frac{\partial^2 x'_k}{\partial x_m \partial x_p} g_{mn} \frac{\partial x'_\ell}{\partial x_n} + \frac{\partial x'_k}{\partial x_m} \frac{\partial g_{mn}}{\partial x_p} \frac{\partial x'_\ell}{\partial x_n} + \frac{\partial x'_k}{\partial x_m} g_{mn} \frac{\partial^2 x'_\ell}{\partial x_n \partial x_p} \right]. \end{aligned} \tag{6.3.29}$$

so that

$$M'^{j}_{k\ell}\dot{q}'^{k}\dot{q}'^{\ell} = \frac{1}{2}\frac{\partial x_p}{\partial x'_j}\left[2g_{m\ell}\frac{\partial^2 x'_k}{\partial x_m \partial x_p}\frac{\partial x_s}{\partial x'_k} + \frac{\partial g_{s\ell}}{\partial x_p}\right]\dot{q}^s\dot{q}^\ell. \tag{6.3.30}$$

It is, however, an identity, using

$$\frac{\partial x'_p}{\partial x_k}\frac{\partial}{\partial x'_j}\frac{\partial x_m}{\partial x'_p} = -\left(\frac{\partial}{\partial x'_j}\frac{\partial x'_p}{\partial x_k}\right)\frac{\partial x_m}{\partial x'_p}, \tag{6.3.31}$$

that

$$\frac{\partial x_p}{\partial x'_j}g_{mq}\frac{\partial^2 x'_\ell}{\partial x_m \partial x_p}\frac{\partial x_s}{\partial x'_\ell}\dot{q}^s\dot{q}^q = -g_{k\ell}\frac{\partial x'_p}{\partial x_k}\frac{\partial^2 x_m}{\partial x'_j \partial x'_p}\dot{q}^\ell\dot{q}^m. \tag{6.3.32}$$

Then, we see that the first term of (6.3.28) cancels with the addition of $M'^{j}_{k\ell}\dot{q}^k\dot{q}^\ell$, so that

$$\ddot{q}'^{j} + M'^{j}_{k\ell}\dot{q}'^{k}\dot{q}'^{\ell} = \frac{\partial x_m}{\partial x'_j}(\ddot{q}^m + M^m_{k\ell}\dot{q}^k\dot{q}^\ell), \tag{6.3.33}$$

i.e., the geodesic equation transforms as a tensor under diffeomorphisms on $\{q\}$ induced by diffeomorphisms on $\{x\}$, the fundamental underlying geometric coordinates. This expression corresponds to what we shall define as a *covariant derivative*, to be discussed further below.

The connection M^ℓ_{mn} is, however, not compatible with the metric g_{ij}, and therefore one cannot construct a locally flat space (i.e., an equivalence principle) in the Hamilton framework, consistent with the fact that there is no equivalence principle in the standard Hamilton mechanics. Equation (6.3.10) is therefore, although covariant, not a geodesic equation in the sense that it could be derived from a minimum path length principle constructed with the metric g_{ij}. It maintains, however, the local diffeomorphism covariance derived from the underlying Gutzwiller coordinates, and therefore the resulting geodesic type equation corresponds to a proper geometric embedding of the Hamiltonian motion.

Although the dynamics represented by (6.3.14) does not have an equivalence principle, parallel transport carried out in the Gutzwiller manifold $\{x\}$, with transformation back to the Hamilton space $\{q\}$ results in precisely the truncated connection form (6.3.15) (Horwitz and Ben Zion 2007). We shall see that this argument applies as well to the covariant derivative to be discussed below. Applying the method of geodesic deviation to the orbits described in terms of the geometric embedding (6.3.14) results in a stability criterion that has been found to be remarkably effective in detecting chaotic behavior in Hamiltonian systems.

6.4 The Nature of the Embedding: Relation Between $\phi(x)$ and $F(q)$

We now discuss the nature of this embedding. Since q and x are related only by the non-integrable condition (6.2.11), i.e. $dq^j = g^{ij} dx_i$ (we have called $x^i \equiv q^i$ to identify it with the variable occurring in the conventional Hamiltonian with the special conformal choice of metric), the two sets of coordinates are not uniquely related. However, we can show systematically that all derivatives of the function $\phi(x)$ are determined by $F(q)$ and its derivatives, and, conversely, that if $\phi(x)$ is known, all derivatives of $F(q)$ are determined by $\phi(x)$ by requiring that the momenta $p^i(t)$ and $\pi^i(t)$ are equal for all times. We shall call this requirement *dynamical equivalence*. In a sufficiently analytic domain, the two functions can therefore be put into correspondence.

In the following, we shall use the explicit form of the metric for the geometric Hamiltonian in the special coordinate choice for which (6.3.4) is valid.[10] Although not necessary, we take the geometric Hamiltonian to be on the same energy shell as (6.3.1), i.e.,

$$H_G = \phi(x) \frac{\pi^2}{2m} = E. \tag{6.4.1}$$

Since then

$$V(q) = E \left(1 - \frac{1}{F}\right), \tag{6.4.2}$$

the Hamilton equation for the time derivative of the momentum is

$$\dot{p}^j = -\frac{\partial V}{\partial q^j} = E \frac{\partial}{\partial q^j} \left(\frac{1}{F}\right). \tag{6.4.3}$$

For the geometric dynamics,

$$\dot{\pi}^j = -\frac{\partial \phi}{\partial x_j} \frac{\pi^2}{2m} = -\frac{\partial \phi}{\partial x_j} \frac{E}{\phi} = -E \frac{\partial}{\partial x_j} \ln \phi. \tag{6.4.4}$$

It follows from (6.4.3) and (6.4.4) that (using the fact that $\phi = F$)

$$\phi \frac{\partial \phi}{\partial x_j} = \frac{\partial F}{\partial q^j}. \tag{6.4.5}$$

For the time derivatives of the coordinates, we have, setting $p^j = \pi^j$,

[10] In this special choice of coordinates, the placement of indices does not always make the covariance evident.

$$\dot{q}^j = \frac{p^j}{m} = \frac{\pi^j}{m} \tag{6.4.6}$$

and[11]

$$\dot{x}_j = \frac{\phi}{m}\pi^j = \frac{\phi}{m}p^j = \phi\dot{q}^j. \tag{6.4.7}$$

We therefore see that the relation

$$\dot{x}_i = g_{ij}\dot{q}^j, \tag{6.4.8}$$

taken to be defined along a geodesic curve in x^j, is a necessary consequence of the dynamical equivalence of the two systems (see Calderon et.al. 2013 for an alternative, but eventually equivalent, approach).

Proceeding to the second derivatives, we have

$$\begin{aligned} \ddot{p}^j &= E\frac{\partial^2}{\partial q^j \partial q^k}\left(\frac{1}{F}\right)\dot{q}^k \\ \ddot{\pi} &= -E\left(\frac{\partial^2}{\partial x_j \partial x_k}\ln\phi\right)\dot{x}^k. \end{aligned} \tag{6.4.9}$$

With (6.3.16) and (6.4.7) we may rewrite these equations as

$$\begin{aligned} \ddot{p}^j &= E\left(\frac{\partial^2}{\partial q^j \partial q^k}\frac{1}{F}\right)\frac{\pi^k}{m} \\ \ddot{\pi}^j &= -E\left(\frac{\partial^2}{\partial x_j \partial x_k}\ln\phi\right)\frac{\phi}{m}\pi^k. \end{aligned} \tag{6.4.10}$$

Comparing coefficients of π^k, we find

$$\frac{\partial^2}{\partial q^j \partial q^k}\left(\frac{1}{F}\right) = -\left(\frac{\partial^2}{\partial x_j \partial x_k}\ln\phi\right)\phi. \tag{6.4.11}$$

This result provides the equivalence to second order in the power series expansions of the functions $F(q)^{-1}$ and $\ln\phi$.

Explicitly, carrying out the derivatives, one obtains

$$\frac{2}{F^3}\frac{\partial F}{\partial q^j}\frac{\partial F}{\partial q^k} - \frac{1}{F^2}\frac{\partial^2 F}{\partial q^j \partial q^k} = \frac{1}{\phi}\frac{\partial \phi}{\partial x_j}\frac{\partial \phi}{\partial x_k} - \frac{\partial^2 \phi}{\partial x_j \partial x_k}. \tag{6.4.12}$$

Substituting (6.4.5) into the first term on the right hand side, one obtains one half of the first term on the left; after cancellation, one obtains

[11] We remark that in this context, the conformal factor plays the role of a local mass scaling.

$$\frac{\partial^2 \phi}{\partial x_j \partial x_k} = \frac{1}{F^2}\left(\frac{\partial^2 F}{\partial q^j \partial q^k} - \frac{1}{F}\frac{\partial F}{\partial q^j}\frac{\partial F}{\partial q^k}\right). \tag{6.4.13}$$

This procedure can evidently be continued, and we therefore conclude that, in a domain of mutual analyticity, the Taylor series expansions of the two functions can be put into correspondence. We derive in the following a general formula.

It is convenient to write (6.4.9) in the form

$$\begin{aligned} \ddot{p}^j &= \frac{E}{m}\frac{\partial^2}{\partial q^j \partial q^k}\left(\frac{1}{F}\right)p^k \\ \ddot{\pi}^j &= -\frac{E}{m}\left(\frac{\partial^2}{\partial x_j \partial x_k}\ln\phi\right)\phi p^k. \end{aligned} \tag{6.4.14}$$

Differentiating with respect to t, one obtains

$$\begin{aligned} \left(\frac{d}{dt}\right)^3 p^j &= \frac{E}{m}\left[\frac{\partial^2}{\partial q^j \partial q^k}\left(\frac{1}{F}\right)\right]p^k + \frac{E}{m}\frac{\partial^2}{\partial q^j \partial q^k}\left(\frac{1}{F}\right)\dot{p}^k \\ \left(\frac{d}{dt}\right)^3 \pi^j &= -\frac{E}{m}\frac{d}{dt}\left(\frac{\partial^2}{\partial x_j \partial x_k}\ln\phi\right)\phi p^k - \frac{E}{m}\left(\frac{\partial^2}{\partial x_j \partial x_k}\ln\phi\right)\phi\dot{p}^k. \end{aligned} \tag{6.4.15}$$

Subtracting the two expressions, the last terms in each cancel by (6.4.11), so that we have, carrying out the derivatives in t to obtain factors of $\dot{q}^\ell = \frac{p^\ell}{m}$ and $\dot{x}_\ell = \phi\frac{p^\ell}{m}$,

$$0 = \left(\frac{d}{dt}\right)^3 p^j - \left(\frac{d}{dt}\right)^3 \pi^j = \frac{E}{m^2}p^k p^\ell\left[\frac{\partial^3}{\partial q^j \partial q^k \partial q^\ell}\left(\frac{1}{F}\right) + \phi\frac{\partial}{\partial x_\ell}\left(\phi\frac{\partial^2}{\partial x_j \partial x_k}\ln\phi\right)\right]. \tag{6.4.16}$$

We therefore have the condition

$$\frac{\partial^3}{\partial q^j \partial q^k \partial q^\ell}\left(\frac{1}{F}\right) = -\phi Sym\frac{\partial}{\partial x_\ell}\left(\phi\frac{\partial^2}{\partial x_j \partial x_k}\ln\phi\right), \tag{6.4.17}$$

where Sym implies symmetrization.

Taking further derivatives with respect to t of (6.4.16) provides higher derivatives of the condition (6.4.17) (with symmetrization), since derivatives of the momenta do not contribute due to the vanishing of the remaining factor (as sufficient conditions). One then finds the condition for the fourth order that

$$\frac{\partial^4}{\partial q^j \partial q^k \partial q^\ell \partial q^m}\left(\frac{1}{F}\right) = -\phi Sym\frac{\partial}{\partial x_m}\left(\phi\frac{\partial}{\partial x_\ell}\left(\phi\frac{\partial^2}{\partial x_j \partial x_k}\ln\phi\right)\right). \tag{6.4.18}$$

It is clear that higher derivatives then obey the relation (as can easily be proved by induction)

$$\frac{\partial^n}{\partial q^{i_1}\partial q^{i_2}\ldots\partial q^{i_n}}\left(\frac{1}{F}\right) = -\phi\, Sym \frac{\partial}{\partial x_{i_n}}\left(\phi \frac{\partial}{\partial x_{i_{n-1}}}\left(\ldots\phi\frac{\partial}{x_{i_3}}\left(\phi\frac{\partial^2}{\partial x_{i_2}\partial x_{i_1}}\ln\phi\right)\ldots\right)\right). \tag{6.4.19}$$

These relations between the coefficients of the Taylor expansion in the neighborhood of a given point in x and q, are not, in general, integrable, but provide a relation between the functions $\phi(x)$ and $F(q)$ in analytic domains. We emphasize that the differentiability in t that we have used was restricted to geodesic curves on x. We remark that the process may alternatively be carried out to provide formulas for the derivatives to all orders of $\phi(x)$ in terms of the function $F(q)$ and its derivatives.

6.5 Poisson Bracket Structure

It follows directly from this analysis that the Poisson bracket structure on the two sets of coordinates is identical. To see this, we start with the time derivative of a function $f(x,\pi)$ on the geometrical coordinates:

$$\frac{d}{dt}f(x,\pi) = \frac{\partial f}{\partial x_i}\dot{x}_i + \frac{\partial f}{\partial \pi^i}\dot{\pi}^i \tag{6.5.1}$$

Since we are assuming Hamilton equations for the geometrical Hamiltonian H_G, this result is a Poisson bracket with the usual symplectic symmetry:

$$\frac{d}{dt}f(x,\pi) = \{f, H_G\}. \tag{6.5.2}$$

On the other hand, from (6.3.17), the Poisson bracket (6.5.1) becomes

$$\frac{d}{dt}f(x,\pi) = \frac{\partial f}{\partial x_i}g_{ij}\frac{\pi^j}{m} + \frac{\partial f}{\partial \pi^i}\dot{\pi}^i \tag{6.5.3}$$

Since the Poisson bracket is formed by following the flow in phase space, the motion in x_i follows a geodesic; therefore $\frac{\partial f}{\partial x_i}$ is projected by $\dot{x}_i$ along a geodesic curve, and we can set

$$\frac{\partial f}{\partial x_i}g_{ij} = \frac{\partial \hat{f}}{\partial q^j} \tag{6.5.4}$$

in Eq. (6.5.3), where $\hat{f}$ is equal to f considered a function of q, p. Furthermore, $\frac{\pi^j}{m} = \frac{p^j}{m} = \dot{q}^j$, so that (6.5.3) becomes, with our requirement that the momenta p and π and all derivatives are equal,

$$\frac{d}{dt}f(x,\pi) = \frac{\partial \hat{f}}{\partial q^i}\dot{q}^i + \frac{\partial \hat{f}}{\partial p^i}\dot{p}^i. \tag{6.5.5}$$

Therefore, we have

$$\frac{d}{dt}f(x,\pi) = \frac{d}{dt}\hat{f}(q,p), \tag{6.5.6}$$

establishing the equivalence of the Poisson brackets in both representations.[12]

We have shown in the above that a geometric Hamiltonian of the form (6.3.2) with conformal metric can be constructed which has properties closely related to the original Hamiltonian motion. The geodesics of the geometrical Hamiltonian H_G, in the special coordinate system for which the correspondence (6.3.4) is constructed, do not follow the orbits of the original Hamiltonian, but the local mapping along the geodesic curve, required by the dynamical equivalence of the two pictures provides the modified geodesic type motion (6.3.14) which, in the special choice of coordinates for which (6.2.7) and (6.2.8) are valid, precisely coincides with the equations of motion generated by the original Hamiltonian, as in (6.3.16). We have, moreover, shown that the modified connection form (6.3.15) is a good connection form, in the sense that the diffeomorphisms induced on the coordinates denoted by $\{q\}$ by diffeomorphisms on $\{x\}$ leave the geodesic type Eq. (6.3.14) invariant in form. In fact, Eq. (6.3.14) can be written as the vanishing of a covariant derivative

$$\frac{D}{Dt}\dot{q}^k = \left(\frac{d}{dt}\delta^k{}_j + M^k{}_{ij}\dot{q}^i\right)\dot{q}^j = 0. \tag{6.5.7}$$

The notion of covariant derivative in terms of parallel transport may be considered as parallel transport in the coordinates $\{x\}$ and transformation back to the coordinates $\{q\}$ in two steps, first from the tangent space of the geometric coordinates to the coordinates $\{x\}$, and then to $\{q\}$. The vanishing of the covariant derivative (6.5.7) of $\dot{q}^k$ carries this geometrical interpretation, i.e., it effectively carries the motion along the tangent space of $\{x\}$. We would therefore expect that the second covariant derivative of $\dot{q}^k$, and in fact, all covariant derivatives, would vanish as well.

It is straightforward to compute the second covariant derivative:

$$\frac{D^2}{Dt^2}\dot{q}^k = \left(\frac{d}{dt}\right)^2\dot{q}^k + \frac{3}{2}\frac{\partial g_{m\ell}}{\partial x_k}\ddot{q}^m\dot{q}^\ell + \frac{1}{2}\dot{q}^p\dot{q}^m\dot{q}^\ell g_{qp}\frac{\partial^2 g_{\ell m}}{\partial x_k \partial x_q} + \frac{1}{4}\frac{\partial g_{ij}}{\partial x_k}\frac{\partial g_{m\ell}}{\partial x_j}\dot{q}^i\dot{q}^m\dot{q}^\ell, \tag{6.5.8}$$

where we have used the replacement $\dot{x}_q = g_{pq}\dot{q}^p$ along the geodesic curve. In the special choice of coordinates, one then obtains

$$\frac{D^2}{Dt^2}\dot{q}^k = \left(\frac{d}{dt}\right)^2\dot{q}^k - \frac{E}{m}\left\{\frac{1}{\phi}\frac{\partial\phi}{\partial x_k}\frac{\partial\phi}{\partial x_\ell} - \frac{\partial^2\phi}{\partial x_k \partial x_\ell}\right\}\dot{q}^\ell. \tag{6.5.9}$$

---

[12] In Chap. 7, as remarked above, we show that the relation between the two Hamiltonian structures we have discussed can be understood in terms of a canonical transformation (which, of course, preserves the Poisson bracket relations).

From (6.4.9), which follows from equating the momenta, one may carry out the derivatives as in (6.3.11)–(6.4.12) and divide by m to show that the expression (6.5.9) for the second covariant derivative vanishes. We infer that all covariant derivatives (for smooth motion) vanish, indicating that the mechanism forming the geodesic motion on the geometrically embedded coordinates $\{q\}$ is effectively associated with parallel transport on the geometric coordinates $\{x\}$.

We emphasize that the vanishing of covariant derivatives is not the trivial result of differentiating a quantity that is identically zero; the covariant derivative is not an ordinary derivative, but contains information on the connection form so that the operation acts essentially on the tangent space. Since, in the geometrical embedding, the connection M^k_{ij} is not compatible with the metric, this result demonstrates that the covariant derivative acts effectively, as we have pointed out before, on the tangent space of the underlying geometrical manifold.

6.6 Variational Correspondence

In order to establish a variational correspondence we now turn to the local relation between the functions $\phi(x)$ and $F(q)$. We wish to study the result of a variation in q due to a variation in x along a geodesic path. The variation δq^i generated locally by the transport of a point x along a geodesic is

$$\delta q^i = \int_0^{\delta q} dq^i = \int_0^{\delta\xi} g^{ij}(x_0 + \xi\eta)\eta_j d\xi, \tag{6.6.1}$$

where we take ξ to be an affine parameter along a small segment of the geodesic curve, which we approximate as a straight line with constant direction given by the unit vector . Since the upper limit on ξ is small, we can expand in Taylor series to obtain

$$\delta q^i|_{x+\delta x} = \int_0^{\delta\xi} \left\{ g^{ij}(x_0)\eta_j + \frac{\partial g^{ij}}{\partial x_k}\eta_j\eta_k\xi + \frac{1}{2}\frac{\partial^2 g^{ij}}{\partial x_k \partial x_\ell}\eta_j\eta_k\eta_\ell\xi^2 + \cdots \right\} d\xi; \tag{6.6.2}$$

carrying out the integral over the affine parameter ξ, the resulting powers of $\delta\xi$ match the occurrence of the unit vectors, so we may write the result as

$$\delta q^i|_{x+\delta x} = g^{ij}(x_0)\delta x_j + \frac{1}{2}\frac{\partial g^{ij}}{\partial x_k}\delta x_j \delta x_k + \frac{1}{3!}\frac{\partial^2 g^{ij}}{\partial x_k \partial x_\ell}\delta x_j \delta x_k \delta x_\ell + \cdots \tag{6.6.3}$$

We remark that since the derivative of g^{ij} occurs symmetrically in j, k in (6.6.3), the second term is essentially complementary to the truncated connection (6.3.15) with respect to a complete connection form of the type (6.3.11). Thus the variation in q^i is not along a geodesic curve in $\{q\}$ corresponding to the geodesic curve in $\{x\}$ on which the variation is generated.

Formally, the structure continues to higher order (with the next term proportional to $\frac{1}{4!}$)), but since we have approximated the geodesic curve by a straight line, we do not expect the result to be accurate to high orders; the second order term is sufficient for our present purposes.[13]

We then substitute this result into $F(q + \delta q)$; in the expansion of this function, one uses (6.6.3), i.e.,

$$\begin{aligned} F(q+\delta q) &= F\left(q+g^{ij}(x_0)\delta x_j + \frac{1}{2}\frac{\partial g^{ij}}{\partial x_k}\delta x_j \delta x_k + \frac{1}{3!}\frac{\partial^2 g^{ij}}{\partial x_k \partial x_\ell}\delta x_j \delta x_k \delta x_\ell + \cdots\right) \\ &= F(q) + \frac{\partial F}{\partial q^i}\left(g^{ij}(x_0)\delta x_j + \frac{1}{2}\frac{\partial g^{ij}}{\partial x_k}\delta x_j \delta x_k + \cdots\right) \\ &\quad + \frac{1}{2}\frac{\partial^2 F}{\partial q^i \partial q^m}\left(g^{ij}(x_0)\delta x_j + \frac{1}{2}\frac{\partial g^{ij}}{\partial x_k}\delta x_j \delta x_k + \cdots\right) \\ &\quad \left(g^{mj}(x_0)\delta x_j + \frac{1}{2}\frac{\partial g^{mj}}{\partial x_k}\delta x_j \delta x_k + \cdots\right) + \cdots \end{aligned} \tag{6.6.5}$$

Comparison of the first order term with the expansion of $\phi(x + \delta x)$ yields (recalling that $g^{ij} = \frac{1}{\phi}\delta_{ij}$ and contracting the δ's) the relation (6.4.5), which also follows from equating the first derivatives of the momenta π and p. Keeping, moreover, second order terms from the first order part of the expansion, and adding them to the second order terms from the second order part of the expansion, one may compare with the second order part of the expansion of $\phi(x + \delta x)$. One finds (contracting the δ's) precisely the result (6.4.13).

The second order result we have obtained is sufficient for our purposes. What we have shown is that the functions $\phi(x)$ and $F(q)$ can be put locally into correspondence pointwise, establishing a relation between the coordinates $\{x\}$ and $\{q\}$ along geodesic curves; this relation is, furthermore, consistent with the condition of dynamical equivalence of the motions generated by the two Hamiltonians. We have therefore established a basis for the correspondence between small variations in the two steps of coordinates.

6.7 Variational Principles of Hamilton–Lagrange

In this section we recast the analytic mechanics of Hamilton and Lagrange in the Hamilton space $\{q\}$ in terms of variations generated in an underlying geometrical set of coodinates $\{x\}$. We start with the Lagrangian obtained from the Legendre

[13] Note that we could have followed the converse to define

$$\int_0^{\delta x} dx_j = \delta x_j|_{q+\delta q} = \int_0^{\delta q} g_{jk}(q_0 + \eta\xi)\eta^k d\xi, \tag{6.6.4}$$

and continue as in (6.6.2).

transform of (6.3.2), i.e.,

$$L_G = \pi^i \dot{x}_i - g_{ij}(x)\frac{\pi^i \pi^j}{2m}. \tag{6.7.1}$$

Since $\pi^i = g^{ij}\dot{x}/m$, one obtains

$$L_G = \frac{m}{2} g^{ij}\dot{x}_i \dot{x}_j \tag{6.7.2}$$

The variation of the action $S = \int dt L$ is then (leaving out the overall factor $m/2$)

$$\delta S = \int dt \left\{ 2g^{ij}\delta\dot{x}_i \dot{x}_j + \frac{\partial g^{ij}}{\partial x_k}\delta x_k \dot{x}_i \dot{x}_j \right\}. \tag{6.7.3}$$

We now replace $\dot{x}_i$ by $g_{i\ell}\dot{q}^\ell$ and integrate by parts the time derivative on $\delta\dot{x}_i$ to obtain an integral with coefficient δx_i:

$$\delta S = \int dt \left\{ -2\frac{\partial g^{ij}}{\partial x_m} g_{mn} g_{j\ell} \dot{q}^n \dot{q}^\ell - 2g^{ij} g_{mn}\frac{\partial g_{j\ell}}{\partial x_m}\dot{q}^n \dot{q}^\ell - 2g^{ij} g_{j\ell}\ddot{q}^\ell - \frac{\partial g_{\ell m}}{\partial x_i}\dot{q}^\ell \dot{q}^m \right\} \delta x_i. \tag{6.7.4}$$

Using the identity

$$\frac{\partial g^{ij}}{\partial x_m} = -g^{ip}\frac{\partial g_{pq}}{\partial x_m} g^{qj} \tag{6.7.5}$$

in the first term of (6.7.4), we see that the first two terms cancel; setting the coefficient of the arbitrary variation δx_i to zero, one obtains the equation for the embedded Hamiltonian dynamics (6.2.18).

The final form of the embedded Lagrangian can now be inferred, using our previous results, in the special coordinate system for which (6.2.8) is valid, that is, writing (we delete an overall sign)

$$\delta S = \int dt \left\{ \ddot{q}^i + \frac{1}{2}\frac{\partial g_{\ell n}}{\partial x_i}\dot{q}^\ell \dot{q}^n \right\} \delta x_i \tag{6.7.6}$$

and using (6.2.7), (6.2.8) for $g_{\ell n}$, we find

$$\frac{\partial \phi}{\partial x_i} = g^{ji}\frac{\partial \phi}{\partial q^j} = \frac{1}{\phi}\frac{\partial \phi}{\partial q^j} = \frac{E-V}{E}\frac{E}{(E-V)^2}\frac{\partial V}{\partial q^i} = \frac{1}{(E-V)}\frac{\partial V}{\partial q^i}. \tag{6.7.7}$$

The result (6.7.4) then becomes, with $\dot{q}^2 = \frac{2}{m}(E-V)$,

$$\delta S = \int dt \left\{ \ddot{q}^i + \frac{1}{m}\frac{\partial V}{\partial q^i} \right\} \delta x_i, \tag{6.7.8}$$

clearly equivalent to taking a Lagrangian of the form

$$L = \frac{m}{2}\dot{q}^2 - V(q). \tag{6.7.9}$$

The variation of the associated action can be performed with respect to $\delta q^i = \delta x^i/\phi$, and is therefore arbitrary (we have assumed implicitly that δx_i is along a geodesic curve; the arbitrariness of δq_i then depends on the existence of a dense set of geodesics through the point x). We have therefore recovered the standard Hamiltonian theory which was embedded in the manifold $\{q\}$.

6.8 Summary and Conclusions

A method was developed in 2007 (Horwitz and Ben Zion 2007) involving the study of a geometric Hamiltonian (Gutzwiller 1990; Curtiss and Miller 1985), which was put into dynamical equivalence with the standard Hamiltonian form by defining a conformal metric. On this level, one finds that the motion generated by the standard Hamiltonian H of (6.2.6) can be transformed to a motion, defined by a connection form, on a geometric embedding of the original Hamiltonian motion. In this representation, the geodesic deviation results in a formula with remarkable predictive capability for the stability of the original Hamiltonian motion.

The local mapping (6.2.11) cannot, clearly, be understood as a relation between closed one-forms, but only as a map in the tangent space along geodesic curves. There is therefore not a direct constructive global relation between the coordinates $\{q\}$ (the standard Hamiltonian coordinates) and $\{x\}$. However, we have shown that the functions $\phi(x)$ and $F(q) \equiv E/(E - V(q))$ are related in suitable analytic domains on $\{q\}$ and $\{x\}$ by well-defined relations between the coefficients of the Taylor series expansions of each of the functions around selected points $\{q\}$ and $\{x\}$. The Poisson bracket is, moreover, invariant under the mapping from the variables x to q.

The Lagrangian formulation of the dynamics of these systems makes explicit an intrinsic equivalence, thus accounting for the relation of the easily demonstrable instability (or stability) of the embedded motion to that induced by the original Hamiltonian.

Chapter 7
Canonical Transformation of Potential Model Hamiltonian Mechanics to Geometrical Form

In this chapter, using the methods of symplectic geometry, we establish the existence of a *canonical transformation* from potential model Hamiltonians of standard form in a Euclidean space to an equivalent geometrical form on a manifold, where the one to one corresponding motions are along geodesic curves. We show that there are many possibilities for this construction, one of which is treated constructively in the preceding chapter. The advantage of this representation is that it admits the computation of geodesic deviation as a test for local stability, as shown in recent previous studies. And discussed in the previous chapter, to be a very effective criterion for determining the stability of the orbits generated by the potential model Hamiltonian. We describe here a general algorithm for finding the generating function for the canonical transformation and describe some of the properties of this mapping under local diffeomorphisms. We give a convergence proof for this algorithm for the one dimensional case. We also provide a formulation of geodesic deviation, closely related to what we have presented in Chap. 5 but with somewhat more emphasis on the geometric structure, which relates the stability of the motion in the geometric form to that of the Hamiltonian standard form.

7.1 Introduction

This chapter is concerned with the development of a new method for embedding the motion generated by a classical Hamiltonian of standard form into a Hamiltonian defined by a bilinear form on momenta with coordinate dependent coefficients (forming an invertible matrix) by means of a canonical transformation constituting a *symplectomorphism*. This type of Hamiltonian, which we shall call *geometric*, results, as in the previous chapter, in equations of motion of geodesic form. The coefficients of the resulting bilinear form in velocities can be considered to be a connection form associated with the coefficients in the momenta in the geometric Hamiltonian considered as a *metric* on the corresponding coordinates. The advantage of this result, which may be considered to be an embedding of the motion induced by

L. Horwitz and Y. Strauss, *Unstable Systems*, Mathematical Physics Studies,
https://doi.org/10.1007/978-3-030-31570-2_7

the original Hamiltonian into an auxiliary space for which the motion is governed by a geodesic structure, is that the deviation of geodesics on such a manifold (involving higher order derivatives than the usual Lyapunov criteria) can provide a very sensitive test of the stability of the original Hamiltonian motion.

In previous chapter, an *ad hoc* construction of a geometrical embedding using a conformal metric (Horwitz and Ben Zion 2008) was introduced. This embedding displays strong sensitivity to instability through geodesic deviation.

Casetti and Pettini (1993), as mentioned in Chap. 6, have investigated the application of the Jacobi metric and the extension of the analysis of the resulting Jacobi equations along a geodesic curve in terms of a parametric oscillator; such a procedure could be applied to the construction we discuss here as well. The relation of the stability of geometric motions generated by metric models previously considered to those of the motion generated by the original Hamiltonian is generally, however, difficult to establish. The transformation that we shall construct here preserves a strong relation with the original motion due to its canonical structure.

The methods we shall use are fundamentally geometric, involving the properties of symplectic manifolds which enable the definition and construction of the canonical transformation without using the standard Lagrangian methods. These geometric methods provide a rigorous framework for this construction, which makes accessible a more complete understanding of the dynamics.

In the theory of symplectic manifolds (da Silva 2006), a well defined mechanism exists for transforming a Hamiltonian of the form (1.1) to that of (1.2) (with a possibly conformal metric) by a rigorous canonical transformation, called a symplectomorphism, admitting the use of geodesic deviation to determine stability, which would then be clearly associated with the original Hamiltonian motion. The remarkable success of the procedure introduced in Chap. 6 can be better understood in terms of this canonical theory, although some insights were provided in (Calderon et.al. 2013). We shall discuss this theory, and describe some of its properties, in this chapter.

We remark that in an analysis (Strauss 2015) of the geodesic deviation treated as a parametric oscillator, a procedure of second quantization was carried out, as described in Chap. 5, providing an interpretation of excitation modes for the instability in a "medium" represented by the background Hamiltonian motion. This interpretation would be applicable to the results of the construction we present here as well.

In the following, we describe this mapping and an algorithm for obtaining solutions. We give a convergence proof for the recurrence relations for the generating function in the one dimensional case which, in more general form, is applicable as well to the general n-dimensional case. Although the algorithm for the construction is clearly effective (and convergent), its realization requires considerable computation for specific applications, planned to be carried out in the future. The resulting programs could then be applied to a wide class of systems to provide stability criteria without exhaustive simulation; the local criteria to be developed could, furthermore, be used for the control of intrinsically chaotic systems (Lewkowics et.al. 2016).

In this chapter we discuss some general properties of the framework. In Sect. 7.2, we give the basic mathematical methods in terms of the geometry of symplectic manifolds.

A central motivation for our construction is to make available the study of stability by means of geodesic deviation. This procedure is studied in Sect. 7.3, in terms of geometric methods, making clear the relation between stability in the geometric manifold and the original Hamiltonian motion.

In Sect. 7.4, an algorithm is described for solving the nonlinear equations for the generating function of the canonical transformation. In Sect. 7.5, we study this algorithm for the one dimensional case, and prove convergence of the series expansions, under certain assumptions in Sect. 7.6. The results can be generalized to arbitrary dimension.

The series expansions that we obtain can be studied by methods of Fourier series representations; the nonlinearity leads to convolutions of analytic functions (see, for example (Hille 1976)) that may offer approximation methods that could be useful in studying specific cases.

Since the iterative expansions for the generating function could be expected to have only bounded domains of convergence, we consider, in Sect. 7.7, the possibility of shifting the origin of the expansion in general dimension. As for the analytic continuation of a function of a complex variable, this procedure can extend the definition of the generating function to a maximal domain.

Since the image space of the symplectomorphism has geometrical structure, it is natural to study its properties under local diffeomorphisms. A local change of variable alters the structure of the symplectomorphism. We study the effect of such diffeomorphims on the generating function (holding the original Euclidean variables fixed) in Sect. 7.8.

Further mathematical implications, such as relations to Morse theory (e.g. Frankel 1997; Milnor 1969), are briefly discussed in Sect. 7.9.

7.2 Symplectic Manifolds and Symplectomorphisms

The notion of a symplectic geometry is well-known in analytic mechanics through the existence of the Poisson bracket of Hamilton–Lagrange mechanics, i.e., for A, B functions of the canonical variables q, p on phase space, the Poisson bracket is defined by

$$\{A, B\}_{PB} = \sum_k \left\{ \frac{\partial A}{\partial q_k} \frac{\partial B}{\partial p^k} - \frac{\partial B}{\partial q_k} \frac{\partial A}{\partial p^k} \right\}. \tag{7.2.1}$$

The antisymmetric bilinear form of this expression has the symmetry of the *symplectic group*, associated with the symmetry of the bilinear form $\xi_i \eta^{ij} \xi_j$, with $i, j = 1, 2, \ldots 2n$ and η^{ij} an antisymmetric matrix (independent of ξ); the $\{q_k\}$ and $\{p^k\}$ can be considered as the coordinatization of a symplectic manifold.

We provide hare a short introduction to the basic elements of the theory of symplectic manifolds that we use in subsequent sections for producing a mapping (symplectomorphism) from standard Hamiltonian form to a geometric type Hamiltonian. The discussion in this section essentially follows (da Silva 2006)).

Let V be an n-dimensional vector space over $\mathbb{R}$ and let V^* be the dual space to V. Let $\Omega : V \times V \mapsto \mathbb{R}$ be a bilinear map. Then Ω defines a mapping $\hat{\Omega} : V \mapsto V^*$ by $(\hat{\Omega}(x))(y) := \Omega(x, y)$ $(x, y \in V)$. A skew symmetric bilinear map $\Omega : V \times V \mapsto \mathbb{R}$ is *symplectic* if $\hat{\Omega}$ is bijective. The pair (V, Ω) is then called a *symplectic vector space*.

A subspace $W \subset V$ of a symplectic vector space (V, Ω) is called *symplectic* if the restriction $\Omega|_{W\times W}$ is non-degenerate. A subspace $W \subset V$ of (V, Ω) is called *isotropic* if $\Omega|_{W\times W} \equiv 0$. Given a linear subspace $W \subset V$ of a symplectic vector space (V, Ω) the *symplectic orthogonal* of W is defined by

$$W^\Omega = \{v \in V : \Omega(v, u) = 0, \ \forall u \in W\}.$$

It can be shown that W is a symplectic subspace of (V, Ω) if and only if $V = W \oplus W^\Omega$. By the definition of an isotropic subspace, a subspace W is isotropic if and only if $W \subseteq W^\Omega$. It can be shown that if W is an isotropic subspace of (V, Ω) then $\dim W \leq \frac{1}{2} \dim V$. A subspace $W \subset V$ of a symplectic vector space (V, Ω) is called *Lagrangian* if W is an isotropic subspace of (V, Ω) and $\dim W = \frac{1}{2} \dim V$.

Definition 7.1 (*Symplectomorphism of vector spaces*) A *symplectomorphism* between two symplectic vector spaces (V_1, Ω_1) and (V_2, Ω_2) is a linear isomorphism $\varphi : V_1 \mapsto V_2$ such that $\varphi^*\Omega' = \Omega$ (where φ^* is the pullback map defined by $(\varphi^*\Omega)'(x, y) := \Omega'(\varphi(x), \varphi(y))$). □

If a symplectomorphism $\varphi : V_1 \mapsto V_2$ exists then the symplectic vector spaces (V_1, Ω) and (V_2, Ω') are said to be *symplectomorphic*.

Let $\mathcal{M}$ of a differentiable manifold and let ω be a 2-form on $\mathcal{M}$. Then $\omega_p : T_p\mathcal{M} \times T_p \mathcal{M} \mapsto \mathbb{R}$, the restriction of ω to a point $p \in \mathcal{M}$, is a skew symmetric bilinear form on $T_p\mathcal{M}$ varying smoothly with p. We say that ω is closed if $d\omega = 0$, where d is exterior derivative. The 2-form ω is called symplectic if ω is closed and ω_p is symplectic as a bilinear map from $T_p\mathcal{M} \times T_p\mathcal{M}$ to $\mathbb{R}$ for all $p \in \mathcal{M}$.

Definition 7.2 (*Symplectic manifold*) A *symplectic manifold* is a pair $(\mathcal{M}, \omega)$ where $\mathcal{M}$ is a manifold and ω is a symplectic 2-form on $T\mathcal{M}$. □

Note that the dimension of a symplectic manifold is necessarily even.

A canonical example of a symplectic manifold, and the focus of our discussion in this chapter, is the cotangent bundle $\mathcal{M} = T^*X$ of an n-dimensional manifold X. Let the manifold structure of X be described by an atlas of coordinate charts $(\mathcal{U}, x_1, x_2, \ldots, x_n)$, with $x^i : \mathcal{U} \mapsto \mathbb{R}$. The differential 1-forms $(dx^1)_x, (dx^2)_x, \ldots,$ $(dx^n)_x$ at a point $x \in X$ form a basis of the cotangent space T_x^*X. Therefore, if $\beta \in T_x^*X$, we can expand $\beta = \sum_i^n \beta_i (dx^i)_x$, for real coefficients β_i, $1 \leq 1 \leq n$. This expansion induces coordinate charts on T^*X as mappings $\Phi_\mathcal{U} : T^*\mathcal{U} \mapsto \mathbb{R}^{2n}$ defined by

$$\Phi(x, \beta) = (x^1, \ldots, x_n, \beta_1, \ldots, \beta_n).$$

The coordinate charts for T^*X are then $(\mathcal{T}^*\mathcal{U}, x^1, \ldots, x^n, \beta_1, \ldots, \beta_n)$ and the coordinates $x^1, \ldots, x^n, \beta_1, \ldots, \beta_n$ are called the *cotangent coordinates* associated to the coordinates $x^1, \ldots, x^n$ on $\mathcal{U}$.

Let $(\mathcal{U}, x^1, \ldots, x^n)$ be a coordinate chart for an n-dimensional manifold X with associated cotangent coordinate chart $(\mathcal{T}^*\mathcal{U}, x^1, \ldots, x^n, \xi_1, \ldots, \xi_n)$. Consider a 1-form α defined on $T^*\mathcal{U}$ defined by

$$\alpha := \sum_i^n \xi_i \, dx^i.$$

By its definition, it appears that α depends on the choice of coordinates on $\mathcal{U}$. However, this is not the case and α is, in fact, defined intrinsically. If $(\mathcal{U}', x'^1, \ldots, x'^m)$ is another chart of X with associated cotangent bundle chart $(\mathcal{U}', x'^1, \ldots, x'^n, \xi_1, \ldots, \xi_n)$ then, for a point $x \in \mathcal{U} \cap \mathcal{U}'$, since

$$\xi_j' = \sum_i \left(\frac{\partial x^i}{\partial x'^j}\right)\xi_i, \quad dx'^j = \sum_i \left(\frac{\partial x'^j}{\partial x^i}\right)dx^i,$$

If we define $\alpha' = \sum_i^n \xi_i' \, dx'^i$, we have

$$\alpha' = \sum_j \xi_j' \, dx'^j = \sum_j \sum_{i,k} \left(\frac{\partial x^i}{\partial x'^j}\right)\left(\frac{\partial x'^j}{\partial x^k}\right)\xi_i dx^k = \sum_i \xi_i dx^k = \alpha.$$

The 1-form α is called the *tautological 1-form* (or Liouville 1-form) of X. The exterior derivative of α

$$\omega = -d\alpha = \sum_i dx^i \wedge d\xi_i,$$

Is a symplectic 2-form on T^*X called the *canonical symplectic 2-from on T^*X*. Thus, the cotangent bundle $\mathcal{M} = T^*X$ is inherently a symplectic manifold.

Definition 7.3 (*Symplectomorphism of manifolds*) Let $(\mathcal{M}_1, \omega_1)$ and $(\mathcal{M}_2, \omega_2)$ be two $2n$-dimensional symplectic manifolds. A diffeomorphism $\varphi : \mathcal{M}_1 \mapsto \mathcal{M}_2$ is called a *symplectomorphism* if $\varphi^*\omega_2 = \omega_1$ (where φ^* is the pullback map). □

Definition 7.4 (*Lagrangian submanifold*) Let $(\mathcal{M}, \omega)$ be a $2n$-dimensional symplectic manifold. A submanifold Y of $\mathcal{M}$ is called a *Lagrangian submanifold* if, at each point $p \in Y$ the tangent space T_pY is a Lagrangian subspace of $T_p\mathcal{M}$ (i.e., $\omega_p|_{T_p\mathcal{M}} \equiv 0$ and $\dim T_pY = \frac{1}{2} \dim T_p\mathcal{M}$). □

Let $(\mathcal{M}_1, \omega_1)$ and $(\mathcal{M}_2, \omega_2)$ be two symplectic manifolds. We would like to determine when a given diffeomorphism $\varphi : \mathcal{M}_1 \mapsto \mathcal{M}_2$ is a symplectomorphism. Define the projection maps $\mathrm{pr}_1 : \mathcal{M}_1 \times \mathcal{M}_2 \mapsto \mathcal{M}_1$ and $\mathrm{pr}_2 : \mathcal{M}_1 \times \mathcal{M}_2 \mapsto \mathcal{M}_2$

respectively by $\text{pr}_1(p_1, p_2) = p_1$ and $\text{pr}_2(p_1, p_2) = p_2$ for $(p_1, p_2) \in \mathcal{M}_1 \times \mathcal{M}_2$. Then,

$$\omega = (\text{pr}_1)^*\omega_1 + (\text{pr}_2)^*\omega_2,$$

and

$$\tilde{\omega} = (\text{pr}_1)^*\omega_1 - (\text{pr}_2)^*\omega_2,$$

are symplectic 2-forms on $\mathcal{M}_1 \times \mathcal{M}_2$. The symplectic 2-form $\tilde{\omega}$ is called the *twisted product form* on $\mathcal{M}_1 \times \mathcal{M}_2$. The graph of a diffemorphism $\varphi : \mathcal{M}_1 \mapsto \mathcal{M}_2$ is a $2n$-dimensional submanifold Γ_φ of $\mathcal{M}_1 \times \mathcal{M}_2$ defined by

$$\Gamma_\varphi = \{(p, \varphi(p)) \in \mathcal{M}_1 \times \mathcal{M}_2 \, : \, p \in \mathcal{M}_1\}$$

We have the following proposition (da Silva 2006):

Proposition 7.4 *A diffeomorphism $\varphi : \mathcal{M}_1 \mapsto \mathcal{M}_2$ is a symplectomorphism if and only if Γ_φ is a Lagrangian submanifold of $(\mathcal{M}_1 \times \mathcal{M}_2, \tilde{\omega})$.* □

Let X_1 and X_2 be two n-dimensional manifolds. Let $\mathcal{M}_1 = T^*X_1$ and $\mathcal{M}_2 = T^*X_2$ be the respective cotangent bundles with tautological 1-forms α_1, α_2 and canonical symplectic forms ω_1 and ω_2. Identifying,

$$\mathcal{M}_1 \times \mathcal{M}_2 = T^*X_1 \times T^*X_2 \simeq T^*(X_1 \times X_2)$$

the tautological 1-form on $T^*(X_1 \times X_2)$ is

$$\alpha = (\text{pr}_1)^*\alpha_1 + \text{pr}_2^*\alpha_2.$$

The canonical symplectic form on $T^*(X_1 \times X_2)$ is then

$$\omega = -d\alpha = -d(\text{pr}_1)^*\alpha_1 + \text{pr}_2^*\alpha_2) = (\text{pr}_1)^*(-d\alpha_1) + \text{pr}_2^*(-d\alpha_2) = (\text{pr}_1)^*\omega_1 + \text{pr}_2^*\omega_2.$$

In order to associate this canonical form to the corresponding twisted product form appearing in Proposition 7.4 we define an involution $\sigma_2 : \mathcal{M}_2 \mapsto \mathcal{M}_2$ by

$$\sigma_2(q, \beta_q) = (q, -\beta_q), \quad (q, \beta_q) \in \mathcal{M}_2 = T^*X_2,$$

with $q \in X_2$ and β_q a restriction of a p-form β on X_2 to the point q. With $\sigma : \mathcal{M}_1 \times \mathcal{M}_2 \mapsto \mathcal{M}_1 \times \mathcal{M}_2$ defined by $\sigma := \text{id}_{\mathcal{M}_1} \times \sigma_2$ we have then

$$\sigma^*\omega = \sigma^*((\text{pr}_1)^*\omega_1 + (\text{pr}_2)^*\omega_2) = (\text{pr}_1)^*\omega_1 - (\text{pr}_2)^*\omega_2 = \tilde{\omega}.$$

Therefore, if Y is a Lagrangian submanifold of $(\mathcal{M}_1 \times \mathcal{M}_2, \omega)$ then $Y^\sigma = \sigma(Y)$ is a Lagrangian submanifold of $(\mathcal{M}_1 \times \mathcal{M}_2, \tilde{\omega})$. By Proposition 7.4, if Y^σ is the graph of a diffeomorphism φ then φ is a symplectomorphism.

Let $f \in C^\infty(X_1 \times X_2)$. Then df is a closed 1-form on $X_1 \times X_2$ and it can be shown that

$$Y = \{(x, y), (df)_{(x,y)}) : (x, y) \in X_1 \times X_2\}$$

is a Lagrangian submanifold of $T^*(X_1 \times X_2)$. Denoting by $d_x f$ the projection of $df_{(x,y)}$ onto $T_x^* X_1 \times \{0\}$ and by $d_y f$ the projection of $df_{(x,y)}$ onto $\{0\} \times T_y^* X_2$ and making the identification

$$Y_f = \{(x, y), (df)_{(x,y)}) : (x, y) \in X_1 \times X_2\} = \{(x, y, d_x f, d_y f)) : (x, y) \in X_1 \times X_2\},$$

we apply the involution σ to obtain a Lagrangian submanifold of $(\mathcal{M}_1 \times \mathcal{M}_2, \tilde{\omega})$

$$Y_f^\sigma = \sigma(Y_f) = \{(x, y, d_x f, -d_y f)) : (x, y) \in X_1 \times X_2\}.$$

Thus, if Y_f^σ is the graph of a diffeomorphism $\varphi : \mathcal{M}_1 \mapsto \mathcal{M}_2$, then φ is a symplectomorphism. We call such a symplectomorphism the *symplectomorphism generated by* f and the function f is then called the generating function for the symplectomorphism φ.

We are left with the task of determining when Y_f^σ is, in fact, the graph Γ_φ of a diffeomorphism φ. Suppose that

$$\varphi(q, p) = (x, \pi),$$

where $(q, p) \in \mathcal{M}_1 = T^* X_1$ (with $q \in X_1$ and $p \in T_q^* X_1$)) and $(x, \pi) \in \mathcal{M}_2 = T^* X_2$ (with $x \in X_2$ and $\pi \in T_x^* X_2$)). If Y_f^σ is to be the graph of φ we should make the identifications

$$p = d_q f, \quad \pi = -d_x f. \tag{7.2.1}$$

Let $(\mathcal{U}_1, q^1, q^2, \ldots, q^n)$ be a coordinate chart on X_1 for which the associated coordinate chart on $T^*\mathcal{U}_1 \subset \mathcal{M}_1 = T^* X_1$ is $(T^*\mathcal{U}_1, q^1, q^2, \ldots q^n, p_1, p_2, \ldots, p_n)$. Similarly, let $(\mathcal{U}_2, x_1, x_2, \ldots, x_n)$ be a coordinate chart on X_2 for which the associated coordinate chart on $T^*\mathcal{U}_2 \subset \mathcal{M}_1 = T^* X_1$ is $(T^*\mathcal{U}_2, x^1, x^2, \ldots x^n, \pi_1, \pi_2, \ldots, \pi_n)$. In this case, we can express (7.2.1) with respect to these coordinates and obtain the equations

$$p_i = \frac{\partial f}{\partial q^i}(q, x), \quad \pi_i = -\frac{\partial f}{\partial x^i}(q, x), \; 1 \le i \le n. \tag{7.2.2}$$

If there exists a solution of the first set of equations in (7.2.2) for x in terms of q and p, i.e., the functions $x^i = x^i(q, p) = x^i(q^1, \ldots, q^n, p_1, \ldots, p_n)$, $1 \le i \le n$, are well defined, then we can plug these functions into the second set of equations in (7.2.2) to obtain $\pi_i = -\frac{\partial f}{\partial x^i}(q, x(q, p))$, $1 \le i \le n$, and thus find the symplectomorphism $\varphi(q, p) = (x(q, p), \pi(q, p))$. By the implicit function theorem, the condition for solving for x in terms of q, p is that

$$\det\left(\frac{\partial^2 f(q,x)}{\partial q^i \partial x^j}\right) \neq 0. \tag{7.2.3}$$

Equations (7.2.2) and (7.2.3) form the basis for the fundamental equation below used for the calculation of the generating function of a symplectomorphism mapping a standard dynamical problems into the form of dynamics described by geodesic motion on a manifold associated with a Hamiltonian of geometric form.

Equations (7.2.2), of the form of the usual canonical transformation derived by adding a total derivative to the Lagrangian in Hamilton–Lagrange mechanics, have been obtained here by a more general and more powerful geometric procedure enabling, as we shall see, a simple formulation of the transformation from the standard Hamiltonian form to a geometrical type Hamiltonian.

7.3 Geodesic Deviation

We have discussed geodesic deviation in Chap. 5 with emphasis on its dynamic character to be able to study quantization. In this section, we emphasize its geometric interpretation for application to the study of stability in a classical framework.

The principal reason for introducing the canonical transformation from Hamiltonian form to the geometric form, as we have pointed out in the introduction, is to make accessible the very sensitive measure of stability provided by geodesic deviation. In this section we develop a rigorous geometrical formulation of this technique which makes clear the relation between stability in the geometric space and stability in the original Hamiltonian space.

Consider a Hamiltonian function of the form

$$H(q,p) = \frac{1}{2}\Sigma_{i=1}^n p_i^2 + V(q), \quad q = (q^1, \ldots, q^n), p = (p_1, \ldots, p_n), \tag{7.3.1}$$

where $V(q)$ is a potential function. The Hamiltonian $H(q,p)$ is defined on phase space, i.e., the symplectic manifold $\mathcal{M}_1 = T^*\mathbf{R}^n \simeq \mathbf{R}^{2n}$ with global coordinates $q_1, \ldots, q_n, \ p_1, \ldots p_n$ and a symplectic 2-form ω which, with respect to the above coordinates, takes the form

$$\omega = \Sigma_{i=1}^n dq^i \wedge dp_i \tag{7.3.2}$$

Assume that $\varphi: \ (\mathcal{M}_1, \omega) \mapsto (\mathcal{M}_2, \tilde{\omega})$ is a symplectomorphism mapping the phase space $(\mathcal{M}_1, \omega) = (T^*\mathbf{R}^n, \omega)$ onto a symplectic manifold $(\mathcal{M}_2, \tilde{\omega}) = (T^*\mathcal{M}, \tilde{\omega})$, where $\mathcal{M}_2 = T^*\mathcal{M}$ is the cotangent bundle of an n-dimensional Riemannian manifold $\mathcal{M}$ and $\tilde{\omega}$ is the canonical symplectic form on $T^*\mathcal{M}$. More specifically we consider $\mathcal{M}$ to be a Riemannian manifold such that the Hamiltonian function H_{geo} on $\mathcal{M}_2$, defined by the relation $H = \varphi^* H_{geo} = H_{geo} \circ \varphi$, is a *geometrical Hamiltonian* as discussed in the previous chapter (of the form $\frac{1}{2m} g^{ij}\pi_i\pi_j$). By this we mean that in each local chart $(\mathcal{U}, x^1, \ldots, x^n, \pi_1, \ldots, \pi_n)$ on $\mathcal{M}_2$, with $x^1, \ldots, x^n$ coordinates on $\mathcal{M}$ and $\pi_1, \ldots, \pi_n$ the fiber coordinates, H_{geo} is of geometric form.

Let $\varphi : (\mathcal{M}_1, \omega) \mapsto (\mathcal{M}_2, \tilde{\omega})$ be a symplectomorphism as described above and let $\mathbf{X}$ be a Hamiltonian vector field in the phase space $\mathcal{M}_1$ corresponding to the Hamiltonian function H of the form (7.3.1). Thus, if ω is the canonical symplectic form on $\mathcal{M}_1$, the vector field $\mathbf{X}$ satisfies the equation

$$i_{\mathbf{X}}\omega = dH \tag{7.3.3}$$

and the integral curves of $\mathbf{X}$, obtained by solving Hamilton's equations for H, are trajectories of the Hamiltonian dynamical system defined by H. Since φ is a symplectomorphism, the pullback by φ of the canonical symplectic form $\tilde{\omega}$ on $\mathcal{M}_2$ satisfies $\varphi^* \tilde{\omega} = \omega$. If $d\varphi_* : T\mathcal{M}_1 \mapsto T\mathcal{M}_2$ is the differential of φ and we define the vector field $\mathbf{X}_{geo} = d\varphi_*(\mathbf{X})$ then we have

$$i_{\mathbf{X}_{geo}}\tilde{\omega} = dH_{geo}, \tag{7.3.4}$$

i.e., $\mathbf{X}_{geo}$ is a Hamiltonian vector field with respect to the Hamiltonian function H_{geo}. The integral curves for $\mathbf{X}_{geo}$, obtained by solving Hamilton's equations for H_{geo}, correspond to geodesics in $\mathcal{M}$. We shall refer to such integral curves of $\mathbf{X}_{geo}$ as $\mathcal{M}_2$ *(or cotangent bundle) geodesics*. Hence, if $\gamma \subset \mathcal{M}_1$ is a trajectory of the original dynamical system then $\gamma^{\varphi} = \varphi(\gamma)$ is an $\mathcal{M}_2$ geodesic and if $\tilde{\pi} : \mathcal{M}_2 \mapsto \mathcal{M}$ is the projection of the cotangent bundle $\mathcal{M}_2 = T^*\mathcal{M}$ on the base manifold $\mathcal{M}$ then $\tilde{\pi}(\gamma^{\varphi})$ is a geodesic in $\mathcal{M}$.

Given a point $u \in \mathcal{M}_1$, denote by $\gamma_u \subset \mathcal{M}_1$ the curve given by $\gamma_u(t) = \phi_t(u)$, where $\phi_t : \mathcal{M}_1 \mapsto \mathcal{M}_1$ is the evolution group of the Hamiltonian dynamical system generated by H, i.e., γ_u is a trajectory of the system such that $\gamma_u(0) = u$. Consider a point $u_0 \in \mathcal{M}_1$ such that $H(u_0) = E_0$ and let $\tilde{E}_0 \subset \mathcal{M}_1$ be an equal energy hypersurface with $u_0 \in \tilde{E}_0$, i.e., a hypersurface in $\mathcal{M}_1$ such that $u_0 \in \tilde{E}_0$ and such that $dH = 0$ on $\tilde{E}_0$. By the conservation of the Hamiltonian along the flow, for every point $u \in \tilde{E}_0$ we have that $\gamma_u \subset \tilde{E}_0$.

Let $W \subset \tilde{E}_0$ be a *surface of section* in $\tilde{E}_0$ passing through u_0, i.e., a $(2n-2)$ dimensional hypersurface in $\tilde{E}_0$ transverse to the trajectories of the dynamical system and defined in some open $\tilde{E}_0$ neighborhood of u_0. By construction, the Hamiltonian H has the same value at all points $u \in W$ and the trajectories of the system are transverse to W at all points of intersection.

Let $u \in W$ be an arbitrary point. Then u is the base point of a trajectory γ_u given by $\gamma_u(t) = \phi_t(u)$. Considering a time interval $0 \leq t \leq T$ $(T > 0)$ we define a submanifold $\mathcal{N}_{u_0} \subset \tilde{E}_0 \subset \mathcal{M}_1$ by

$$\mathcal{N}_{u_0} = \{\phi_t(u) \; : \; \forall u \in W, \; \forall t \in [0, T]\} \tag{7.3.5}$$

By construction $\mathcal{N}_{u_0}$ is parametrized by (u, t), $u \in W$, $t \in [0, T]$ and $\mathcal{N}_{u_0}$ consists of trajectories of the dynamical system corresponding to all possible initial points $u \in W$. In particular, $\mathcal{N}_{u_0} \cap \gamma_{u_0} = \{\phi_t(u_0) \; : \; \forall t \in [0, T]\}$.

Next, consider variations of γ_{u_0} in $\mathcal{N}_{u_0}$. Let $\gamma_{var} \subset W$ be a smooth curve parametrized by a parameter α and based at the point $u_0 \in W$. Thus, for some interval $I \subset \mathbf{R}$, with $0 \in I$, γ_{var} is given by a smooth function $u(\alpha)$, such that $u(\alpha) \in W$, $\forall \alpha \in I$ and $u(0) = u_0$. To the curve γ_{var} corresponds a two dimensional surface $S_{var} \subset \mathcal{N}_{u_0}$ via the definition

$$S_{var} = \{\phi_t(u(\alpha)) \, : \, \alpha \in I, \; t \in [0, T]\} \, . \tag{7.3.6}$$

By construction, (α, t), $t \in [0, T]$, $\alpha \in I$ are coordinates on S_{var}. We call S_{var} the *variational surface* of γ_{u_0} corresponding to the *variation curve* γ_{var}. Note that γ_{var} is carried by the flow ϕ_t to a variation curve $\gamma_{var}(t)$ at time t defined by $\gamma_{var}(t) = \phi_t(\gamma_{var})$ and given explicitly by the function $\gamma_{var}(\alpha, t) = \phi_t(u(\alpha))$, where $u(\alpha)$ is the function defining γ_{var}.

We now go back to the symplectomorphism φ. Denoting by $\mathcal{M}_3 = T\mathcal{M}$ the tangent bundle for $\mathcal{M}$ then, by the Riemannian nature of $\mathcal{M}$, there is a natural bijective mapping $G \, : \, \mathcal{M}_3 \mapsto \mathcal{M}_2$ defined by

$$G(x, \mathbf{v}) := (x, g(\mathbf{v}, \cdot)), \quad (x, \mathbf{v}) \in \mathcal{M}_3, \;\; \mathbf{v} \in T_x\mathcal{M} \tag{7.3.7}$$

where $g(\cdot, \cdot)$ is the metric tensor on $\mathcal{M}$ and x is a point in $\mathcal{M}$. Given the symplectomorphism $\varphi \, : \, \mathcal{M}_1 \mapsto \mathcal{M}_2$, we define a mapping

$$Q \, : \, \mathcal{M}_1 \mapsto \mathcal{M}_3, \qquad Q := G^{-1} \circ \varphi \, .$$

Thus, by the application of the mapping Q to a trajectory γ_u of the original dynamical system we obtain a $\mathcal{M}_3$ *(or tangent bundle) geodesic* $\gamma_u^Q = Q(\gamma_u) = G^{-1}(\gamma_u^\varphi) = (G^{-1} \circ \varphi)(\gamma_u)$. If $\pi \, : \, \mathcal{M}_3 \mapsto \mathcal{M}$ is the projection of the tangent bundle $\mathcal{M}_3 = T\mathcal{M}$ on the base manifold $\mathcal{M}$ then $\pi(\gamma_u^Q) = \tilde{\pi}(\gamma_u^\varphi)$ is a geodesic in $\mathcal{M}$.

Next, apply the mapping Q to $\mathcal{N}_{u_0}$ to obtain a submanifold $\mathcal{N}_{u_0}^Q \subset \mathcal{M}_3$ given by

$$\mathcal{N}_{u_0}{}^Q = Q\left(\mathcal{N}_{u_0}\right) = \{Q[\phi_t(u)] \, : \, \forall u \in W, \; \forall t \in [0, T]\} \, . \tag{7.3.8}$$

Again, by construction, $\mathcal{N}_{u_0}{}^Q$ is parametrized by (u, t), $u \in W$, $t \in [0, T]$. In fact, for each $u \in W$ the curve $\gamma_u{}^Q = Q(\gamma_u)$ is an $\mathcal{M}_3$ geodesic curve given by $\gamma_u{}^Q(t) = Q[\phi_t(u)]$ and $\mathcal{N}_{u_0}{}^Q$ consists of all such geodesic curves corresponding to all possible initial points $u \in W$. Considering the variational surface S_{var} we observe that, since $S_{var} \subset \mathcal{N}_{n_0}$ then $QS_{var} \subset \mathcal{N}_{u_0}^Q$. Hence, by the application of Q to S_{var} we obtain a two dimensional surface in $\mathcal{M}_3$

$$\begin{aligned} S_{var}^Q = Q(S_{var}) = \{Q(\phi_t(u(\alpha))) \, : \, \alpha \in I, \; t \in [0, T]\} = \\ = \{Q(\gamma_{var}(\alpha, t)) \, : \, \alpha \in I, \; t \in [0, T]\}. \end{aligned} \tag{7.3.9}$$

Note that (α, t), $t \in [0, T]$, $\alpha \in I$ are coordinates on S_{var}^Q and that, if we denote $\gamma_{u_0}{}^Q = Q(\gamma_{u_0})$, then S_{var}^Q is a *surface of variation for* $\gamma_{u_0}^Q$ *consisting of* $\mathcal{M}_3$ *geodesics.*

Our goal is now to investigate the deviation of nearby trajectories of the original Hamiltonian system by considering the deviation of the corresponding geodesics in M_3. We quantify the deviation of nearby trajectories from the base trajectory γ_0 in N_{u_0}, i.e., on the variational surface $S_{var}(\gamma_{var})$, by studying the evolution along γ_0 of the tangent vector to the variation curve ${\gamma_{var}}'^{t}$. The tangent vector, which we call the *phase space trajectory deviation vector* is formally given by

$$\mathbf{v}_{dev}(t) = \left[\frac{\partial}{\partial\alpha}{\gamma_{var}}'^{t}(\alpha.t)\right]|_{\alpha=0} = \left[\frac{\partial}{\partial\alpha}\phi_t(u(\alpha))\right]|_{\alpha=0} \quad , \mathbf{v}_{dev}(t) \in T\mathcal{M}_1. \tag{7.3.10}$$

The deviation vector $\mathbf{v}_{dev}(t)$ is mapped by the differential of the mapping Q into a deviation vector in $T\mathcal{M}_3$, formally given by

$$\begin{aligned}\mathbf{J}_{dev}(t) = \left[\frac{\partial}{\partial\alpha}Q(\gamma_{var}(\alpha,t))\right]_{\alpha=0} = \left[\frac{\partial}{\partial\alpha}Q[\phi_t(u(\alpha))]\right]_{\alpha=0} = \\ = dQ\left(\left[\frac{\partial}{\partial\alpha}\phi_t(u(\alpha))\right]_{\alpha=0}\right) = dQ_*(\mathbf{v}_{dev}(t)), \quad \mathbf{J}_{dev}(t) \in T[\mathcal{M}_3].\end{aligned} \tag{7.3.11}$$

where $dQ_* : T\mathcal{M}_1 \mapsto T\mathcal{M}_3$ is the differential of the map Q.

In order to obtain a more explicit expression for $\mathbf{J}_{dev}(t)$ we will need a more explicit expression for the points in ${S^Q}_{var}$. Recall the fact that (t, α), $t \in [0, T]$, $\alpha \in I$ serve as coordinates in ${S^Q}_{var}$. If $\gamma_{u(\alpha)}$ is the phase space trajectory corresponding to the initial point $u(\alpha) \in W$ (this trajectory is given by $\gamma_{u(\alpha)}(t) = \phi_t(u(\alpha))$) and $\gamma^Q_{u(\alpha)} = Q(\gamma_{u(\alpha)})$ is its mapping to an $\mathcal{M}_3$ geodesic, then the point in $\mathcal{M}_3$ corresponding to the pair (α, t) is $\gamma^Q_{u(\alpha)}(t) = Q\left[\phi_t(u(\alpha))\right] = (x(\alpha, t), \mathbf{T}(\alpha, t))$, where $x(\alpha, t) = \pi\left(\gamma^Q_{u(\alpha)}(t)\right) \in \mathcal{M}$ is a point on the geodesic $\pi\left(\gamma^Q_{u(\alpha)}\right) \subset \mathcal{M}$ and $\mathbf{T}(\alpha, t) \in T_{x(\alpha,t)}\mathcal{M}$ is the tangent vector to $\pi\left(\gamma^Q_{u(\alpha)}\right)$ at the point $x(\alpha, t)$. Now, the collection of all tangent vectors at all points on all of the geodesics in $\pi({N^Q}_{u_0})$ forms a vector field defined on $\pi({N^Q}_{u_0})$ and, in particular, along the variation curve ${\gamma^{Q,t}}_{var} = Q(\gamma_{var}(t))$, its α derivative is given by the covariant derivative $\frac{\nabla\mathbf{T}(t,\alpha)}{\partial\alpha}$. Thus, we find that

$$\mathbf{J}_{dev}(t) = \left[\frac{\partial}{\partial\alpha}Q[\phi_t(u(\alpha))]\right]_{\alpha=0} = \left(\frac{\partial x(t,\alpha)}{\partial\alpha})|_{\alpha=0}, \frac{\nabla\mathbf{T}(t,\alpha)}{\partial\alpha}|_{\alpha=0}\right)^T. \tag{7.3.12}$$

Note that $\mathbf{J}_{dev}(t) \in T_{x(t,0)}\mathcal{M} \oplus T_{x(t,0)}\mathcal{M} = T\mathcal{M}_3$.

The standard definition of the geodesic deviation vector for geodesics in $\mathcal{M}$ is

$$\mathbf{J}(t) = \left(\frac{\partial x(t,\alpha)}{\partial\alpha}\right)|_{\alpha=0}, \quad \mathbf{J}(t) \in T_{x(t,0)}\mathcal{M}. \tag{7.3.13}$$

Furthermore, it can be shown (see Theorem 10 of Frankel (1997)) that

$$\left(\frac{\nabla \mathbf{T}(t,\alpha)}{\partial \alpha}\right)|_{\alpha=0} = \frac{\nabla \mathbf{J}(t)}{\partial t}. \tag{7.3.14}$$

Hence we have

$$\mathbf{J}_{dev}(t) = \left(\mathbf{J}(t), \frac{\nabla \mathbf{J}(t)}{\partial t}\right)^T, \tag{7.3.15}$$

where t is the affine parameter parametrizing $\pi(\gamma_{u_0})$.

The equation of evolution of $\mathbf{J}_{dev}(t)$, i.e. the *dynamical system representation* of the geodesic deviation equation, was studied in Chap. 5.

Let $\mathbf{X}, \mathbf{Y}, \mathbf{Z} \in T_p\mathcal{M}$ be (n-dimensional) vectors and let $R_p(\mathbf{X}, \mathbf{Y}) : T_p\mathcal{M} \mapsto T_p\mathcal{M}$ be the curvature transformation at the point $p \in \mathcal{M}$ i.e., the linear transformation with matrix elements $[R_p(\mathbf{X}, \mathbf{Y})]_j{}^i = R^i{}_{jk\ell} X^i Y^j$ so that

$$R_p(\mathbf{X}, \mathbf{Y})\mathbf{Z} = (R^i{}_{jk\ell} X^k Y^\ell Z^j)\partial_i, \tag{7.3.16}$$

where ∂_i are coordinate vectors at p and (X^k, Y^k, Z^k, $1 \le k \le n$ are the components of $\mathbf{X}$, $\mathbf{Y}$, $\mathbf{Z}$ with respect to the basis $\{\partial_k\}_{k=1}{}^n$). The quantities $R^i{}_{jk\ell}$ are the components of the Riemann curvature tensor at the point p.

Furthermore, if $< \cdot, \cdot >_{T_p\mathcal{M}}$ denotes the inner product defined on $T_p\mathcal{M}$ with the metric $g(\cdot, \cdot)$ on $\mathcal{M}$, then for $\mathbf{W} \in T_p\mathcal{M}$ we have

$$< R_p(\mathbf{X}, \mathbf{Y})\mathbf{Z}, \mathbf{W} >_{T_p\mathcal{M}} = R^i{}_{jk\ell} X^k Y^\ell Z^j W_i, \tag{7.3.17}$$

where $W_i = g_{ij} W^j$. For the geodesic $\gamma_0{}^{\mathcal{Q}} (\equiv \gamma_{u_0}{}^{\mathcal{Q}}) \in \mathcal{M}$, given in terms of the function $\gamma_0{}^{\mathcal{Q}}(t) = \mathcal{Q}[\phi_t(u_0)]$, using the above notation for the curvature transformation, the geodesic deviation equation along $\gamma_0{}^{\mathcal{Q}}$ is

$$\frac{\nabla^2 \mathbf{J}(t)}{dt^2} + R_{\gamma_0{}^{\mathcal{Q}}(t)}(\mathbf{J}(t), \mathbf{T}(t))(\mathbf{T}(t)) = 0. \tag{7.3.18}$$

where $\mathbf{J}(t)$ is the geodesic deviation vector defined above, $\mathbf{T}(t) \equiv \mathbf{T}_{\gamma_0{}^{\mathcal{Q}}(t)}$ is the tangent vector to $\gamma_0{}^{\mathcal{Q}}$ at the point $\gamma_0{}^{\mathcal{Q}}(t)$ and $R_{\gamma_0{}^{\mathcal{Q}}(t)}$ is the curvature tensor at the point $\gamma_0{}^{\mathcal{Q}}(t)$. The dynamical system representation of the geodesic deviation equation corresponds to putting (7.3.18) into the form

$$\frac{\nabla}{dt}\begin{pmatrix} \mathbf{J}(t) \\ \frac{\nabla \mathbf{J}(t)}{dt} \end{pmatrix} = \begin{pmatrix} 0 & I \\ -R_{\gamma_0{}^{\mathcal{Q}}(t)}(\cdot, \mathbf{T}(t))\mathbf{T}(t)) & 0 \end{pmatrix} \begin{pmatrix} \mathbf{J}(t) \\ \frac{\nabla \mathbf{J}(t)}{dt} \end{pmatrix}. \tag{7.3.19}$$

Denoting

$$\hat{R}_{\gamma_0{}^{\mathcal{Q}}(t)} = \begin{pmatrix} 0 & I \\ -R_{\gamma_0{}^{\mathcal{Q}}(t)}(\cdot, \mathbf{T}(t))\mathbf{T}(t)) & 0 \end{pmatrix} \tag{7.3.20}$$

and using (7.3.15), we may write (7.3.19) in the shorter form

$$\frac{\nabla \mathbf{J}_{dev}}{dt} = \hat{R}_{\gamma_0{}^Q(t)} \mathbf{J}_{dev}, \tag{7.3.21}$$

The behavior of the solution $\mathbf{J}_{dev}$ of the Eq. (7.3.21) determines the deviation properties of geodesics near $\gamma_0{}^Q$ as a function of t and, through the relation $\mathbf{V}_{trj}(t) = dQ^{-1}(\mathbf{J}_{dev}(t))$ obtained from (7.3.11), as well as the deviation of trajectories of the original dynamical system near γ_0 over time. The deviation of trajectories of the original system near γ_0 is therefore governed by the curvature transformation $R_{\gamma_0{}^Q(\cdot)}$ along the geodesic $\gamma_0{}^Q(\cdot)$.

7.4 Formulation of the Algorithm

The purpose of the canonical transformation we have discussed above is to construct a Hamiltonian of the geometrical form by means of a canonical transformation from a Hamiltonian of the standard potential model form. As above, we label the coordinates and momenta of the image space by $\{x^i\}$ and $\{\pi_i\}$ (we do not require that p_i and π_i are necessarily simply related for all t here; the equivalence of the dynamics is assured by the canonical nature of the transformation). We must therefore find the generating function $f(q, x)$ and the metric $g^{ij}(x)$ from the statement

$$\frac{p^2}{2m} + V(q) = \frac{1}{2m} g^{ij}(x) \pi_i \pi_j \tag{7.4.1}$$

Substituting (7.2.2) for the momenta, the problem is to solve (note that the left hand side treats the indices as Euclidean since it does not carry the local coordinate transformations available to the geometric form on the right hand side)

$$V(q) + \frac{1}{2m} \frac{\partial f(q,x)}{\partial q^i} \frac{\partial f(q,x)}{\partial q^i} = \frac{1}{2m} g^{ij}(x) \frac{\partial f(q,x)}{\partial x^i} \frac{\partial f(q,x)}{\partial x^j} \tag{7.4.2}$$

This nonlinear system appears to be difficult to solve, but we provide an algorithm capable of yielding solutions in the following.

Assuming analyticity in the neighborhood of the origin of the coordinates $\{q\}$, and in the potential term $V(q)$, one can write a power series expansion of the generating function and the potential, and identify the resulting powers of $q_i, q_j \ldots$ and their products. This procedure provides an effective recursive algorithm for a system of nonlinear first order equations in the expansion coefficients since the powers of q on the right hand side occurring in the expansion of $f(q, x)$ are higher by one order than the expansions on the left hand side, which contain derivatives with respect to q. Assuming analyticity in $\{x\}$ as well near the origin (as for Riemann normal coordinates), one can find a recursion relation for the resulting coefficients.

For example, in two dimensions, one may expand, into some radius of convergence,

$$f(q^1, q^2, x^1, x^2) = \Sigma_{k,\ell=0}^{\infty} C_{k,\ell}(x^1, x^2)(q^1)^k (q^2)^\ell \tag{7.4.3}$$

and expand $V(q^1, q^2)$ in power series

$$V(q^1, q^2) = \Sigma_{k,\ell=0}^{\infty} v_{k,\ell}(q^1)^k (q^2)^\ell \tag{7.4.4}$$

Substituting into the relation (7.4.2) (in two dimensional form), and equating coefficients of powers of q^1 and q^2, one finds the following recursion relations:

$$\begin{aligned}
v_{k,\ell} + \Sigma_{k=0}^{m} \Sigma_{\ell=0}^{n} \Big[& (k+1)(m-k+1) C_{(k+1),\ell}(x^1, x^2) C_{(m-k+1),(n-1)}(x^1, x^2) + \\
& (\ell+1)(n-\ell+1) C_{k,(\ell+1)}(x^1, x^2) C_{(m-k),(n-\ell+1)}(x^1, x^2) + 2v_{n,m} \Big] \\
= \Sigma_{k=0}^{m} \Sigma_{\ell=0}^{n} \Big[& g^{11}(x^1, x^2) \frac{\partial C_{k,\ell}}{\partial x^1}(x^1, x^2) \frac{\partial C_{m-k,n-1}}{\partial x^1}(x^1, x^2) \\
& + 2g^{12}(x^1, x^2) \frac{\partial C_{k,\ell}}{\partial x^1}(x^1, x^2) \frac{\partial C_{m-k,n-1}}{\partial x^2}(x^1, x^2) \\
& + g^{22}(x^1, x^2) \frac{\partial C_{k,\ell}}{\partial x^2}(x^1, x^2) \frac{\partial C_{m-k,n-1}}{\partial x^2}(x^1, x^2) \Big]
\end{aligned} \tag{7.4.5}$$

It is clear that this algorithm can easily be extended to n dimensions. Our initial investigations indicate reasonable behavior, with strong indications of convergence, for some simple cases.

Although the physically interesting cases are in two or more dimensions, where curvature generated by the geometric Hamiltonian plays an important role in the formation of geodesic curves and for many practical problems, we shall describe the general structure of the calculation in one dimension below and give as well a convergence proof for this case, which can be extended to arbitrary dimension. Some basic properties of the higher dimensional structure are discussed below.

7.5 One Dimensional Study

In one dimension, Eq. (7.4.2) becomes

$$V(q) + \frac{1}{2m}\left(\frac{\partial f(q,x)}{\partial q}\right)^2 = \frac{1}{2m} g(x) \left(\frac{\partial f(q,x)}{\partial x}\right)^2 \tag{7.5.1}$$

The recursion relation for the one dimensional case for

$$f = \Sigma q^\ell C_\ell(x)$$
$$V(q) = \Sigma_\ell V^{(\ell)} q^\ell \tag{7.5.2}$$

becomes

$$\Sigma_{m=0}^{\ell}\{(\ell+1-m)(m+1)C_{\ell+1-m}C_{m+1} - g(x)C'_{\ell-m}C'_m\} + V^{(\ell)} = 0 \tag{7.5.3}$$

Now, taking

$$C_\ell(x) = \Sigma_0^\infty b_{\ell m} x^m$$
$$g(x) = \Sigma_0^\infty g_n x^n \tag{7.5.4}$$

we find (for coefficients of x^r)
$r = 0$:

$$\Sigma_{m=0}^{\ell}\{(\ell+1-m)(m+1)b_{\ell+1-m,0}b_{m+1,0} - g_0 b_{\ell-m,1}b_{m,1}\} + V^{(\ell)} = 0 \tag{7.5.5}$$

and for
$r \geq 1$:

$$\Sigma_{m=0,0\leq p\leq r}^{\ell}(\ell+1-m)(m+1)b_{\ell+1-m,p}b_{m+1,r-p}$$
$$- \Sigma_{n,1\leq p\leq r+1}^{\ell} g_n b_{\ell-m,p} b_{m,r-n-p+2} \times p(r-n-p+2) = 0. \tag{7.5.6}$$

Note that for the case $r \geq 1$, the potential does enter explicitly since it has no x dependence. The relations (7.5.5) and (7.5.6) provide the basis for a systematic recursion.

One can easily work out several terms to see how the algorithm develops. It is clear that it is iteratively closed, but it is difficult to draw detailed conclusions on the solutions without extensive computations, as well as specification of potential models.

We give in the next section a proof, however, for one dimension, that, with some reasonable assumptions, such a computation converges. The method of proof can be generalized to n dimensions.

7.6 Convergence of the Algorithm in One Dimension

Now, in (7.5.3), define

$$D_m = mC_m, \tag{7.6.1}$$

and note that the first term in (7.5.3) can then be written as

$$\Sigma_{m=0}^{\ell} D_{\ell+1-m} D_{m+1} = \Sigma_{m=1}^{\ell+1} D_n A^{(\ell)}{}_{nm} D_m, \tag{7.6.2}$$

where the symmetric matrices $A^{(\ell)}{}_{nm}$ consist of completely skew diagonal 1's, a reflection of the combinatorial origin of the coefficients. The trace is zero for even and unity for odd ℓ's, and the eigenvalues are ± 1. They can occur in any order, but the orthogonal matrices that diagonalize $A^{(\ell)}$ may be constructed so that the eigenvalues alternate (this is convenient for our proof of convergence but not necessary). Let us call these orthogonal matrices $u^{(\ell)}{}_{nm}$ and represent the "vectors" D_m in terms of the eigenvectors d_n^{ℓ} as

$$D_m = \Sigma_{n=1}^{\ell+1} u^{(\ell)}{}_{mn} d_n{}^{\ell}, \tag{7.6.3}$$

where

$$\Sigma_{n=1}^{\ell+1} u^{(\ell)}{}_{mn} u^{(\ell)}{}_{m'n} = \delta_{mm'}. \tag{7.6.4}$$

We then obtain

$$\Sigma_{m=0}^{\ell} D_{\ell+1-m} D_{m+1} = \Sigma_{m=1}^{\ell+1} D_n A^{(\ell)}{}_{nm} D_m = \Sigma_{m=1}^{\ell+1} \lambda^{(\ell)}{}_m (d_m{}^{\ell})^2. \tag{7.6.5}$$

Now, consider the sum in the second term of (7.6.3):

$$\Sigma_{m=0}^{\ell} C'_{\ell-m}(x) C'_m(x) = \Sigma_{m=0}^{\ell} C'_m B^{(\ell)}{}_{mn} C'_n, \tag{7.6.6}$$

where $B^{(\ell)}{}_{mn} = A^{(\ell)}{}_{m+1,n+1}$, the same set of matrices as $A^{(\ell)}$, occurring here with indices $1, \ldots \ell + 1$ as well. By shifting the indices in the vectors C'_n by unity, one obtains the same structure as for the left hand side, i.e. for $m = 0, \ldots \ell$, and f the eigenvectors constructed from C',

$$C'_{m-1} = \Sigma_{n=1}^{\ell+1} u^{(\ell)}{}_{mn} f_n{}^{\ell}. \tag{7.6.7}$$

We then have

$$\Sigma_{m=0}^{\ell} C'_m B^{(\ell)}{}_{mn} C'_n = \Sigma_{m=1}^{\ell+1} \lambda^{(\ell)}{}_m (f_m{}^{\ell})^2 \tag{7.6.8}$$

so that our condition for a solution to the Eqs. (7.5.3) becomes

$$V^{(\ell)} + \Sigma_{m=1}^{\ell+1} \lambda^{(\ell)}{}_m [(d_m{}^{\ell})^2 - g(x)(f_m{}^{\ell})^2] = 0. \tag{7.6.9}$$

We now study the convergence of the d and f sums as $\ell \to \infty$. Inverting (7.6.3) and (7.6.7), we obtain

$$d_m{}^{\ell} = \Sigma_{n=1}^{\ell+1} n C_n u^{(\ell)}{}_{nm} \tag{7.6.10}$$

and

$$f_m{}^{\ell} = \Sigma_{n=1}^{\ell+1} C'_{n-1} u^{(\ell)}{}_{nm}. \tag{7.6.11}$$

Since $u^{(\ell)}{}_{nm}$ is an orthogonal matrix, it follows that

$$\Sigma_{n=1}^{\ell+1}(f_m{}^{\ell})^2 = \Sigma_{n=1}^{\ell+1}C'_{n-1}{}^2 \tag{7.6.12}$$

and

$$\Sigma_{n=1}^{\ell+1}(d_m{}^{\ell})^2 = \Sigma_{n=1}^{\ell+1}n^2 C_n{}^2 \tag{7.6.13}$$

It is sufficient to argue that the sequences in these sums are decreasing. The alternating (due to the $\lambda_m{}^{\ell}$) series appearing in (7.6.9) then converges.

We first remark that the generating function $f(q, X)$ is C^∞ in both variables, so that all orders of derivative with respect to q exist. We seek solutions that can be represented as power series in q. Suppose that this series converges for all values of $q < q_0(x)$ (the radius of convergence can depend on x), and call D_ϵ the domain of x such that $|q_0(x)| \geq \epsilon > 0$, The ratio test prescribes that, for each such x,

$$|\frac{C_{\ell+1}}{C_\ell}| < \frac{1}{|q_0(x)|} \tag{7.6.14}$$

The series (7.5.2) corresponds to the Taylor expansion

$$f(q, x) = \Sigma_0^\infty \frac{1}{\ell!} f^\ell, \tag{7.6.15}$$

where

$$f^\ell = \frac{\partial^\ell f}{\partial q^\ell}; \tag{7.6.16}$$

The ratio condition then becomes

$$|\frac{f^{\ell+1}}{f^\ell}| < |\frac{\ell+1}{q_0}|. \tag{7.6.17}$$

If the derivatives do not grow faster than linearly, this condition should be satisfied for sufficiently large ℓ. Taking $|q_0| = \epsilon$, the convergence would be uniform in D_ϵ.

Now, consider the decreasing property. As for any series depending on a dimensional variable, we may scale the dimension, for $|q_0| > 0$, so that $|q_0(x)| > 1$ for all $x \in D_\epsilon$ (the ratio $C_{\ell+1}/C_\ell$ scales with $1/q$ as well). This choice of scale is adequate for all $x \in D_\epsilon$ for a scale such that $\epsilon > 1$. Then, uniformly, the $|C_\ell(x)|$ forms a decreasing sequence, leading to convergence of the d series in (7.6.9) (the factor m in (7.6.1) does not affect the convergence for large m). A similar argument can be followed for the f series following the convergence of the series in q for $\partial f(q, x)/\partial x$.

This completes our proof of convergence.

Such nonlinear expansions can be studied by means of Fourier series representations in terms of (upper half place) analytic functions (see, for example (Hille 1976)), which may provide useful approximation techniques in specific cases. We leave such a study for future work.

7.7 Shift of Origin for Expansion

We now return to arbitrary dimension. The algorithm proposed in Sect. 7.3 contains an expansion of the potential function $V(q)$ around some point $q = 0$; for a polynomial potential or some other entire function, there would be no question of convergence of this expansion, but the algorithm itself may have only a finite domain of convergence. To extend the range of the resulting functions, it would then be necessary to carry out the expansions around some new origin at, e.g., $q = q_0$.

Therefore, let us now consider expanding $V(q)$ around q_0, and carry out the same procedure. We then rewrite (7.3.2) for the modified problem with a new potential function

$$V'(q) = V(q + q_0) \tag{7.7.1}$$

as

$$V'(q) + \delta_{ij}\frac{1}{2m}\frac{\partial \tilde{f}(q,x')}{\partial q^i}\frac{\partial \tilde{f}(q,x')}{\partial q^j} = g^{ij}(x')\frac{\partial \tilde{f}(q,x')}{\partial x'^i}\frac{\partial \tilde{f}(q,x')}{\partial x'^j}, \tag{7.7.2}$$

where we observe that the solutions $\tilde{f}(q, x')$ and the manifold which we label x' will be different from $f(q, x)$ on the manifold x since the potential function $V'(q)$ is different; however, the variable q on the original space is still designated by q since it is the argument of $V'(q)$.

The assumptions underlying (7.7.2) imply that in the generating function $\tilde{f}(q, x')$, q and x' are independent variables; we may then proceed by recognizing that, as a result of the solution algorithm, x' can only be a function of x in the mapping $q, x \to q, x'$.

We can now use the chain rule of derivatives for the right hand side and consider $\tilde{f}(q, x')$ as a function of q, x, at least locally under this map. Calling this function $h(q + q_0, x)$, we can rewrite (7.7.2) as

$$V'(q) + \delta^{ij}\frac{1}{2m}\frac{\partial h(q+q_0,x)}{\partial q^i}\frac{\partial h(q+q_0,x)}{\partial q^j} = \tilde{g}^{ij}(x)\frac{\partial h(q+q_0,x)}{\partial x^i}\frac{\partial h(q+q_0,x)}{\partial x^j}, \tag{7.7.3}$$

where

$$\tilde{g}^{ij}(x) = g^{k\ell}(x')\frac{\partial x^i}{\partial x'^k}\frac{\partial x^j}{\partial x'^\ell}. \tag{7.7.4}$$

Replacing as a change of variables $q + q_0 \to q$, $V'(q)$ becomes $V(q)$, and (7.7.3) becomes

$$V(q) + \delta^{ij}\frac{1}{2m}\frac{\partial h(q,x)}{\partial q^i}\frac{\partial h(q,x)}{\partial q^j} = \tilde{g}^{ij}(x)\frac{\partial h(q,x)}{\partial x^i}\frac{\partial h(q,x)}{\partial x^j}, \tag{7.7.5}$$

Since this equation has a solution (among others) of the form for which

$$\tilde{g}^{ij}(x) = g^{ij}(x), \tag{7.7.7}$$

by applying the same algorithm, we may choose this solution with the consequence that

$$g^{k\ell}(x')\frac{\partial x^i}{\partial x'^k}\frac{\partial x^j}{\partial x'^\ell} = g^{ij}(x). \tag{7.7.8}$$

With this choice we may follow shifts from $q \to q_0 \to q_1 \dots$ within the domains of convergence choosing the same algorithm for solution at every step, building a set of overlapping neighborhoods which construct a manifold, on which covariance is maintained through the canonical transformation.

7.8 Change in Generating Function Induced by Diffeomorphisms in the Geometric Space

The structure of the image space has the property of supporting local diffeomorphisms. However, our construction concerns a mapping from the coordinates $\{q, p\}$ to $\{x, \pi\}$; therefore a diffeomorphism of the latter set of variables necessarily involves a change in the generating function of the transformation.

In this section, we calculate the effect of an infinitesimal coordinate transformation on the geometrical space, holding the Hamiltonian variables $\{q, p\}$ unchanged, on the generating function of the canonical transformation, i.e., $f \to \tilde{f}$.

On the original choice of coordinates, for which

$$\begin{aligned} p_i &= \frac{\partial f(q,x)}{\partial q^i} \\ \pi_i &= -\frac{\partial f(q,x)}{\partial x^i} \end{aligned} \tag{7.8.1}$$

we now consider a new mapping from q, p to x', π' differing infinitesimally from x, π according to

$$x'^i = x^i + \lambda^i(x), \tag{7.8.2}$$

where $\lambda^i(x)$ is small.

After this mapping, we can write

$$\begin{aligned} p_i &= \frac{\partial \tilde{f}(q,x')}{\partial q^i} \\ \pi'_i &= -\frac{\partial \tilde{f}(q,x')}{\partial x'_i} \end{aligned} \tag{7.8.3}$$

To study $\tilde{f}(q, x')$, let us define

$$g^i(q, x') = \frac{\partial \tilde{f}(q, x')}{\partial x'^i} = -\pi'_i. \tag{7.8.4}$$

Then,

$$g_i(q, x+\lambda) \cong \frac{\partial \tilde{f}(q, x)}{\partial x^i} + \frac{\partial^2 \tilde{f}(q, x)}{\partial x^i \partial x^j}\lambda^j(x) \tag{7.8.5}$$

so that

$$-\pi'_i \cong \frac{\partial \tilde{f}(q, x)}{\partial x^i} + \frac{\partial^2 \tilde{f}(q, x)}{\partial x^i \partial x^j}\lambda^j(x). \tag{7.8.6}$$

This result could have been obtained directly from (7.8.3) but it is perhaps helpful to define the function $g^i(q, x')$ to clarify the computation.

We now impose invariance of

$$\pi'_i dx'^i = \pi_i dx^i, \tag{7.8.7}$$

which leads, through the Hamilton–Lagrange construction, to invariance of the Hamiltonian. We now write out

$$\begin{aligned}
-\pi'_i dx'^i &\cong \left[\frac{\partial \tilde{f}(q, x)}{\partial x^i} + \frac{\partial^2 \tilde{f}(q, x)}{\partial x^i \partial x^j}\lambda^j(x)\right] \times \left[dx^i + \frac{\partial \lambda^i}{\partial x^k}dx^k\right] \\
&= \frac{\partial \tilde{f}(q, x)}{\partial x^i}dx^i + \frac{\partial^2 \tilde{f}(q, x)}{\partial x^i \partial x^j}\lambda^j(x)dx^i + \frac{\partial \tilde{f}(q, x)}{\partial x^i}\frac{\partial \lambda^i}{\partial x^k}dx^k + \frac{\partial^2 \tilde{f}(q, x)}{\partial x^i \partial x^j}\lambda^j(x)\frac{\partial \lambda^i}{\partial x^k}dx_k \\
&= -\pi_i dx^i.
\end{aligned} \tag{7.8.8}$$

Therefore, to order λdx,

$$\begin{aligned}
dx^i \frac{\partial f(q, x)}{\partial x^i} &= dx^i \left\{\frac{\partial \tilde{f}(q, x)}{\partial x^i} + \frac{\partial^2 \tilde{f}(q, x)}{\partial x^i \partial x^j}\lambda^j(x) + \frac{\partial \tilde{f}(q, x)}{\partial x^k}\frac{\partial \lambda^k}{\partial x^i}\right\} \\
&= dx^i \left\{\frac{\partial \tilde{f}(q, x)}{\partial x^i} + \frac{\partial}{\partial x^i}\left[\frac{\partial \tilde{f}(q, x)}{\partial x^k}\lambda^k\right]\right\},
\end{aligned} \tag{7.8.9}$$

so that

$$dx_i \frac{\partial f(q, x)}{\partial x^i} = dx_i \frac{\partial}{\partial x^i}\left[\tilde{f}(q, x) + \lambda_k \frac{\partial \tilde{f}(q, x)}{\partial x^k}\right] \tag{7.8.10}$$

If we write (say, integrate up to some x^i)

$$f(q, x) = \tilde{f}(q, x) + \lambda^k \frac{\partial \tilde{f}(q, x)}{\partial x^k}, \tag{7.8.11}$$

we may approximately invert to get

$$\tilde{f}(q,x) \cong f(q,x) - \lambda^k \frac{\partial f(q,x)}{\partial x^k}. \tag{7.8.12}$$

This corresponds to a conformal-like local transformation. The algebra of such generators is

$$\left[\lambda^{ia}\frac{\partial}{\partial x^i}, \lambda^{jb}\frac{\partial}{\partial x^j}\right] = \left(\lambda^{ia}\frac{\partial \lambda^{jb}}{\partial x^i} - \lambda^{ib}\frac{\partial \lambda^{ja}}{\partial x^i}\right)\frac{\partial}{\partial x^j} \tag{7.8.13}$$

Thus the algebra is of a conformal type, but the coefficients may run on, so that the group may not be finite dimensional.

Example Suppose $\lambda_i{}^a = \epsilon_i{}^j(a)x_j$, such as a rotation generator (we may factor out the infinitesimal scale), for $\epsilon_i{}^j(a)$ antisymmetric constants. Then,

$$\left[\lambda_i{}^a\frac{\partial}{\partial x_i}, \lambda_j{}^b\frac{\partial}{\partial x_j}\right] = x_j M_i{}^j(b,a)\frac{\partial}{\partial x_i}, \tag{7.8.14}$$

where

$$M_i{}^j(b,a) = \epsilon_i{}^k(b)\epsilon_k{}^j(a) - \epsilon_i{}^k(a)\epsilon_k{}^j(b). \tag{7.8.15}$$

For the rotation group, these form a finite Lie algebra. The group acts on the generating function, which then forms a representation.

7.9 Mapping of Bounded Submanifolds

Since the mapping that we have constructed carries a Euclidean phase space into a geometrical form, it is natural to study possibly non-trivial topological properties that this geometrical space could have. As a simple example, consider a potential in the Euclidean space in two dimensions which contains two identical finite depth potential wells with lower bound E_0, and centers spaced along the x-axis. Above a certain energy, say E_1, there is just one connected region of motion, and between E_1 and E_0 there are two separated regions. The total energy serves as a *height* function, in the terminology of Morse theory (Milnor 1969).

Let us first consider a particle with energy $E_0 < E < E_1$. A particle in one of these wells has an orbit that is confined to this well. If it reaches the boundary where $E = V$, the momentum (and velocity) vanishes, and the orbit necessarily then retraces its path as under time reversal. Under the symplectomorphism, this orbit is mapped into a geodesic curve, and by the property of $1:1$ mapping, the corresponding geodesic curve must stop and retrace its path as under time reversal as well. The family of all such orbits for a given value of E defines a boundary in the geometric space, and is therefore a closed submanifold with boundary.

It is clear that such orbits associated with each well (at a given value of E) separately are disjoint since they are disjoint in the original space. Increasing the energy above the value E_1 would result in a single connected region for the geometric orbits. Therefore the homotopy classes of the possible orbits change as a function of the height function E. We shall explore the consequences, in particular, of the existence of topological invariants, in this context, in a later publication.

7.10 Summary and Conclusions

In this chapter we have constructed a canonical transformation from a Hamiltonian of the usual form to a geometric form.

We have given the basic mathematical formulation in terms of the geometry of symplectic manifolds.

For the central purpose of our construction, we formulate the process of studying stability by means of geodesic deviation in terms of geometric methods, making clear the relation between stability in the geometric manifold and the original Hamiltonian motion.

We then give an algorithm for solving the nonlinear equations for the generating function of the canonical transformation. This algorithm was studied for the simple case of one dimension, and we proved convergence of the recursive scheme under certain reasonable assumptions.

Since the series expansions generated by the algorithm for finding the solutions for the generating function may have a bounded domain of convergence, we studied (in general dimension) the possibility of shifting the origin in order to carry out the expansions based on a new origin. As for the analytic continuation of a function of a complex variable, this procedure can extend the solutions for the generating function to a maximal domain.

Since the image space of the symplectomorphism has geometrical structure, it is natural to study its properties under local diffeomorphisms. A local change of variables $\{x, \pi\} \rightarrow \{x', \pi'\}$ (leaving the variables of the original space unchanged) alters the structure of the mapping from the original variables $\{q, p\}$ to the new variables $\{x', \pi'\}$; we studied the effect of infinitesimal diffeomorphims of this type on the generating function.

We finally discussed briefly the mapping of bounded closed submanifolds, created by potential wells in the Hamiltonian space, corresponding to closed submanifolds in the geometric space, where Morse theory may be applied, to open the possibility of obtaining a new class of conserved quantities associated with homotopies of the image space.

Chapter 8
Summary

In the first part of this book, we have described an approach to the treatment of unstable quantum systems from a fundamentally new point of view, although the methods we have used have their source in the older work of Nagy and Foias (1970), for which a system with exact semigroup evolution on a Hilbert space is embedded in a larger space in which the evolution is unitary. Projection back to the smaller space, a subspace of the larger space, then has exact semigroup evolution. In the second part of the book, we study new methods for determining the stability of classical Hamiltonian systems based on embedding the motion generated by a potential model Hamiltonian of usual type into a system described geometrically, evolving along geodesic orbits, for which stability can be measured by geodesic deviation.

Gamow (1928) was the first to apply the quantum theory to the description of the unstable nucleus in 1928, in particular, to alpha decay. It was observed that the nucleus decays exponentially; Gamow therefore wrote a Schrödinger equation with complex energy eigenvalue $E - i\frac{\Gamma}{2}$. The solution, $\psi(t) \propto e^{-i(E-i\frac{\Gamma}{2})}\psi(0)$ then has an exponential decaying form $|\psi(t)|^2 \propto e^{-\Gamma t}$. In their description of K meson decay, Wu and Yang (1964) generalized Gamow's formula to two dimensions, corresponding to the short (2π) and long lived (3π) decay modes, providing a phenomenologically useful description of the K meson decay. However, Gamow's method did not provide a basic understanding of how a quantum system could display an instability described by a decay law of this type. In 1930 Weisskopf and Wigner (1930) formulated a basic theory that could provide, for times not too short or too long, an explanation for this semigroup type behavior. The Wigner Weisskopf formulation, however, does not provide an actual semigroup behavior; applying it to the two dimensional K decay problem results in a deviation from semigroup behavior due to the non-orthogonality of the pole residues contributing to (almost) exponential decay of each mode that is in disagreement with the semigroup property observed experimentally (Cohen and Horwitz 2001).

Lax and Phillips (1967), based on the fundamental mathematical ideas of Nagy and Foias, developed a theory which resulted in exact semigroup evolution in a

L. Horwitz and Y. Strauss, *Unstable Systems*, Mathematical Physics Studies,
https://doi.org/10.1007/978-3-030-31570-2_8

Hilbert space for hyperbolic wave equations, as occur in electromagnetic radiation, but its application to quantum mechanical evolution was only discovered (Strauss and Horwitz 2000) in 2000. This method is explained in detail in Chap. 2 of this book. The method developed there is very efficient for problems involving Hamiltonians with spectrum on the whole real line, but for half line spectrum (positive energy), it is much more difficult to apply. To deal with this problem an "approximate" Lax–Phillips theory is worked out in Chap. 4, based on the development of the theory of quantum Lyapunov functions, detecting the *direction of time* in Chap. 3.

In the second part of the book, we concentrate on classical Hamiltonian systems. A very powerful method to deal with such systems is to establish a relation between usual type of potential model Hamiltonian system and a corresponding *geometrical* form of the type discussed by Gutzwiller (1990) and Curtiss and Miller (1985). The motion in the geometrical form is described by geodesic curves on a manifold, where stability can be tested by examining geodesic deviation. Chapter 5 is devoted to the study of geodesic deviation, and it is shown that the oscillator emerging in the Jacobi equations for the stability of the geodesic deviation can be quantized as done by Maassen (1989), where the quanta are interpreted as excitations of the dynamical medium. This forms a transition between the quantum treatment of instability and the classical treatment in the second part of the book. A rigorous discussion of geodesic deviation is given in this chapter as well.

In our discussion of classical unstable systems, we show that there is a simple correspondence between a classical potential type Hamiltonian and a geometric type Hamiltonian with conformal metric (Horwitz and Ben Zion 2007). The geodesic deviation criteria have been shown to provide a remarkable predictive power (Yahalom et. al. 2015; Ben Zion and Horwitz 2007, 2008) for the instabilities of the corresponding potential model Hamiltonian motion. This predictive power can be understood, as we show in Chap. 7, to follow from the existence of a *canonical transformation* relating the potential model Hamiltonian to the geometric form, for which many possibilities for the metric become available through a constructive algorithm. We study here some properties of this construction; this development has opened a topic for future research as well.

Appendix A
Hardy Spaces

A.1 Hardy Spaces of Scalar Valued Functions on the Unit Disc

The use of Hardy spaces of functions analytic in the upper, respectively lower, half-plane for the description of irreversible evolution appears as a central theme throughout various chapters of this book. This is no surprise, since the analytic properties of Hardy spaces incorporates in a natural way the causality structure of irreversible processes, such as semigroup evolution. For the convenience of the reader, we provide in this appendix a short discussion of the structure and properties of Hardy spaces. Of course, this is but a small fraction of a full discussion of these mathematical constructions and we have tried to include only the basic facts concerning Hardy space which are directly relevant for the mathematical frameworks considered in the present text. For more complete presentations of the subject of H^p spaces (in particular Hardy spaces) we refer the reader to Duren (1970), Hoffman (1962), Rosenblum and Rovnyak (1985) or any other reference from the wide literature on this subject.

We start our discussion with the definition of Hardy spaces of scalar valued functions on the open unit disc D:

Definition A.10 (*Hardy spaces in the unit disk (scalar case)*) For $0 < p \leq \infty$ we denote by $\mathcal{H}^p(D)$ the class of all functions $f : D \mapsto \mathbb{C}$ analytic in the unit disc such that the functions $f_r(\theta) := f(re^{i\theta})$ satisfy $\sup_{0<r<1} \|f_r\|_p < \infty$. □

Here $\|\cdot\|_p$ denotes the L^p norm on the unit circle T, i.e.,

$$\|f\|_p = \left(\frac{1}{2\pi} \int_0^{2\pi} |f(e^{it})|^p dt \right)^{1/p}, \quad f \in L^p(T),\ 1 \leq p < \infty,$$

where $dt/2\pi$ is normalized Lebesgue measure on the unit circle. The L^∞ norm is given by

L. Horwitz and Y. Strauss, *Unstable Systems*, Mathematical Physics Studies,
https://doi.org/10.1007/978-3-030-31570-2

$$\|f\|_\infty = \sup_{t\in[0,2\pi]} |f(e^{it})|, \quad f \in L^\infty(T).$$

Each function $f \in \mathcal{H}^p(D)$ has a non-tangential boundary value $f(e^{it}), t \in [0, 2\pi)$ almost everywhere with respect to Lebesgue measure on the unit circle T. For $1 \leq p \leq \infty$ the Hardy space $\mathcal{H}^p(D)$ is, in fact, a Banach space under a norm defined by

$$\|f\| = \lim_{r\to 1^-} \|f_r\|_p, \quad f \in \mathcal{H}^p(D). \tag{A.1}$$

Denote by $\mathcal{H}^p(T)$ the set of boundary value functions $f(e^{it})$ on T of functions in $\mathcal{H}^p(D)$. By the existence of the limit in Eq. (A.1) the set $\mathcal{H}^p(T)$ is a vector subspace of $L^p(T)$. In fact, $\mathcal{H}^p(T)$ is a closed subspace of $L^p(T)$ and we have

$$\|f\|_p = \|f\| = \lim_{r\to 1^-} \|f_r\|_p, \tag{A.2}$$

where on the left hand side of Eq. (A.2) f is the boundary valued function.

By their isometric isomorphism we often consider $\mathcal{H}^p(D)$ and $\mathcal{H}^p(T)$ as identical function spaces. Since the functions in $\mathcal{H}^p(D)$ are analytic in the unit disc each such function has a Taylor expansion

$$f(z) = \sum_{n=0}^{\infty} a_n z^n$$

It is clear that every Hardy space $\mathcal{H}^p(D)$ contains all finite sums of the form $p(z) = \sum_{k=0}^{n} a_k z^k$, i.e., all of the polynomials. The boundary value of such a polynomial on T is a function of the form $p(e^{it}) = \sum_{k=0}^{\infty} a_k e^{int}$, i.e., a polynomial in e^{it}. In fact, we have

Theorem A.12 *For $0 < p < \infty$ the Hardy space $\mathcal{H}^p(T)$ is the L^p closure of the set of polynomials in e^{it}.* □

For $1 \leq p \leq \infty$ there is another characterization of $\mathcal{H}^p(T)$ which is more useful for our purposes. We have the following theorem:

Theorem A.13 *For $1 \leq p \leq \infty$ the Hardy space $\mathcal{H}^p(T)$ is exactly the closed subspace of $L^p(T)$ consisting of functions whose Fourier coefficients vanish for all $n < 0$.* □

Thus, for $1 \leq p \leq \infty$ the Hardy space $\mathcal{H}^p(T)$ consists of all functions in $f \in L^p(T)$ such that

$$\int_{-\pi}^{\pi} f(t)e^{int}dt = 0, \quad n = 1, 2, \ldots.$$

A.1.1 Canonical Factorization of $\mathcal{H}^p$ Functions

A.1.1.1 Blaschke Products

If $f(z)$ is a non-zero analytic function in D then the zeros of $f(z)$ cannot have an accumulation point in D. Of course, no difficulty arises if the number of zeros of f is finite. If the number of zeros of f is infinite then any accumulation point of these zeros must be on the unit circle T. In fact, we have the following theorem:

Theorem A.14 *Let $f : D \mapsto \mathbb{C}$ be a bounded analytic function. Assume furthermore that $f(0) \neq 0$. Let $(\alpha_n)_{n=1}^{\infty}$ be the sequence of zeros of f in D, each repeated according to its multiplicity as a zero of f. Then we have $\sum_{n=1}^{\infty}(1 - |\alpha_n|) < \infty$, i.e., the product $\prod_{n=1}^{\infty} |\alpha_n|$ converges to a positive real number.* □

Complementing the above result, we have:

Theorem A.15 *Let $(\alpha_n)_{n=1}^{\infty}$, $0 < |\alpha_1| \leq |\alpha_2| \leq |\alpha_3| \leq \cdots < 1$ be a sequence of complex numbers in D. Then the infinite product*

$$B(z) = \prod_{n=1}^{\infty} \frac{\overline{\alpha_n}}{|\alpha_n|} \cdot \frac{\alpha_n - z}{1 - \overline{\alpha_n} z} \tag{A.3}$$

converges uniformly on each closed disc $\overline{D}$, $0 < r < 1$, if and only if $\sum_{n=1}^{\infty}(1 - |\alpha_n|) < \infty$ (i.e., the infinite product $\prod_{n=1}^{\infty} |\alpha_n|$ converges to a positive real number). The zeros of $B(z)$ are exactly the numbers α_n, $n \in \mathbb{N}$ with multiplicity equal to the number of times α_n appears in the sequence $(\alpha_n)_{n=1}^{\infty}$. Furthermore, we have

$$|B(z)| < 1, \quad \forall z \in D, \qquad |B(e^{it})| = 1, \quad t \in [0, 2\pi]$$

□

Theorems A.14 and A.15 motivate the following definition of a Blaschke product:

Definition A.11 (*Blaschke product*) A Blaschke product is a function $B(z)$ analytic in the unit disk D of the form

$$B(z) = z^{m_0} \prod_{n=1}^{\infty} \left[\frac{\overline{\alpha_n}}{|\alpha_n|} \cdot \frac{\alpha_n - z}{1 - \overline{\alpha_n} z} \right]^{m_n}.$$

where

1. $m_0, m_1, m_2, m_3, \ldots$ are non-negative integers;
2. The constants $\alpha_n \in D$ are non-zero distinct complex numbers;
3. The product $\prod_{n=1}^{\infty} |\alpha_n|^{m_n}$ converges.

□

We have seen in Theorem A.14 that the infinite product defining $B(z)$ converges uniformly on compact subsets of the unit disk D and that the only zeros of $B(z)$ are the numbers α_n.

A.1.1.2 Canonical Factorization of $\mathcal{H}^p$ Functions

The next two theorems, concerning canonical factorizations, provide key information regarding the structure of functions in $\mathcal{H}^p(D)$. First we need some definitions:

Definition A.12 (*inner function*) An inner function is a function $f(z)$ analytic in the unit disk, such that $|f(z)| \leq 1$, $\forall z \in D$ and $|f(e^{it})| = 1$, a.e. with respect to Lebesgue measure on the unit circle T. □

It is readily seen, from our discussion of Blaschke products above, that a Blaschke product is an inner function.

Definition A.13 (*outer function*) An outer function for the class $\mathcal{H}^p$ is a function of the form

$$F(z) = e^{i\gamma} \exp\left(\frac{1}{2\pi}\int_{-\pi}^{\pi} \frac{e^{it}+z}{e^{it}-z} \log\psi(t)dt\right), \tag{A.4}$$

where $\gamma \in \mathbb{R}$, $\psi(t) \geq 0$, $\log\psi(t) \in L^1(T)$ and $\psi(t) \in \mathcal{L}^p(T)$. □

We can now state the first factorization theorem:

Theorem A.16 *Let $f(z)$ be a function in $\mathcal{H}^p(D)$. Then f can be expressed in the form $f(z) = g(z)F(z)$ where $g(z)$ is an inner function and $F(z)$ is an outer function for the class $\mathcal{H}^p(D)$. This factorization is unique up to a constant factor of modulus 1. The outer function in this factorization is given by*

$$F(z) = e^{i\gamma} \exp\left(\frac{1}{2\pi}\int_{-\pi}^{\pi} \frac{e^{it}+z}{e^{it}-z} \log|f(e^{it})|dt\right).$$

□

The inner-outer factorization of an $\mathcal{H}^p(D)$ function can be refined by further factorization of the inner part. For this we need to define one more class of functions:

Definition A.14 (*singular inner function*) A singular inner function is an inner function without zeros which is positive at the origin. □

Theorem A.17 *Let $S(z)$ be a singular inner function. Then there exists a unique positive singular measure μ on the unit circle such that*

$$S(z) = \exp\left(-\int_{-\pi}^{\pi} \frac{e^{it}+z}{e^{it}-z} d\mu(t)\right)$$

□

With the introduction of singular inner functions we can state the canonical factorization theorem for $\mathcal{H}^p(D)$ functions:

Theorem A.18 (Canonical factorization theorem) *Let $f(z)$ be an $\mathcal{H}^p(D)$ class function ($p > 0$). Assuming that $f(z)$ is not identically zero there exists a unique factorization of f of the form $f(z) = B(z)S(z)F(z)$, where $B(z)$ is a Blaschke product, $S(z)$ is a singular inner function, and $F(z)$ is an outer function of class $\mathcal{H}^p(D)$. Conversely, each product $B(z)S(z)F(z)$, with $F(z) \in \mathcal{H}^p(D)$, is an $\mathcal{H}^p(D)$ function.* □

Let us summarize the canonical factorization of an $\mathcal{H}^p(D)$ function. Given $f(z) \in \mathcal{H}^p(D)$, let m_0 be the order of the zero of f at $z = 0$, let $\alpha_1, \alpha_2, \ldots$ be the list of zeros of f other than the origin and let $m_1, m_2, \ldots$ be their corresponding multiplicities. Then

$$B(z) = z^{m_0} \prod_{n=1}^{\infty} \left[\frac{\overline{\alpha_n}}{|\alpha_n|} \cdot \frac{\alpha_n - z}{1 - \overline{\alpha_n} z}\right]^{m_n}.$$

$$F(z) = e^{i\gamma} \exp\left(\frac{1}{2\pi} \int_{-\pi}^{\pi} \frac{e^{it}+z}{e^{it}-z} \log \psi(t) dt\right),$$

and

$$S(z) = \frac{f(z)}{B(z)F(z)} = \exp\left(-\int_{-\pi}^{\pi} \frac{e^{it}+z}{e^{it}-z} d\mu(t)\right)$$

where μ is a positive singular measure.

A.1.2 Hardy Spaces on the Half-Plane

It is seen from the discussion above that a natural domain for the definition of a Hardy space is the unit disc D. However, for many applications, and in particular most application considered in this book, a more appropriate domain is the upper-half of the complex plane. For this reason we consider in this section Hardy spaces of the upper half-plane $\mathbb{C}^+ = \{z \in \mathbb{C} : \operatorname{Im} z > 0\}$. We define two types of Hardy spaces over $\mathbb{C}^+$:

Definition A.15 (*Hardy space of the first kind over* $\mathbb{C}^+$) For $0 < p < \infty$ we denote by $\mathcal{H}^p(\mathbb{C}^+)$ the space of functions analytic in $\mathbb{C}^+$ and such that the L^p norms

$$M_p(y, f) := \left(\int_{-\infty}^{\infty} |f(x + iy)|^p \, dx \right)^{1/p}$$

are bounded for $0 < y < \infty$. We shall denote by $\mathcal{H}^\infty(\mathbb{C}^+)$ the space of all bounded analytic functions in $\mathbb{C}^+$. The function spaces $\mathcal{H}^p(\mathbb{C}^+)$ are called Hardy spaces of the first kind. □

Definition A.16 (*Hardy space of the second kind over* $\mathbb{C}^+$) Consider the conformal mapping $\varphi(z) = i\frac{1+z}{1-z}$ of the unit disc onto the upper half-plane $\mathbb{C}^+$. For $0 < p \leq \infty$ the Hardy space of the second kind is denoted by $H^p(\mathbb{C}^+)$ and is defined to be the set of all function holomorphic on $\mathbb{C}^+$ such that $f \circ \varphi \in H^p(D)$. □

The next two theorems characterize the boundary behavior of functions in $\mathcal{H}^p(\mathbb{C}^+)$. First, we have

Theorem A.19 *Let* $f \in \mathcal{H}^p(\mathbb{C}^+)$, $0 < p < \infty$. *Then* $f(z) \to 0$ *as* $z \to \infty$ *in each fixed half-plane* $Im\, z \geq \delta > 0$.

With regard to the boundary values on the real axis we first have:

Theorem A.20 *Let* $1 \leq p \leq \infty$ *and let* $f \in \mathcal{H}^p(\mathbb{C}^+)$. *Then:*

1. $f(z)$ *has non-tangential boundary value a.e on the real axis. Denote the boundary value function thus obtained by* $f(t)$, $t \in \mathbb{R}$.
2. *The boundary value function of* $f(z)$ *on* $\mathbb{R}$ *is in* $L^p(\mathbb{R})$ *and we have*

$$f(z) = \frac{1}{\pi} \int_{-\infty}^{\infty} f(t) \frac{y}{(x - t)^2 + y^2} dt, \quad z = x + iy, \ y > 0$$

3. $\lim_{y \to 0^+} \int_{-\infty}^{\infty} |f(x + iy)|^p dx = \int_{-\infty}^{\infty} |f(t)|^p dt$

□

The last theorem can be partially extended to the case $0 < p < 1$. In fact, we have the following theorem:

Theorem A.21 *Let* $f \in \mathcal{H}^p(\mathbb{C}^+)$, $0 < p \leq \infty$. *Then the boundary function*

$$f(x) = \lim_{y \to 0^+} f(x + iy), \quad x \in \mathbb{R}$$

exists a.e. on $\mathbb{R}$, *the boundary value function* $f(t)$ *is in* $L^p(\mathbb{R})$ *and*

$$\int_{-\infty}^{\infty} \frac{\log |f(t)|}{1 + t^2} dt > -\infty.$$

□

Another important result states that the integral in item (2) of Theorem A.20, reconstructing $f(z)$ from its boundary value function $f(x)$, can, in fact, be written in a Cauchy integral like form. We have

Theorem A.22 *Let $f \in \mathcal{H}^p(\mathbb{C}^+)$, $1 \le p \le \infty$, Then*

$$\frac{1}{2\pi i}\int_{-\infty}^{\infty} \frac{f(t)}{t-z}dt = \begin{cases} f(z), & Im\, z > 0 \\ 0, & Im\, z < 0 \end{cases}$$

Conversely, for a function $h \in L^p(\mathbb{R})$, $1 \le p \le \infty$, such that

$$\frac{1}{2\pi i}\int_{-\infty}^{\infty} \frac{h(t)}{t-z}dt = 0, \quad \forall z \in \mathbb{C}^-,$$

the function

$$f(z) = \frac{1}{2\pi i}\int_{-\infty}^{\infty} \frac{h(t)}{t-z}dt, \quad Im\, z > 0,$$

is an $\mathcal{H}^p(\mathbb{C}^+)$ function with boundary value $f(x) = h(x)$, a.e., $x \in \mathbb{R}$. □

Of the various function spaces $\mathcal{H}^p(\mathbb{C}^+)$, the Hilbert space $\mathcal{H}^2(\mathbb{C}^+)$ is particularly important for our purposes. A central fact concerning the Hardy space $\mathcal{H}^2(\mathbb{C}^+)$ is the Paley–Wiener theorem, which is the half-plane analogue of the fact that $\mathcal{H}^2(D)$ consists of power series $\sum_{n=0}^{\infty} a_n z^n$ such that $\sum_{n=0}^{\infty} |a_n|^2 < \infty$:

Theorem A.23 (Paley–Wiener theorem) *A complex-valued function $f(z)$, analytic in the upper half-plane, belongs to the class $\mathcal{H}^2(\mathbb{C}^+)$ if and only if it has the form*

$$f(z) = \frac{1}{\sqrt{2\pi}}\int_{0}^{\infty} \hat{f}(t)e^{izt}dt \tag{A.5}$$

for some function $\hat{f} \in L^2(\mathbb{R}^+)$. □

The Fourier integral in Eq. (A.5) is analogous to the power series $\sum_{n=0}^{\infty} a_n z^n$.

Corollary A.3 *If $f \in \mathcal{H}^2(\mathbb{C}^+)$ and $\hat{f}$ is the Fourier transform of the boundary value function of f on $\mathbb{R}$ then $\hat{f}(\omega) = 0$, $\forall \omega < 0$.* □

A.1.2.1 Canonical Factorization of $\mathcal{H}^p(\mathbb{C}^+)$ Functions

Functions in $\mathcal{H}^p(\mathbb{C}^+)$ have canonical factorizations similar to functions in $\mathcal{H}^p(D)$. Before we state the relevant theorem we introduce the various factors in this factorization:

Definition A.17 (*Blaschke product on* $\mathbb{C}^+$) A Blaschke product for the upper half-plane $\mathbb{C}^+$ is a function of the form

$$B(z) = \left(\frac{z-i}{z+i}\right)_0^m \prod_n \frac{|z_n^2+1|}{z_n^2+1} \cdot \frac{z-z_n}{z-\overline{z_n}},$$

where $(z_n)_{n=1}^{\infty}$ is a sequence of numbers $z_n \in \mathbb{C}^+$, $z_n \neq i$, satisfying $\sum_n \frac{\operatorname{Im} z_n}{1+|z_n|^2} < \infty$ (the sequence might be finite; also, the values z_n may be repeated, i.e., there may be multiplicities for the zeros of the factors in the product). □

Definition A.18 (*Outer function on* $\mathbb{C}^+$) An outer function for the class $\mathcal{H}^p(\mathbb{C}^+)$ is a function of the form

$$F(z) = e^{i\gamma} \exp\left(\frac{1}{\pi i} \int_{-\infty}^{\infty} \frac{(1+tz)\log \omega(t)}{(t-z)(1+t^2)} dt\right),$$

for some $\gamma \in \mathbb{R}$ and a measurable function $\omega(t) \geq 0$, $t \in \mathbb{R}$ satisfying

$$\int_{-\infty}^{\infty} \frac{\log \omega(t)}{1+t^2} dt > -\infty, \qquad \int_{-\infty}^{\infty} \frac{(\omega(t))^p}{1+t^2} dt < \infty$$

□

Definition A.19 (*Singular inner function for* $\mathbb{C}^+$) A singular inner function for $\mathcal{H}^p(\mathbb{C}^+)$ is a function of the form

$$S(z) = \exp\left(i \int_{-\infty}^{\infty} \frac{1+tz}{t-z} d\mu(t)\right)$$

where μ is a singular positive measure on $\mathbb{R}$. □

Following the introduction of Blaschke products and outer and singular inner functions for $\mathcal{H}^p(\mathbb{C}^+)$ we may state the canonical factorization theorem:

Theorem A.24 (Canonical factorization of $\mathcal{H}^p(\mathbb{C}^+)$ functions) *A function $f \in \mathcal{H}^p(\mathbb{C}^+)$, $0 < p < \infty$, such that $f(z) \not\equiv 0$ has a unique factorization of the form*

$$f(z) = e^{i\alpha z} B(z) S(z) F(z),$$

where $\alpha \geq 0$, $B(z)$ is a Blaschke product on $\mathbb{C}^+$ such that the sequence $(z_n)_{n=1}^{\infty}$ defining $B(z)$ is the sequence of zeros of f in $\mathbb{C}^+$, $F(z)$ is an outer function for $\mathcal{H}^p(\mathbb{C}^+)$ with

$$F(z) = e^{i\gamma} \exp\left(\frac{1}{\pi i} \int_{-\infty}^{\infty} \frac{(1+tz)\log|f(t)|}{(t-z)(1+t^2)} dt \right),$$

and $S(z) = f(z)/(e^{i\alpha z}B(z)F(z))$ *is a singular inner function for* $\mathcal{H}^p(\mathbb{C}^+)$. □

In particular, by the definition of an outer function, for $f \in \mathcal{H}^p(\mathbb{C}^+)$ we have

$$\int_{-\infty}^{\infty} \frac{\log|f(t)|}{1+t^2} dt > -\infty, \tag{A.6}$$

and hence $f \in \mathcal{H}^p(\mathbb{C}^+)$ cannot vanish on any non-zero measure subset in $\mathbb{R}$.

A.1.3 Relation Between $\mathcal{H}^p$ Spaces of the Unit Disc and Upper Half-Plane

In this subsection we investigate the relation between the function spaces $\mathcal{H}^p(D)$ and $\mathcal{H}^p(\mathbb{C}^+)$ and $H^p(\mathbb{C}^+)$. Observe that the linear fractional transformations

$$w = \psi(z) = \frac{z-i}{z+i}, \qquad z = \varphi(w) = i\frac{1+w}{1-w}, \tag{A.7}$$

map, respectively, $\mathbb{C}^+$ onto D and D onto $\mathbb{C}^+$. Our first result is:

Theorem A.25 *If* $0 < p < \infty$ *then* $\mathcal{H}^p(\mathbb{C}^+) \subseteq H^p(\mathbb{C}^+)$. □

By the definition of Hardy spaces of the second kind this last theorem states that for $f \in \mathcal{H}^p(\mathbb{C}^+)$ the function g obtained by

$$g(z) = (f \circ \varphi)(z) = f\left(i\frac{1+z}{1-z}\right) \tag{A.8}$$

belongs to $H^p(D)$. By the above theorem the transformation defined in Eq. (A.8) maps $\mathcal{H}^p(\mathbb{C}^+)$ into a subspace of $\mathcal{H}^p(D)$. We would like to characterize this subspace. Using the transformations in Eq. (A.7) we find a mapping of the boundary $\mathbb{R}$ of $\mathbb{C}^+$ to the boundary ∂D of D of the form

$$e^{i\theta} = \frac{t-i}{t+i}, \quad \theta \in [0, 2\pi), \; t \in \mathbb{R}.$$

Hence we have

$$ie^{i\theta}d\theta = \frac{(t+i)-(t-i)}{(t+i)^2}dt \Rightarrow i\frac{t-i}{t+i}d\theta = \frac{2i}{(t+i)^2}dt \Rightarrow d\theta = \frac{2}{t^2+1}dt,$$

and so, if a function $g(e^{i\theta})$ is integrable on the unit circle, then

$$\frac{1}{2\pi}\int_{-\pi}^{\pi} g(e^{i\theta})d\theta = \frac{1}{\pi}\int_{-\infty}^{\infty} \frac{f(t)}{t^2+1}dt\,,$$

where

$$f(t) = g(e^{i\theta}) = g\left(\frac{t-i}{t+i}\right).$$

In particular, we have in this case

$$\frac{1}{2\pi}\int_{-\pi}^{\pi} |g(e^{i\theta})|^p d\theta = \frac{1}{\pi}\int_{-\infty}^{\infty} \frac{|f(t)|^p}{t^2+1}dt\,,$$

Therefore, by the definition of $H^p(\mathbb{C}^+)$, this Hardy space contains all of functions f that are analytic in $\mathbb{C}^+$ and such that their boundary value function on $\mathbb{R}$ satisfies

$$\int_{-\infty}^{\infty} \frac{|f(t)|^p}{1+t^2}dt < \infty$$

Take a branch of $\log(z+i)$ which is analytic in the upper half-plane, so that the function $(z+i)^{2/p}$ is well defined and analytic in the $\mathbb{C}^+$. Let $f \in \mathcal{H}^p(\mathbb{C}^+)$. Then the function $\tilde{f}(z) = (z+i)^{2/p} f(z)$ is analytic in $\mathbb{C}^+$ and its boundary value function on $\mathbb{R}$ satisfies

$$\int_{-\infty}^{\infty} \frac{|(t+i)^{2/p} f(t)|^p}{1+t^2}dt = \int_{-\infty}^{\infty} \frac{|t+i|^2 |f(t)|^p}{1+t^2}dt = \int_{-\infty}^{\infty} |f(t)|^p dt < \infty$$

Thus, if $f \in \mathcal{H}^p(\mathbb{C}^+)$ then the function $\tilde{f}(z) = (z+i)^{2/p} f(z)$ is in $H^p(\mathbb{C}^+)$. In this case there is a function $\tilde{g} \in H^p(D)$ such that

$$f(z) = \frac{1}{(z+i)^{2/p}}\tilde{g}\left(\frac{z-i}{z+i}\right), \quad z \in \mathbb{C}^+.$$

Transforming back to the unit disc we get that

$$g(z) = f\left(i\frac{1+z}{1-z}\right) = (2i)^{-2/p}(1-z)^{2/p}\tilde{g}(z), \quad z \in D$$

Hence every function $f \in \mathcal{H}^p(\mathbb{C}^+)$ corresponds to a function $g(z) \in H^p(D)$ such that

$$\tilde{g}(z) = (2i)^{2/p} \frac{g(z)}{(1-z)^{2/p}} \in H^p(D).$$

Conversely, if $g \in H^p(D)$ is such that $\tilde{g}(z) = \frac{g(z)}{(1-z)^{2/p}} \in H^p(D)$ and we define a function $\tilde{f}(z)$, $z \in \mathbb{C}^+$, by

$$\tilde{f}(z) = (\tilde{g} \circ \psi)(z) = (2i)^{-2/p}(z+i)^{2/p} f(z), \quad f(z) = g\left(\frac{z-i}{z+i}\right)$$

then we get that

$$\infty > \frac{1}{2\pi} \int_0^{2\pi} |\tilde{g}(e^{i\theta})|^p \, d\theta = \frac{1}{\pi} \int_{-\infty}^{\infty} \frac{|\tilde{f}(t)|^p}{1+t^2} dt = \frac{1}{\pi} \int_{-\infty}^{\infty} \frac{|(2i)^{-2/p}(t+i)^{2/p} f(t)|^p}{1+t^2} dt$$
$$= \frac{1}{4\pi} \int_{-\infty}^{\infty} |f(t)|^p \, dt$$

so that $f \in \mathcal{H}^p(\mathbb{C}^+)$. Thus the Hardy space $\mathcal{H}^p(\mathbb{C}^+)$ corresponds to the subspace of $H^p(D)$ consisting of functions $g \in H^p(D)$ such that $g(z)/(1-z)^{2/p}$ is also in $H^p(D)$. Furthermore, $f \in \mathcal{H}^p(\mathbb{C}^+)$ if and only if $\tilde{f}(z) = (z+i)^{2/p} f(z) \in H^p(\mathbb{C}^+)$. We conclude this section by noting that the mapping

$$g(z) \to f(z) = \pi^{-1/2}(z+i)^{-1} g\left(\frac{z-i}{z+i}\right)$$

is an isometric isomorphism of $H^2(D)$ onto $\mathcal{H}^2(\mathbb{C}^+)$.

A.1.4 Hardy Spaces of Vector and Operator Valued Functions

Let $\mathcal{H}$ be a separable Hilbert space and let $\mathcal{B}(\mathcal{H})$ be the space of bounded linear operators on $\mathcal{H}$. In the following X will denote either $\mathcal{H}$ or $\mathcal{B}(\mathcal{H})$. If D is the unit disc we define:

Definition A.20 (*X valued Hardy spaces on D*) The Hardy space $H_X^p(D)$ is defined to be the space of X valued functions holomorphic on D and such that

$$\sup_{0<r<1} \frac{1}{2\pi} \int_0^{2\pi} \|f(re^{i\theta})\|_X^p d\theta < \infty.$$

□

For each $0 < p < \infty$ and $f \in H_X^p(D)$ we define the p norm of f to be

$$\|f\|_{H_X^p(D)} = \sup_{0<r<1} \left(\frac{1}{2\pi} \int_0^{2\pi} \|f(re^{i\theta})\|_X^p d\theta \right)^{1/p},$$

and it can be shown that, as in the scalar case,

$$\|f\|_{H_X^p(D)} = \lim_{r\to 1^-} \left(\frac{1}{2\pi} \int_0^{2\pi} \|f(re^{i\theta})\|_X^p d\theta \right)^{1/p}$$

For $f \in H_X^\infty(\mathbb{C}^+)$ we define

$$\|f\|_{H_X^\infty(D)} = \sup_{z\in D} \|f(z)\|_X.$$

The next result concerns the existence of boundary values on ∂D for functions in $H_X^p(D)$, with $X = \mathcal{H}$ or $X = \mathcal{B}(\mathcal{H})$:

Theorem A.26 *We have the following limits:*

(i) For $0 < p \le \infty$ let $f \in H_{\mathcal{H}}^p(D)$. Then f has a nontangential limit

$$f(e^{i\theta}) = \lim_{z\to e^{i\theta}} f(z),$$

a.e. on ∂D in the strong topology $\mathcal{H}$.

(ii) For $0 < p \le \infty$ let $f \in H_{\mathcal{B}(\mathcal{H})}^p(D)$. Then f has a nontangential limit

$$f(e^{i\theta}) = \lim_{z\to e^{i\theta}} f(z),$$

a.e. on ∂D in the strong operator topology of $\mathcal{B}(\mathcal{H})$.

(iii) Let $X = \mathcal{H}$ or $X = \mathcal{B}(\mathcal{H})$ and, for $0 < p \le \infty$ let $f \in H_{\mathcal{H}}^p(D)$. Then the non-tangential limit

$$\lim_{z\to e^{i\theta}} \|f(z)\|_X = \|f(e^{i\theta})\|_X$$

exists a.e on ∂D. □

The last theorem ensures the existence of boundary value functions defined in some appropriate sense. Given these boundary value functions, we have

Theorem A.27 *Let $X = \mathcal{H}$ or $X = \mathcal{B}(\mathcal{H})$ then:*

(i) For $0 < ep < \infty$ and $f \in H_X^p(D)$ we have

$$\|f\|_{H^p_X(D)} = \lim_{r\to 1^-} \frac{1}{2\pi}\int_0^{2\pi} \|f(re^{i\theta})\|_X^p\, d\theta = \frac{1}{2\pi}\int_0^{2\pi} \|f(e^{i\theta})\|_X^p\, d\theta.$$

(ii) For $f \in H^\infty_X(D)$ we have

$$\|f\|_{H^\infty_X(D)} = \lim_{r\to 1^-} \left(\max_{|z|=r} \|f(z)\|_X\right) = ess\, sup_{|z|=1} \|f(e^{i\theta})\|_X$$

□

We turn now to consider Hardy spaces on the upper Half-plane. As in the scalar case, in this case we define two classes of hardy spaces:

Definition A.21 (*X valued Hardy spaces on $\mathbb{C}^+$—First kind)* For $0 < p < \infty$, the Hardy space of X-valued functions on $\mathbb{C}^+$, of the first kind, is denoted by $\mathcal{H}^p_X(\mathbb{C}^+)$ and is defined to be the set of all holomorphic X-valued functions f on $\mathbb{C}^+$ such that

$$\sup_{y>0}\left(\int_{-\infty}^{\infty} \|f(x+iy)\|_X^p dx\right)^{1/p} < \infty$$

□

In addition we define the Hardy space $\mathcal{H}^\infty_X(\mathbb{C}^+)$ to be the space of all X valued functions holomorphic on $\mathbb{C}^+$ and such that $\sup_{\operatorname{Im} z>0} \|f(z)\|_X < \infty$. For $0 < p < \infty$ the norm of $f \in \mathcal{H}^p_X(\mathbb{C}^+)$ is defined to be

$$\|f\|_{\mathcal{H}^p_X(\mathbb{C}^+)} = \sup_{y>0}\left(\int_{-\infty}^{\infty} \|f(x+iy)\|_X^p dx\right)^{1/p}.$$

For $p = \infty$ we define

$$\|f\|_{\mathcal{H}^\infty_X(\mathbb{C}^+)} = \sup_{\operatorname{Im} z>0} \|f(z)\|_X.$$

Definition A.22 (*X valued Hardy spaces on $\mathbb{C}^+$—Second kind*) Let $\varphi(z) = i\frac{1+z}{1-z}$ be the conformal mapping of the unit disc onto the upper half-plane $\mathbb{C}^+$. For $0 < p \le \infty$, the Hardy space of X-valued functions on $\mathbb{C}^+$, of the second kind, is denoted by $H^p_X(\mathbb{C}^+)$ and is defined to be the set of all holomorphic X-valued functions f on $\mathbb{C}^+$ such that $f \circ \varphi \in H^p_X(D)$. □

The p-norm of a function $f \in H^p_X(\mathbb{C}^+)$, $0 < p < \infty$, is defined by

$$\|f\|_{H^p_X(\mathbb{C}^+)} = \|f\circ\varphi\|_{H^p_X(D)} = \lim_{r\to 1^-}\left(\frac{1}{2\pi}\int_0^{2\pi} \|f(\varphi(re^{i\theta}))\|_X^p\, d\theta\right)^{1/p}.$$

The functions in $H_X^p(\mathbb{C}^+)$ are characterized by the following theorem:

Theorem A.28 *Let $0 < p < \infty$. If f is a holomorphic X-valued function on $\mathbb{C}^+$, then the following statements are equivalent:*

(i) $f \in H_X^p(\mathbb{C}^+)$.
(ii) $\|f\|_X^p$ *has a harmonic majorant on* $\mathbb{C}^+$.
(iii) $\sup_{y>0} \int_{-\infty}^{\infty} \frac{\|f(x+iy)\|_X^p}{x^2+(y+1)^2} dx < \infty$.

□

In a manner similar to the scalar case, for $X = \mathcal{H}$ and $X = \mathcal{B}(\mathcal{H})$ we have $\mathcal{H}_X^p(\mathbb{C}^+) \subseteq H_X^p(\mathbb{C}^+)$. For vector and operator valued Hardy space functions we have important Poisson and Cauchy type formula relating members of the function space to their boundary value functions:

Theorem A.29 *Let $X = \mathcal{H}$ or $X = \mathcal{B}(\mathcal{H})$. Then for $1 \le p \le \infty$ and $g \in H_X^p(\mathbb{C}^+)$ we have*

$$g(z) = \frac{1}{\pi} \int_{-\infty}^{\infty} \frac{y}{(t-x)^2 + y^2} g(t) dt, \quad y > 0 \tag{A.9}$$

□

Theorem A.30 *Let $X = \mathcal{H}$ or $X = \mathcal{B}(\mathcal{H})$ and $1 \le p < \infty$. Then*

(i) If $g \in \mathcal{H}_X^p(\mathbb{C}^+)$ then the boundary function $g(x)$ satisfies $g \in L_X^p(\mathbb{R})$ and

$$\frac{1}{2\pi i} \int_{-\infty}^{\infty} \frac{1}{t-z} g(t) dt = \begin{cases} g(z), & Im\, z > 0 \\ 0, & Im\, z < 0 \end{cases} \tag{A.10}$$

(ii) Let $g \in L_X^p(\mathbb{R})$ be a function satisfying $\frac{1}{2\pi i} \int_{-\infty}^{\infty} \frac{1}{t-z} g(t) dt = 0$ for $Im\, z < 0$. Then the integral

$$g(z) = \frac{1}{2\pi i} \int_{-\infty}^{\infty} \frac{1}{t-z} g(t) dt, \quad Im\, z > 0 \tag{A.11}$$

defines a function $g \in \mathcal{H}_X^p(\mathbb{C}^+)$ and the boundary value of this function is g. □

We note also that for $g \in L_X^p(\mathbb{R})$ the functions defined by Eqs. (A.11) and (A.9) are the same.

Theorem A.31 *Let $X = \mathcal{H}$ or $X = \mathcal{B}(\mathcal{H})$ and let $1 \le p \le \infty$. Then the Hardy space $\mathcal{H}_X^p(\mathbb{C}^+)$ and the space of boundary value functions $\mathcal{H}_X^p(\mathbb{R})$ are Banach spaces and the mapping of $g(z) \in \mathcal{H}_X^p(\mathbb{C}^+)$ to its boundary value function $g(x) \in \mathcal{H}_X^p(\mathbb{R})$ is an isometry from $\mathcal{H}_X^p(\mathbb{C}^+)$ onto $\mathcal{H}_X^p(\mathbb{R})$ (the norm of $\mathcal{H}_X^p(\mathbb{R})$ is that of $L_X^p(\mathbb{R})$).* □

A result similar to the last theorem holds for Hardy spaces of the second kind $H^p_X(\mathbb{C}^+)$ and the space boundary value functions $H^p_X(\mathbb{R})$, provided that for $1 \leq p < \infty$ the norm in $H^p_X(\mathbb{R})$ is defined by

$$g \in H^p_X(\mathbb{R}), \quad \|g\|_{H^p_X(\mathbb{R})} := \left(\frac{1}{\pi} \int_{-\infty}^{\infty} \frac{\|g(t)\|^p_X}{1+t^2} dt \right)^{1/p}$$

and for $p = \infty$ the norm is defined by

$$g \in H^\infty_X(\mathbb{R}), \quad \|g\|_{H^\infty_X(\mathbb{R})} = \text{ess sup}_{x \in \mathbb{R}} \|g(x)\|_X.$$

Amongst the Hardy spaces presented above the case most interesting for us is when $p = 2$ for which $\mathcal{H}^2_{\mathcal{H}}(\mathbb{C}^+)$ and $\mathcal{H}^2_{\mathcal{H}}(\mathbb{R})$ are Hilbert spaces. We have

Theorem A.32 *(i) For any $f \in \mathcal{H}^2_{\mathcal{H}}(\mathbb{C}^+)$ the integral*

$$\int_{-\infty}^{\infty} \|f(x+iy)\|^2_{\mathcal{H}} \, dx$$

is a non-increasing function of y ($y > 0$).
(ii) For any $f, g \in \mathcal{H}^2_{\mathcal{H}}(\mathbb{C}^+)$,

$$\langle f, g \rangle_{\mathcal{H}^2_{\mathcal{H}}(\mathbb{C}^+)} = \lim_{y \to 0^+} \int_{-\infty}^{\infty} \langle f(x+iy), g(x+iy) \rangle_{\mathcal{H}} dx = \int_{-\infty}^{\infty} \langle f(x, g(x) \rangle_{\mathcal{H}} dx = \langle f, g \rangle_{\mathcal{H}^2_{\mathcal{H}}(\mathbb{R})}$$

where $f(x), g(x) \in \mathcal{H}^2_{\mathcal{H}}(\mathbb{R})$ are the boundary value functions.
(iii) If $f(x) \in \mathcal{H}^2_{\mathcal{H}}(\mathbb{R})$ is the boundary value function of $f(z) \in \mathcal{H}^2_{\mathcal{H}}(\mathbb{C}^+)$ then

$$\lim_{y \to 0^+} \int_{-\infty}^{\infty} \|f(x+iy) - f(x)\|^2_{\mathcal{H}} \, dx = 0$$

We may also apply Fourier transforms and obtain the very important Paley–Wiener theorem, so useful in the Lax–Phillips formalism and other frameworks developed in this book. We have

Theorem A.33 (Plancherel theorem) *The Fourier transform operator is an isometry of $L^2_{\mathcal{H}}(\mathbb{R})$ onto itself and for $f \in L^p_{\mathcal{H}}(\mathbb{R})$ we have*

$$\hat{f}(t) = \lim_{L \to \infty} \frac{1}{\sqrt{2\pi}} \int_{-L}^{L} f(t) e^{-i\omega t} dt,$$

$$f(t) = \lim_{L \to \infty} \frac{1}{\sqrt{2\pi}} \int_{-L}^{L} \hat{f}(\omega) e^{i\omega t} d\omega,$$

where convergence is in the norm of $L^2_{\mathcal{H}}(\mathbb{R})$. □

and

Theorem A.34 (Paley–Wiener) *For a function* $f \in L^2_{\mathcal{H}}(\mathbb{R}^+)$ *set*

$$F(z) = \frac{1}{\sqrt{2\pi}} \int_{0}^{\infty} f(t) e^{izt} dt, \quad Im\, z > 0.$$

then the mapping $f \to F$ *is an isometric isomorphism from* $L^2_{\mathcal{H}}(\mathbb{R}^+)$ *onto* $\mathcal{H}^p_{\mathcal{H}}(\mathbb{C}^+)$ *and we have*

$$\lim_{L \to \infty} \frac{1}{\sqrt{2\pi}} \int_{-L}^{L} F(t+iy) e^{-i(t+iy)x} dt = \begin{cases} f(x), & x > 0 \\ 0, & x < 0 \end{cases}$$

where convergence is in the norm of $L^2_{\mathcal{H}}(\mathbb{R})$.

Appendix B
Contraction Operators and Their Dilations

A common thread throughout the chapters of the present book is the premise that continuous, contractive semigroups are the basic mathematical constructions for the description of irreversible evolution. Given this assumption, the study of contraction operators and continuous contractive semigroups, their properties, spectral analysis and canonical representations, becomes the cornerstone in the development of both general mathematical formalisms, as well as specific models, for the description of the evolution of unstable physical systems undergoing irreversible evolution. In quantum mechanics, the study of self-adjoint operators is absolutely essential for the description of the evolution of systems undergoing unitary, reversible, evolution and the spectral analysis, centered on the spectral theorem for normal operators, of such operators is a highly developed theory. We are very fortunate to have a well developed mathematical theory, including properties, spectral analysis, model operators and more, of general contraction operators and contractive semigroups. This theory is discussed in length by Sz.-Nagy and Foias in their book on the subject (Sz.-Nagy and Foias 1970) (see references therein for the many contributors to the development of the theory of contractions). For the convenience of the reader we describe in this appendix the parts of the Sz.-Nagy–Foias theory which are essential to understanding of the structures developed in the present book. This, of course, leaves out many subjects discussed by Sz.-Nagy and Foias in their book and we urge the reader to explore it in order to grasp the strength of the formalism presented there. The presentation of the material in this appendix follows closely the books of Sz.-Nagy and Foias (1970) and Rosenblum and Rovnyak (1985).

B.1 Isometries, Unilateral Shifts and the Wold Decomposition

Definition B.23 (*Contraction Operator*) Let $\mathcal{H}$ be a Hilbert space. An operator $T \in \mathcal{B}(\mathcal{H})$ is called a contraction if

L. Horwitz and Y. Strauss, *Unstable Systems*, Mathematical Physics Studies,
https://doi.org/10.1007/978-3-030-31570-2

$$\forall \mathbf{x} \in \mathcal{H}, \quad \|T\mathbf{x}\| \leq \|\mathbf{x}\|$$

i.e., if $\|T\| \leq 1$. □

An important class of contraction operators consists of partial isometries,

Definition B.24 (*Partial Isometry*) Let $\mathcal{H}$ be a Hilbert space. An operator $T \in \mathcal{B}(\mathcal{H})$ is called a partial isometry if

$$\forall \mathbf{x}, \mathbf{y} \in (Ker(T))^{\perp}, \quad \langle T\mathbf{x}, T\mathbf{y}\rangle = \langle \mathbf{x}, \mathbf{y}\rangle$$

□

If T is a partial isometry on $\mathcal{H}$ then $\mathcal{W} = (Ker(T))^{\perp}$ is called the *initial space* for T. The subspace $\mathcal{U} = T\mathcal{W}$ is called the *final space* for T.

Proposition B.5 *An operator $T \in \mathcal{B}(\mathcal{H})$ is a partial isometry if and only if T^*T is a projection operator and in this case T^*T is the orthogonal projection on the initial space of T.* □

Proof Let T be a partial isometry on $\mathcal{H}$ and let $\mathcal{W} = (Ker(T))^{\perp}$ be the initial space for T. Then

$$\forall \mathbf{x} \in \mathcal{W}, \ \forall \mathbf{y} \in \mathcal{H}, \quad \langle T^*T\mathbf{x}, \mathbf{y}\rangle = \langle T\mathbf{x}, T\mathbf{y}\rangle = \langle T\mathbf{x}, TP_{\mathcal{W}}\mathbf{y}\rangle = \langle \mathbf{x}, P_{\mathcal{W}}\mathbf{y}\rangle = \langle \mathbf{x}, \mathbf{y}\rangle$$

Since $\mathbf{y} \in \mathcal{H}$ is arbitrary we conclude that for all $\mathbf{x} \in \mathcal{W}$ we have $T^*T\mathbf{x} = \mathbf{x}$. Furthermore, for every $\mathbf{x} \in \mathcal{W}^{\perp}$ we have $T\mathbf{x} = 0$. Hence T^*T is an orthogonal projection on $\mathcal{W}$. Conversely, assume that T^*T is an orthogonal projection on a subspace $\mathcal{W}$ in $\mathcal{H}$. Then,

$$\forall \mathbf{x} \in \mathcal{W}^{\perp}, \quad 0 = \langle \mathbf{x}, T^*T\mathbf{x}\rangle = \|T\mathbf{x}\|^2 \Rightarrow T\mathbf{x} = \mathbf{0} \Rightarrow \mathcal{W}^{\perp} \subseteq Ker(T).$$

Obviously, we also have $Ker(T) \subseteq \mathcal{W}^{\perp}$ and hence $\mathcal{W}^{\perp} = Ker(T))$. Finally, since T^*T is an orthogonal projection on $\mathcal{W}$, then for all $\mathbf{x}, \mathbf{y} \in \mathcal{W}$ we have $\langle T\mathbf{x}, T\mathbf{y}\rangle = \langle \mathbf{x}, T^*T\mathbf{y}\rangle = \langle \mathbf{x}, \mathbf{y}\rangle$. Thus T is a partial isometry. ■

Corollary B.4 *If T is a partial isometry with initial space $\mathcal{W}$ and final space $\mathcal{U}$ then T^* is a partial isometry with initial space $\mathcal{U}$ and final space $\mathcal{W}$ and TT^* is an orthogonal projection on $\mathcal{U}$.* □

Proof Assume that T is a partial isometry with initial space $\mathcal{W}$ and final space $\mathcal{U}$. Then, by the fact that $T^*(T\mathbf{x}) = \mathbf{x}$ we have $T^*\mathcal{U} = \mathcal{W}$. Letting $\mathbf{y} \in \mathcal{U}^{\perp}$, for every $\mathbf{x} \in \mathcal{H}$ we have $0 = \langle \mathbf{y}, T\mathbf{x}\rangle = \langle T^*\mathbf{y}, \mathbf{x}\rangle$ so that $T^*\mathbf{y} = \mathbf{0}$ and we conclude that $\mathcal{U}^{\perp} \subseteq Ker(T^*)$. Moreover, for $\mathbf{y} \in Ker(T^*)$ we again have $0 = \langle T^*\mathbf{y}, \mathbf{x}\rangle = \langle \mathbf{y}, T\mathbf{x}\rangle$ which implies that $Ker(T^*) \subseteq \mathcal{U}^{\perp}$ and hence $\mathcal{U}^{\perp} = Ker(T^*)$. Finally, for $\mathbf{x}, \mathbf{y} \in \mathcal{U}$ we have

$$\langle T^*\mathbf{x}, T^*\mathbf{y}\rangle = \langle T^*(T\tilde{\mathbf{x}}), T^*(T\tilde{\mathbf{y}})\rangle = \langle \tilde{\mathbf{x}}, \tilde{\mathbf{y}}\rangle = \langle T\tilde{\mathbf{x}}, T\tilde{\mathbf{y}}\rangle = \langle \mathbf{x}, \mathbf{y}\rangle$$

where $\mathbf{x} = T\tilde{\mathbf{x}}, \mathbf{y} = T\tilde{\mathbf{y}}$. Thus T^* is a partial isometry and, by Proposition B.5 above, TT^* is an orthogonal projection on its initial space, i.e., on $\mathcal{U}$. ■

An *isometry* T on $\mathcal{H}$ is a partial isometry satisfying $Ker(T) = \{\mathbf{0}\}$. Note that in this case the initial space of T is equal to $\mathcal{H}$ and we have $T^*T = I$. A *unitary* operator U on $\mathcal{H}$ is an isometry on $\mathcal{H}$ whose final space is $\mathcal{H}$. Observe that in this case we have

$$U^*U = I, \quad UU^* = I$$

Indeed, this is a defining relation of a unitary operator on $\mathcal{H}$. Thus a unitary operator on $\mathcal{H}$ is a partial isometry whose initial and final spaces are both equal to $\mathcal{H}$.

Proposition B.6 *Let T be an isometry on $\mathcal{H}$ and let $\mathcal{W} \subseteq \mathcal{H}$ be a subspace such that $T\mathcal{W} = \mathcal{W}$. Then $\mathcal{W}$ reduces T and $T_{\mathcal{W}} := T|_{\mathcal{W}}$ (the restriction of T to $\mathcal{W}$), is unitary on $\mathcal{W}$* □

Proof Assume that T is an isometry on $\mathcal{H}$ and that $T\mathcal{W} = \mathcal{W}$. Then for an arbitrary $\mathbf{y} \in \mathcal{W}$ there exists $\mathbf{x} \in \mathcal{W}$ such that $\mathbf{y} = T\mathbf{x}$ and we have

$$T^*\mathbf{y} = T^*T\mathbf{x} = \mathbf{x} \in \mathcal{W}$$

hence $T^*\mathcal{W} \subseteq \mathcal{W}$. Thus $\mathcal{W}$ is invariant for both T and T^* and is therefore a reducing subspace for T. Note that

$$\forall \mathbf{x}, \mathbf{y} \in \mathcal{W}, \quad \langle \mathbf{x}, T_{\mathcal{W}}\mathbf{y} \rangle = \langle \mathbf{x}, T\mathbf{y} \rangle = \langle T^*\mathbf{x}, \mathbf{y} \rangle = \langle T_{\mathcal{W}}^*\mathbf{x}, \mathbf{y} \rangle \Rightarrow T_{\mathcal{W}}^* = T^*|_{\mathcal{W}}$$

and hence

$$\forall \mathbf{x} \in \mathcal{W}, \quad T_{\mathcal{W}}^* T_{\mathcal{W}}\mathbf{x} = T^*T\mathbf{x} = \mathbf{x} \Rightarrow T_{\mathcal{W}}^* T_{\mathcal{W}} = I_{\mathcal{W}}$$

Let $\mathbf{y} \in \mathcal{W}$ be an arbitrary element. Since $T\mathcal{W} = \mathcal{W}$ there exists $\mathbf{x} \in \mathcal{W}$ such that $\mathbf{y} = T\mathbf{x}$ and we have

$$T_{\mathcal{W}} T_{\mathcal{W}}^*\mathbf{y} = TT^*\mathbf{y} = (TT^*)T\mathbf{x} = T(T^*T)\mathbf{x} = T\mathbf{x} = \mathbf{y} \Rightarrow T_{\mathcal{W}} T_{\mathcal{W}}^* = I_{\mathcal{W}}$$

Hence $T_{\mathcal{W}}$ is unitary on $\mathcal{W}$. ■

At this point we introduce two basic notions of the theory of isometric operators on Hilbert spaces, i.e., unilateral shift operators and wandering subspaces:

Definition B.25 (*Unilateral shift*) An operator $S \in \mathcal{B}(\mathcal{H})$ is called a unilateral shift operator if S is an isometry on $\mathcal{H}$ and if $s - \lim_{n\to\infty}(S^*) = 0$, i.e., for any $\mathbf{x} \in \mathcal{H}$ we have $\lim_{n\to\infty} \|(S^*)^n\mathbf{x}\| = 0$. □

Definition B.26 (*Wandering subspace*) Let T be an isometry on a Hilbert space $\mathcal{H}$. A subspace $\mathcal{L}$ is called a *wandering subspace* for T if for any integers $n_1, n_2 \geq 0$, $n_1 \neq n_2$ we have $T^{n_2}\mathcal{L} \perp T^{n_1}\mathcal{L}$. □

Note that if T is an isometry and if we assume $T^n\mathcal{L} \perp \mathcal{L}$, $\forall n > 0$ then for $n_2 > n_1$ we have

$$T^{(n_2-n_1)}\mathcal{L} \perp \mathcal{L}, \Rightarrow T^{n_1}T^{(n_2-n_1)}\mathcal{L} \perp T^{n_1}\mathcal{L} \Rightarrow T^{n_2}\mathcal{L} \perp T^{n_1}\mathcal{L}$$

Thus, to show that $\mathcal{L}$ is wandering for T it is enough to show that $T^n\mathcal{L} \perp \mathcal{L}$, $\forall n > 0$.

Assume that $\mathcal{L}$ is a wandering subspace for an isometry T. Since $T^{n_2}\mathcal{L} \perp T^{n_1}\mathcal{L}$, for all integers $\forall n_1, n_2 \geq 0$, $n_1 \neq n_2$, we may form the orthogonal direct sum

$$M_+(\mathcal{L}) = \bigoplus_{n=0}^{\infty} T^n\mathcal{L}.$$

Observe that

$$TM_+(\mathcal{L}) = T\left(\bigoplus_{n=0}^{\infty} T^n\mathcal{L}\right) = \bigoplus_{n=1}^{\infty} T^n\mathcal{L}$$

and so

$$M_+(\mathcal{L}) = \mathcal{L} \oplus TM_+(\mathcal{L}).$$

The central result regarding isometric operators on Hilbert spaces is the Wold decomposition:

Theorem B.35 (Wold decomposition) *Let T be an isometry on a Hilbert space $\mathcal{H}$. Then:*

(i) $\mathcal{H}$ decomposes into an orthogonal sum $\mathcal{H} = \mathcal{H}_0 \oplus \mathcal{H}_1$ such that both $\mathcal{H}_0$ and $\mathcal{H}_1$ reduce T. The restriction $U := T|_{\mathcal{H}_0}$ of T to $\mathcal{H}_0$ is a unitary operator on $\mathcal{H}_0$ and the restriction $S := T|_{\mathcal{H}_1}$ of T to $\mathcal{H}_1$ is a unilateral shift on $\mathcal{H}_1$.

(ii) Denote $\mathcal{L} = \mathcal{H} \ominus T\mathcal{H}$. Then $\mathcal{L}$ is a wandering subspace for T such that

$$\mathcal{H}_1 = \bigoplus_{n=0}^{\infty} T^n\mathcal{L} = \{\mathbf{x} \in \mathcal{H} : \lim_{n\to\infty} \|(T^*)^n\mathbf{x}\| = 0\}, \qquad \mathcal{H}_0 = \cap_{n=0}^{\infty} T^n\mathcal{H}$$

The subspaces $\mathcal{H}_0$ or $\mathcal{H}_1$ may be equal to $\{\mathbf{0}\}$. □

Proof We prove an operator version of the Wold decomposition equivalent to Theorem B.35. Let T be an isometry on $\mathcal{H}$. Then:

(i) The operator $P_{\mathcal{L}} = I - TT^*$ is an orthogonal projection operator on $\mathcal{L} = \mathcal{H} \ominus T\mathcal{H}$.

(ii) We have $s - \lim_{n\to\infty} T^n(T^*)^n = P_0$ where P_0 is an orthogonal projection operator on $\mathcal{H}$.

(iii) We have $P_0\mathcal{H} = \cap_{n=0}^{\infty} T^n\mathcal{H}$.

(iv) The operator sum $\sum_{n=0}^{\infty} T^n P_{\mathcal{L}}(T^*)^n$ converges strongly to an orthogonal projection operator P_1 on $\mathcal{H}$.

(v) We have $P_1\mathcal{H} = \{\mathbf{x} \in \mathcal{H} : \lim_{n\to\infty} \|(T^*)^n\mathbf{x}\| = 0.$
(vi) Both $P_0\mathcal{H}$ and $P_1\mathcal{H}$ are reducing subspaces for T.
(vii) The restriction $T|P_0\mathcal{H}$ is a unitary operator on $\mathcal{H}_0 = P_0\mathcal{H}$.
(viii) The restriction $S = T|P_1\mathcal{H}$ is a unilateral shift operator.
(ix) We have $I = P_0 + \sum_{n=0}^{\infty} T^n P_{\mathcal{L}}(T^*)^n$ and $\mathcal{H} = P_0\mathcal{H} \oplus \left(\bigoplus_{n=0}^{\infty} T^n P_{\mathcal{L}}(T^*)^n\right) = \mathcal{H}_0 \oplus \mathcal{H}_1$ where $\mathcal{H}_0 = P_0\mathcal{H}$ and $\mathcal{H}_1 = \bigoplus_{n=0}^{\infty} T^n P_{\mathcal{L}}(T^*)^n\mathcal{H}$.

To show (i) we note that, by Corollary B.4, TT^* is the projection on the final space of T i.e. on $T\mathcal{H}$. To show (ii) and (iii) we observe that

$$T^{n+1}(T^*)^{n+1} = T^n(TT^*)(T^*)^n \leq T^n(T^*)^n$$

since TT^* is a projection. To show the validity of (iv) note that

$$s - \lim_{m\to\infty}[\sum_{n=0}^{m} T^n P_{\mathcal{L}}(T^*)^n] = s - \lim_{m\to\infty}[\sum_{n=0}^{m} T^n(I - TT^*)(T^*)^n] =$$

$$= s - \lim_{m\to\infty}[\sum_{n=0}^{m}(T^n(T^*)^n - T^{n+1}(T^*)^{n+1})] = s - \lim_{m\to\infty}[I - T^{m+1}(T^*)^{m+1})] = I - P_0 = P_1 .$$

To see (v) we observe that

$$\forall \mathbf{x} \in \mathcal{H}, \quad \lim_{n\to\infty} \|(T^*)^n\mathbf{x}\|^2 = \lim_{n\to\infty}\langle T^n(T^*)^n\mathbf{x}, \mathbf{x}\rangle = \langle P_0\mathbf{x}, \mathbf{x}\rangle = \|P_0\mathbf{x}\|^2 .$$

and hence $\mathbf{x} \in P_1\mathcal{H}$ we have $\lim_{n\to\infty} \|(T^*)^n\mathbf{x}\|^2 = 0$ if and only if $\mathbf{x} \in P_1\mathcal{H}$. To show (vi) we note that for $n \geq 0$ we have $T[T^n(T^*)^n] = T^{n+1}(T^*)^n T^* T = T^{n+1}(T^*)^{n+1}T$, take the strong limit as n goes to infinity, and get that $TP_0 = P_0T$. In a similar manner we get that $TP_0T^* = P_0$ so that $(P_0TP_0)(P_0TP_0)^* = (P_0TP_0)(P_0T^*P_0) = (P_0TP_0T^*P_0) = P_0$ and hence $T|P_0\mathcal{H}$ is a unitary operator. To see that validity of (viii) first set $S = T|P_1\mathcal{H}$. Then, since $P_1\mathcal{H}$ reduces T we get that $S^* = T^*|P_1\mathcal{H}$ and hence, by (v), we get that $s - \lim_{n\to\infty}(S^*)^n P_1\mathcal{H} = 0$ and hence S is a unilateral shift on $P_1\mathcal{H}$. The validity of the first relation in (ix) is immediate from (iv). Furthermore, since

$$T^n P_{\mathcal{L}}(T^*)^n = T^n(I - TT^*)(T^*)^n = T^n(T^*)^n - T^{n+1}(T^*)^{n+1}$$

then $T^n P_{\mathcal{L}}(T^*)^n$ is the orthogonal projection on the subspace $T^n\mathcal{H} \ominus T^{n+1}\mathcal{H} = T^n(\mathcal{H} \ominus T\mathcal{H}) = T^n\mathcal{L}$. The subspaces $T^n P_{\mathcal{L}}\mathcal{H}$ are then orthogonal to each other and, by (iv), also to $P_1\mathcal{H}$. With this observation, application of the first relation in (ix) completes the proof. ■

Corollary B.5 *An isometry $T \in \mathcal{B}(\mathcal{H})$ is a shift operator if and only if $\cap_{n=0}^{\infty} T^n\mathcal{H} = \{0\}$.* □

Corollary B.6 *Let $S \in \mathcal{B}(\mathcal{H})$ be a shift operator and set $\mathcal{L} = Ker(S^*)$. Then*

$$\mathcal{L} = \mathcal{H} \ominus S\mathcal{H}, \qquad \mathcal{H} = \bigoplus_{n=0}^{\infty} S^n \mathcal{L},$$

and for each $\mathbf{x} \in \mathcal{H}$ *there exists a unique sequence of vectors* $\{\mathbf{x}_n\}_{n=0^\infty}$, $\mathbf{x}_n \in \mathcal{L}$, *such that*

$$\mathbf{x} = \sum_{n=0}^{\infty} S^n \mathbf{x}_n, \quad \|\mathbf{x}\|^2 = \sum_{n=0}^{\infty} \|\mathbf{x}_n\|^2, \quad \mathbf{x}_n = P_0(S^*)^n \mathbf{x}, \tag{B.1}$$

where $P_0 = I - SS^*$ *is the projection in* $\mathcal{H}$ *on* $\mathcal{L}$. □

Corollaries B.5 and B.6 are immediate consequences of Theorem B.35. In particular, note that (i) implies that $\mathcal{L} = Ker(S^*) = \mathcal{H} \ominus S\mathcal{H}$ and that (ix) implies that

$$S^*\mathbf{x} = S^* \sum_{n=0}^{\infty} S^n \mathbf{x}_n = S^* \sum_{n=0}^{\infty} S^n P_0 (S^*)^n \mathbf{x} =$$
$$= \sum_{n=1}^{\infty} S^{(n-1)} P_0 (S^*)^n \mathbf{x} = \sum_{n=0}^{\infty} S^n P_0 (S^*)^{(n+1)} \mathbf{x} = \sum_{n=0}^{\infty} S^n \mathbf{x}_{n+1} \,. \tag{B.2}$$

Furthermore, we have

$$S\mathbf{x} = S \sum_{n=0}^{\infty} S^n \mathbf{x}_n = \sum_{n=0}^{\infty} S^{(n+1)} \mathbf{x}_n = \sum_{n=1}^{\infty} S^n \mathbf{x}_{n-1} \,.$$

Given a Hilbert space $\mathcal{H}$, a subspace $\mathcal{L}$ is called *cyclic* for an operator $A \in \mathcal{B}(\mathcal{H})$ if $\mathcal{H} = \overline{\bigvee_{n=0}^{\infty} A^n \mathcal{L}}$. Thus, if S is a unilateral shift operator on $\mathcal{H}$ then $\mathcal{L} = Ker(S^*)$ is a cyclic subspace for S. We then say that $\mathcal{L}$ is a *generating subspace* for S and its dimension, $\dim(\mathcal{L})$, is called the *multiplicity* of S.

A unilateral shift is uniquely determined by its multiplicity up to an isometric isomorphism. We have the following result, which we state without proof:

Lemma B.1 *Let S and S' be, respectively, unilateral shifts on the Hilbert spaces $\mathcal{H}$ and $\mathcal{H}'$. Let $\mathcal{L}$ and $\mathcal{L}'$ be the generating spaces, respectively, for S and for S' and assume that the multiplicities of S and S' are the same, i.e.,* $\dim(\mathcal{L}) = \dim(\mathcal{L}')$. *Then there is an isometric isomorphism $U : \mathcal{H} \mapsto \mathcal{H}'$ such that $US = S'U$ (i.e., $S' = USU^*$)* □

B.2 Bilateral Shifts

Let $U \in \mathcal{B}(\mathcal{H})$ be a unitary operator on a Hilbert space $\mathcal{H}$ and let $\mathcal{L}$ be a wandering subspace for U. Since $U^{-1} = U^*$ is also an isometry on $\mathcal{H}$ we get that

$$U^{n_1}\mathcal{L} \perp U^{n_2}\mathcal{L}, \quad \forall n_1, n_2 \in \mathbb{Z},\ n_1 \neq n_2$$

This leads us to define a subspace $M(\mathcal{L})$ as an orthogonal sum

$$M(\mathcal{L}) = \bigoplus_{n=-\infty}^{\infty} U^n \mathcal{L}$$

It is immediate that $M(\mathcal{L})$ is a reducing subspace for U. Recall that in the case of an isometry T which is not a unitary operator the wandering subspace is uniquely given by $\mathcal{L} = \mathcal{H} \ominus T\mathcal{H}$. This situation is to be contrasted with the case of a unitary operator U in which the operator U does not uniquely define the wandering space. As a simple example for this fact note that if $\mathcal{L}$ is wandering for a unitary operator U then $U\mathcal{L}$ is also wandering for U and we have $M(\mathcal{L}) = M(U\mathcal{L})$. A particular type of unitary operator is the bilateral shift:

Definition B.27 (*Bilateral shift*) An Operator $U \in \mathcal{B}(\mathcal{H})$ on a Hilbert space $\mathcal{H}$ is called a *bilateral shift* if U is unitary and if there exists a subspace $\mathcal{L} \subset \mathcal{H}$ such that $\mathcal{L}$ is wandering for U and $\mathcal{H} = M(\mathcal{L})$.

A bilateral shift is determined by its multiplicity up to unitary equivalence. The proof of this fact is analogous to the one given in the case of a unilateral shift. Next, we make the following observation regarding bilateral shifts

Lemma B.2 *A bilateral shift U an a Hilbert space $\mathcal{H}$ has no eigenvalues.* □

Proof Since U is a bilateral shift on $\mathcal{H}$, then every element $\mathbf{x} \in \mathcal{H}$ can be written in the form

$$\mathbf{x} = \sum_{n=-\infty}^{\infty} U^n \mathbf{x}_n,\ \mathbf{x}_n \in \mathcal{L}, \qquad \|\mathbf{x}\|^2 = \sum_{n=-\infty}^{\infty} \|U^n \mathbf{x}_n\|^2 = \sum_{n=-\infty}^{\infty} \|\mathbf{k}_n\|^2$$

Therefore,

$$U\mathbf{x} = \sum_{n=-\infty}^{\infty} U^{n+1}\mathbf{x}_n = \sum_{n=-\infty}^{\infty} U^n \mathbf{x}_{n-1}$$

Now assume that λ is an eigenvalue of U and that $\mathbf{x} \in \mathcal{H}$ is a corresponding eigenvector. In this case we have

$$U\mathbf{x} = \sum_{n=-\infty}^{\infty} U^n \mathbf{x}_{n-1} = \lambda \sum_{n=-\infty}^{\infty} U^n \mathbf{x}_n \Rightarrow \mathbf{x}_{n-1} = \lambda \mathbf{x}_n$$

and since U is unitary and $|\lambda| = 1$ we get that the norm of $\mathbf{x}$ diverges. ■

We have the following important result:

Proposition B.7 *Every unilateral shift $S \in \mathcal{B}(\mathcal{H})$ on a Hilbert space $\mathcal{H}$ can be extended to a bilateral shift U of the same multiplicity on a Hilbert space $\tilde{\mathcal{H}}$ containing $\mathcal{H}$ as a subspace.* □

Proof Let S be a unilateral shift on $\mathcal{H}$. Setting $\mathcal{L} = \mathcal{H} \ominus S\mathcal{H}$, we have that $\mathcal{L}$ is a wandering subspace for S and that $\mathcal{H} = M_+(\mathcal{L}) = \bigoplus_{n=0}^{\infty} S^n\mathcal{L}$. Let $\tilde{\mathcal{H}} = \ell^2(\mathcal{L})$ be a Hilbert space of square summable, $\mathcal{L}$ valued, sequences

$$\tilde{\mathcal{H}} = \ell^2(\mathcal{L}) = \left\{(\mathbf{x}_n)_{n=-\infty}^{\infty},\ \mathbf{x}_n \in \mathcal{L},\ \sum_{n=-\infty}^{\infty} \|\mathbf{x}_n\|^2 < \infty\right\}.$$

equipped with the inner product

$$\langle \mathbf{x}, \mathbf{y}\rangle_{\ell^2(\mathcal{L})} = \sum_{n=-\infty}^{\infty} \langle \mathbf{x}_n, \mathbf{y}_n\rangle, \quad \mathbf{x} = (\mathbf{x}_n)_{n=-\infty}^{\infty},\ \mathbf{y} = (\mathbf{x}_n)_{n=-\infty}^{\infty}.$$

Let $\ell^2_+(\mathcal{L})$ be the subspace of $\ell^2(\mathcal{L})$ defined by

$$\ell^2_+ = \left\{(\mathbf{x}_n)_{n=0}^{\infty},\ \mathbf{x}_n \in \mathcal{L},\ \sum_{n=0}^{\infty} \|\mathbf{x}_n\|^2 < \infty\right\}$$

By Eq. (B.1), if S is a unilateral shift defined on $\mathcal{H}$, we may define an isometric isomorphism $W_1 : \mathcal{H} \mapsto \ell^2_+(\mathcal{L})$ by

$$\mathbf{x} \in \mathcal{H},\ \mathbf{x} = \sum_{n=0}^{\infty} S^n\mathbf{x}_n,\ \mathbf{x}_n \in \mathcal{L} \Rightarrow (W_1\mathbf{x})_n = \mathbf{x}_n,\ n = 0, 1, , 2, \dots \tag{B.3}$$

Indeed, we have

$$\|W_1\mathbf{x}\|^2_{\ell^2_+(\mathcal{L})} = \sum_{n=0}^{\infty} \|\mathbf{x}_n\|^2 = \|\mathbf{x}\|^2$$

so that W_1 is an isometry.

On $\ell^2_+(\mathcal{L})$ define an operator $S_1 : \ell^2_+(\mathcal{L}) \mapsto \ell^2_+(\mathcal{L})$ by

$$\mathbf{x} = (\mathbf{x}_n)_{n=0}^{\infty} \in \ell^2_+(\mathcal{L}), \quad (S_1\mathbf{x})_n = \begin{cases} 0 & n = 0 \\ \mathbf{x}_{n-1} & n = 1, 2, \dots \end{cases} \tag{B.4}$$

Then S_1 is the canonical unilateral shift on $\ell^2_+(\mathcal{L})$ and we have

$$[W_1 S\mathbf{x}]_n = \left[W_1 S \sum_{k=0}^{\infty} S^k \mathbf{x}_k\right]_n = \left[W_1 \sum_{k=0}^{\infty} S^{(k+1)} \mathbf{x}_k\right]_n =$$

$$= \left[W_1 \sum_{k=1}^{\infty} S^k \mathbf{x}_{k-1}\right]_n = \mathbf{x}_{n-1} = [S_1 W_1 \mathbf{x}]_n \Rightarrow W_1 S = S_1 W_1 .$$

Thus, via the mapping by W_1, we obtain a canonical representation of the action of S on $\mathcal{H}$ in terms of the action of the unilateral shift S_1 on $\ell^2_+(\mathcal{L})$. Now, observe that the operator $U_1 : \ell^2(\mathcal{L}) \mapsto \ell^2(\mathcal{L})$, defined by

$$(U_1\mathbf{x})_n = \mathbf{x}_{n-1}, \quad \mathbf{x} = (\mathbf{x}_n)_{n=-\infty}^{\infty} \in \ell^2(\mathcal{L}) = \tilde{\mathcal{H}}$$

is a bilateral shift on $\tilde{\mathcal{H}}$. Moreover, the restriction $U_1|\ell^2_+(\mathcal{L})$ is identical to the unilateral shift S_1 in Eq. (B.4) which is unitarily equivalent to S. Thus the unilateral shift S on $\mathcal{H}$ is extended to the bilateral shift U_1 on $\tilde{\mathcal{H}}$. ■

Proposition B.8 *Every isometry T on a Hilbert space $\mathcal{H}$ can be extended to a unitary operator on a Hilbert space $\tilde{\mathcal{H}}$ containing $\mathcal{H}$ as a subspace.* □

Proof By the wold decomposition $\mathcal{H}$ has an orthogonal decomposition $\mathcal{H} = \mathcal{H}_0 \oplus \mathcal{H}_1$ with $T_0 := T|\mathcal{H}_0$ a unitary operator on $\mathcal{H}_0$ and $T_1 := T|\mathcal{H}_1$ a unilateral shift on $\mathcal{H}_1$. Extend T_1 to a bilateral shift U_1 on a Hilbert space $\tilde{\mathcal{H}}_1$ containing $\mathcal{H}_1$ as a subspace. Then the Hilbert space $\tilde{\mathcal{H}} = \mathcal{H}_0 \oplus \tilde{\mathcal{H}}_1$ contains $\mathcal{H} = \mathcal{H}_0 \oplus \mathcal{H}_1$ as a subspace and the unitary operator $U = T_0 \oplus U_1$ is an extension of T to $\tilde{\mathcal{H}}$ ■

B.3 Canonical Representations of Unilateral and Bilateral Shifts

Let S be a unilateral shift on a Hilbert space $\mathcal{H}$ and let $\mathcal{L} = \mathcal{H} \ominus S\mathcal{H}$. Construct an isometric isomorphism $W_1 : \mathcal{H} \mapsto \ell^2_+(\mathcal{L})$ as in Eq. (B.3) in the proof of Proposition B.7 and recall that $W_1 S = S_1 W_1$, where S_1 is the canonical shift operator on $\ell^2_+(\mathcal{L})$ defined in Eq. (B.4). Next, consider the Hardy space $\mathcal{H}^2_{\mathcal{L}}(D)$ of all $\mathcal{L}$-valued holomorphic functions $f(z) = \sum_{n=0}^{\infty} \mathbf{a}_n z^n$ on the unit disk D. $\mathcal{H}^2_{\mathcal{L}}(D)$ is a Hilbert space with inner product

$$\langle f, g\rangle_{\mathcal{H}^2_{\mathcal{L}}(D)} = \lim_{r\to 1^-} \frac{1}{2\pi} \int_0^{2\pi} \langle f(re^{i\theta}), g(re^{i\theta})\rangle_{\mathcal{L}} d\theta = \sum_{j=0}^{\infty} \langle \mathbf{a}_j, \mathbf{b}_j\rangle_{\mathcal{L}}, \tag{B.5}$$

for any $f(z) = \sum_{n=0}^{\infty} \mathbf{a}_n z^n$ and $g(z) = \sum_{n=0}^{\infty} \mathbf{b}_n z^n$ in $\mathcal{H}^2_{\mathcal{L}}(D)$. From Eq. (B.5) it is clear that $\mathcal{H}^2_{\mathcal{L}}(D)$ is isomorphic to $\ell^2_+(\mathcal{L})$ through a mapping $W_2 : \ell^2_+(\mathcal{L}) \mapsto \mathcal{H}^2_{\mathcal{L}}(D)$ defined by

$$f = \{\mathbf{k}_n\}_{n=0}^{\infty} \in \ell_+^2(\mathcal{L}), \qquad W_2 f = \sum_{n=0}^{\infty} \mathbf{k}_n z^n . \tag{B.6}$$

If S_1 is the canonical unilateral shift on $\ell_+^2(\mathcal{L})$ defined above we have

$$f = \{\mathbf{k}_n\}_{n=0}^{\infty} \in \ell_+^2(\mathcal{L}), \quad (W_2 S_1 f)(z) = \sum_{n=1}^{\infty} (S_1 f)_n \, z^n = \sum_{n=1}^{\infty} \mathbf{k}_{n-1} z^n = z \left(\sum_{n=0}^{\infty} \mathbf{k}_n z^n \right) =$$
$$= z(W_2 f)(z) = (S_2 W_2 f)(z) \Rightarrow W_2 S_1 = S_2 W_2 , \tag{B.7}$$

where S_2 is the operator of multiplication by the independent variable on $\mathcal{H}_{\mathcal{L}}^2(D)$. S_2 is then a unilateral shift operator on $\mathcal{H}_{\mathcal{L}}^2(D)$. In fact, we have

Proposition B.9 *The operator S_2 on $\mathcal{H}_{\mathcal{N}}^2(D)$ is a unilateral shift of multiplicity $\mathcal{N}$. Its adjoint S_2^* is given by $(S_2^* f)(z) = [f(z) - f(0)]/z$.* □

The combined mapping $W_2 W_1$ is then an isometric isomorphism of $\mathcal{H}$ onto $\mathcal{H}_{\mathcal{L}}^2(D)$ and we conclude that

Corollary B.7 *Every unilateral shift operator S on a Hilbert space $\mathcal{H}$ is unitarily equivalent to multiplication by the independent variable z on $\mathcal{H}_{\mathcal{L}}^2(D)$ for some subspace $\mathcal{L}$ of $\mathcal{H}$.* □

The last transformation we consider makes use of a change of variables through a linear fractional transformation (Möbius transform) of the form $z \to \sigma = i(1+z)/(1-z)$, sending the punctured circumference of the unit disk $C\backslash\{1\}$ onto $\mathbb{R}$ and the unit disk D onto the upper half-plane $\mathbb{C}^+$. The inverse map is then given by $\sigma \to z = (\sigma - i)/(\sigma + i)$. We define a mapping $W_3 : \mathcal{H}_{\mathcal{L}}^2(D) \mapsto \mathcal{H}^2(\mathbb{C}^+; \mathcal{L})$,

$$(W_3 f)(\sigma) = \frac{1}{\sqrt{\pi}} \frac{1}{\sigma + i} f\left(\frac{\sigma - i}{\sigma + i} \right), \quad \sigma \in \mathbb{C}^+. \tag{B.8}$$

The map W_3 is an isometric isomorphism of $\mathcal{H}_{\mathcal{L}}^2(D)$ onto $\mathcal{H}_+^2(\mathbb{C}^+; \mathcal{L})$. If S_2 is the canonical unilateral shift on $\mathcal{H}_{\mathcal{L}}^2(D)$, i.e., multiplication by the independent variable z, then under the mapping W_3 we have

$$\forall f \in \mathcal{H}_{\mathcal{L}}^2(D), \quad (W_3 S_2 f)(\sigma) = \frac{\sigma - i}{\sigma + i} (W_3 f)(\sigma) = (S_3 W_3 f)(\sigma) . \tag{B.9}$$

where S_3 is the canonical unilateral shift on $\mathcal{H}^2(\mathbb{C}^+; \mathcal{L})$. We see that, if S is a unilateral shift operator on a Hilbert space $\mathcal{H}$ a subsequent application of the three mappings from above defines an isometric isomorphism $W : \mathcal{H} \mapsto \mathcal{H}^2(\mathbb{C}^+; \mathcal{L})$, $W = W_3 W_2 W_1$, providing a representation of $\mathcal{H}$ in terms of the Hardy space $\mathcal{H}^2(\mathbb{C}^+; \mathcal{L})$, where $\mathcal{L} = Ker\ S^* = \mathcal{H} \ominus S\mathcal{H}$. The subspace $\mathcal{L}$ is cyclic for S and the multiplicity of S is $\dim(\mathcal{L})$. In addition we have

$$(W\mathbf{x})(\sigma) = \sum_{j=0}^{\infty} \frac{\pi^{-1/2}\mathbf{x}_j}{\sigma + i}\left(\frac{\sigma - i}{\sigma + i}\right)^j, \quad \mathbf{x} = \sum_{j=0}^{\infty} S^j \mathbf{x}_j,\ \mathbf{x}_j \in \mathcal{L}$$

and

$$(W S\mathbf{x})(\sigma) = \frac{\sigma - i}{\sigma + i}(W\mathbf{x})(\sigma) = (S_3 W\mathbf{x})(\sigma).$$

Let U be a bilateral shift on a Hilbert space $\mathcal{H}$. Then, by definition of a bilateral shift, there exists a wandering subspace $\mathcal{L}$ for U such that

$$\mathcal{H} = \bigoplus_{n=-\infty}^{\infty} U^n \mathcal{L}.$$

Let $\ell^2(\mathcal{L})$ be the Hilbert space of square summable, $\mathcal{L}$ valued, sequences. We may now define an isometric isomorphism $W_1' : \mathcal{H} \mapsto \ell^2(\mathcal{L})$ by

$$\mathbf{x} \in \mathcal{H},\ \mathbf{x} = \sum_{n=0}^{\infty} U^n \mathbf{x}_n,\ \mathbf{x}_n \in \mathcal{L} \Rightarrow (W_1'\mathbf{x})_n = \mathbf{x}_n,\ n \in \mathbb{Z}, \tag{B.10}$$

$$\|W_1'\mathbf{x}\|^2_{\ell^2(\mathcal{L})} = \sum_{n=-\infty}^{\infty} \|\mathbf{x}_n\|^2 = \|\mathbf{x}\|^2.$$

On $\ell^2(\mathcal{L})$ define an operator $U_1 : \ell^2(\mathcal{L}) \mapsto \ell^2(\mathcal{L})$ by

$$\mathbf{x} = (\mathbf{x}_n)_{n=-\infty}^{\infty} \in \ell^2(\mathcal{L}), \quad (U_1\mathbf{x})_n = \mathbf{x}_{n-1}. \tag{B.11}$$

Then U_1 is a bilateral shift on $\ell^2(\mathcal{L})$ and we have

$$[W_1'U_1\mathbf{x}]_n = \left[W_1'U_1 \sum_{k=-\infty}^{\infty} U_1^k \mathbf{x}_k\right]_n = \left[W_1 \sum_{k=-\infty}^{\infty} U_1^{(k+1)}\mathbf{x}_k\right]_n =$$
$$= \left[W_1 \sum_{k=-\infty}^{\infty} U_1^k \mathbf{x}_{k-1}\right]_n = \mathbf{x}_{n-1} = [U_1 W_1'\mathbf{x}]_n \Rightarrow W_1'U_1 = U_1 W_1'.$$

Thus, via the mapping W_1', we obtain a canonical representation of the action of U on $\mathcal{H}$ in terms of the action of U_1 on $\ell^2(\mathcal{L})$. Note that if we consider $\ell^2_+(\mathcal{L})$ to be a subspace of $\ell^2(\mathcal{L})$ then the restriction $S_1 = U_1|\ell^2_+(\mathcal{L})$ is the canonical unilateral shift on $\ell^2_+(\mathcal{L})$ and U_1 is a unitary extension of S_1 to a bilateral shift.

Consider the Hilbert space $L^2_{\mathcal{L}}(C)$ of all $\mathcal{L}$-valued Lebesgue square integrable functions on the unit circle C. $L^2_{\mathcal{L}}(C)$ is a Hilbert space with inner product

$$\langle f, g\rangle_{L^2_{\mathcal{L}}(C)} = \frac{1}{2\pi}\int_0^{2\pi} \langle f(e^{i\theta}), g(e^{i\theta})\rangle_{\mathcal{L}} d\theta = \sum_{n=-\infty}^{\infty} \langle \mathbf{a}_n, \mathbf{b}_n\rangle_{\mathcal{L}}, \tag{B.12}$$

for any $f(e^{i\theta}) = \sum_{n=-\infty}^{\infty} \mathbf{a}_n e^{in\theta}$ and $g(e^{i\theta}) = \sum_{n=-\infty}^{\infty} \mathbf{b}_n e^{in\theta}$ in $L^2_{\mathcal{L}}(C)$. From Eq. (B.12) it is clear that $L^2_{\mathcal{L}}(C)$ is isomorphic to $\ell^2(\mathcal{L})$ through a mapping $W_2' : \ell^2(\mathcal{L}) \mapsto L^2_{\mathcal{L}}(C)$ defined by

$$f = \{\mathbf{x}_n\}_{n=-\infty}^{\infty} \in \ell^2(\mathcal{L}), \qquad W_2'(f) = \sum_{n=-\infty}^{\infty} \mathbf{x}_n e^{in\theta}. \tag{B.13}$$

If U_1 is the bilateral shift on $\ell^2(\mathcal{L})$ defined above we have

$$\begin{aligned}(W_2'U_1 f)(e^{i\theta}) &= \sum_{n=-\infty}^{\infty} (U_1 f)_n\, e^{in\theta} = \sum_{n=-\infty}^{\infty} \mathbf{x}_{n-1} e^{in\theta} = e^{i\theta}\left(\sum_{n=-\infty}^{\infty} \mathbf{x}_n e^{in\theta}\right) = \\ &= e^{i\theta}(W_2 f)(e^{i\theta}) = (U_2 W_2 f)(e^{i\theta}) \Rightarrow W_2'U_1 = U_2 W_2',\end{aligned} \tag{B.14}$$

where U_2 is the operator of multiplication by $e^{i\theta}$ on $L^2_{\mathcal{L}}(C)$. U_2 is then the canonical bilateral shift operator on $L^2_{\mathcal{L}}(C)$. Note that $\mathcal{H}^2_{\mathcal{L}}(D)$ is a closed subspace of $L^2_{\mathcal{L}}(C)$ and that the canonical unilateral shift S_2 on $\mathcal{H}^2_{\mathcal{L}}(D)$ is the restriction $S_2 = U_2|\mathcal{H}^2_{\mathcal{L}}(D)$, i.e., U_2 is the extension of S_2 to a bilateral shift on $L^2_{\mathcal{L}}(C)$.

Finally, we consider the mapping $W_3' : L^2_{\mathcal{L}}(C) \mapsto L^2(\mathbb{R}; \mathcal{L})$ defined by

$$(W_3' f)(\sigma) = \frac{1}{\sqrt{\pi}} \frac{1}{\sigma + i} f\left(\frac{\sigma - i}{\sigma + i}\right), \quad \sigma \in \mathbb{R}. \tag{B.15}$$

The map W_3' is an isometric isomorphism of $L^2_{\mathcal{L}}(C)$ onto $L^2(\mathbb{R}; \mathcal{L})$. If U_2 is the canonical bilateral shift on $L^2_{\mathcal{L}}(C)$, then under the mapping W_3' we have

$$\forall f \in L^2_{\mathcal{L}}(C), \quad (W_3' U_2 f)(\sigma) = \frac{\sigma - i}{\sigma + i}(W_3' f)(\sigma) = (U_3 W_3' f)(\sigma) \tag{B.16}$$

where U_3 is a bilateral shift on $L^2(\mathbb{R}, \mathcal{L})$. Moreover, we have $W_3'\mathcal{H}^2_{\mathcal{L}}(D) = \mathcal{H}^2(\mathbb{C}^+; \mathcal{L})$ and $S_3 = U_3|\mathcal{H}^2(\mathbb{C}^+; \mathcal{L})$. We see that, if U is a unilateral shift operator on a Hilbert space $\mathcal{H}$, then then mapping $\hat{W} : \mathcal{H} \mapsto L^2(\mathbb{R}; \mathcal{L})$ defined by $\hat{W} = W_3' W_2' W_1'$ is an isometric isomorphism such that

$$(\hat{W}\mathbf{x})(\sigma) = \sum_{n=-\infty}^{\infty} \frac{\pi^{-1/2}\mathbf{x}_n}{\sigma + i}\left(\frac{\sigma - i}{\sigma + i}\right)^j, \quad \mathbf{x} = \sum_{n=-\infty}^{\infty} U^n \mathbf{x}_n, \ \mathbf{x}_n \in \mathcal{L},$$

and $\hat{W}U = U_3\hat{W}$. In addition, we have

$$\hat{W}(U|(\hat{W}^*\mathcal{H}^2(\mathbb{C}^+; \mathcal{L}))) = U_3\hat{W}|(\hat{W}^*\mathcal{H}^2(\mathbb{C}^+; \mathcal{L})) = U_3|\mathcal{H}^2(\mathbb{C}^+; \mathcal{L}) = S_3\,.$$

Hence $U|(\hat{W}^*\mathcal{H}^2(\mathbb{C}^+; \mathcal{L}))$ is a unilateral shift on $\mathcal{H}$ and U is its extension to a bilateral shift.

B.4 Contraction Operators and Their Dilations

The previous section has centered on the structure theory of isometric operators on Hilbert spaces, with the main result being the Wold decomposition theorem. In the present section we initiate a discussion of more general types of contraction operators. As we shall see isometric operators, and in particular unilateral and bilateral shifts are particularly important within this context. We start with a standard classification of contractions:

Definition B.28 (*Classification of contractions*) Define classes $C_{..}$ of contraction operators on a Hilbert space $\mathcal{H}$ as follows:

1. $T \in C_{0\cdot}$ if $\lim_{n\to\infty} \|T^n\mathbf{x}\| = 0, \quad \forall \mathbf{x} \in \mathcal{H},\ \mathbf{x} \neq \mathbf{0}$.
2. $T \in C_{1\cdot}$ if $\lim_{n\to\infty} \|T^n\mathbf{x}\| \neq 0, \quad \forall \mathbf{x} \in \mathcal{H},\ \mathbf{x} \neq \mathbf{0}$.
3. $T \in C_{\cdot 0}$ if $\lim_{n\to\infty} \|(T^*)^n\mathbf{x}\| = 0, \quad \forall \mathbf{x} \in \mathcal{H},\ \mathbf{x} \neq \mathbf{0}$.
4. $T \in C_{\cdot 1}$ if $\lim_{n\to\infty} \|(T^*)^n\mathbf{x}\| \neq 0, \quad \forall \mathbf{x} \in \mathcal{H},\ \mathbf{x} \neq \mathbf{0}$.

In addition, we set

$$C_{\alpha\beta} := C_{\alpha\cdot} \cap C_{\cdot\beta}$$

□

Next, we introduce the important notion of a *completely non-unitary operator*:

Definition B.29 (*Completely non-unitary contraction*) A contraction T on a Hilbert space $\mathcal{H}$ is called *completely non-unitary* (c.n.u.) if there is no reducing subspace $\mathcal{H}_0 \subseteq \mathcal{H}$ for T such that the restriction $T|\mathcal{H}_0$ is unitary. □

A unilateral shift is a c.n.u. contraction. Indeed, we have seen that if S is a unilateral shift on a Hilbert space $\mathcal{H}$ then for every $\mathbf{x} \in \mathcal{H}$ we have $\lim_{n\to\infty} \|(S^*)^n\mathbf{x}\| = 0$. On the other hand, if S is not c.n.u. there exists a subspace $\mathcal{H}_0 \subset \mathcal{H}$ such that $S|\mathcal{H}_0$ is unitary on $\mathcal{H}_0$ and in this case for $\mathbf{x} \in \mathcal{H}_0$ we would have $\|(S^*)^n\mathbf{x}\| = \|\mathbf{x}\|$, $n = 0, 1, 2, \ldots$. Therefore, S must be completely non-unitary. For more general contractions we have the following theorem, which we state without proof (see Sz.-Nagy and Foias 1970):

Theorem B.36 *Let T be a contraction on a Hilbert space $\mathcal{H}$. Then there is a decomposition of $\mathcal{H}$ into an orthogonal sum $\mathcal{H} = \mathcal{H}_0 \oplus \mathcal{H}_1$, where the subspaces $\mathcal{H}_0$, $\mathcal{H}_1$ are reducing subspaces for T, $T_0 := T|\mathcal{H}_0$ is unitary and $T_1 := T|\mathcal{H}_1$ is c.n.u. . Moreover, we have*

$$\mathcal{H}_0 = \{\mathbf{x} : \mathbf{x} \in \mathcal{H},\ \|T^n\mathbf{x}\| = \|(T^*)^n\mathbf{x}\| = \|\mathbf{x}\|,\ n = 1, 2, \ldots\}$$

□

As an immediate corollary of Theorem B.36 we have:

Corollary B.8 *If T be an isometry on a Hilbert space $\mathcal{H}$. Then the canonical decomposition of Theorem B.36 corresponds to the Wold decomposition.* □

Let T be a contraction operator on a Hilbert space $\mathcal{H}$. Then

$$\forall \mathbf{x} \in \mathcal{H},\ \|T\mathbf{x}\|^2 \leq \|\mathbf{x}\|^2 \Rightarrow \forall \mathbf{x} \in \mathcal{H},\ \langle \mathbf{x}, T^*T\mathbf{x}\rangle \leq \langle \mathbf{x}, \mathbf{x}\rangle \Rightarrow I - T^*T \geq 0\,.$$

Since $\|T^*\| = \|T\|$ we also have

$$\forall \mathbf{x} \in \mathcal{H},\ \|T^*\mathbf{x}\|^2 \leq \|\mathbf{x}\|^2 \Rightarrow \forall \mathbf{x} \in \mathcal{H},\ \langle \mathbf{x}, TT^*\mathbf{x}\rangle \leq \langle \mathbf{x}, \mathbf{x}\rangle \Rightarrow I - TT^* \geq 0$$

Therefore, the operators

$$D_T := (I - T^*T)^{1/2}, \qquad D_{T^*} := (I - TT^*)^{1/2} \tag{B.17}$$

are well defined, non-negative and contractive. In addition it is possible to prove that (Sz.-Nagy and Foias 1970)

$$TD_T = D_T^*T, \qquad D_TT^* = TD_T^*$$

Note that

$$\forall \mathbf{x} \in \mathcal{H},\quad \|D_T\mathbf{x}\|^2 = \langle D_T\mathbf{x}, D_T\mathbf{x}\rangle = \langle \mathbf{x}, D_T^2\mathbf{x}\rangle = \langle \mathbf{x}, (I - T^*T)\mathbf{x}\rangle = \|\mathbf{x}\|^2 - \|T\mathbf{x}\|^2$$

so that the null space $\mathcal{N}_{D_T}$ of D_T is given by

$$\mathcal{N}_{D_T} = \{\mathbf{x} : \mathbf{x} \in \mathcal{H},\ D_T\mathbf{x} = \mathbf{0}\} = \{\mathbf{x} : \mathbf{x} \in \mathcal{H},\ \|T\mathbf{x}\| = \|\mathbf{x}\|\}. \tag{B.18}$$

Similarly,

$$\forall \mathbf{x} \in \mathcal{H},\quad \|D_{T^*}\mathbf{x}\|^2 = \langle D_{T^*}\mathbf{x}, D_{T^*}\mathbf{x}\rangle = \langle \mathbf{x}, D_{T^*}^2\mathbf{x}\rangle = \langle \mathbf{x}, (I - TT^*)\mathbf{x}\rangle = \|\mathbf{x}\|^2 - \|T^*\mathbf{x}\|^2$$

so that the null space $\mathcal{N}_{D_{T^*}}$ of D_{T^*} is given by

$$\mathcal{N}_{D_{T^*}} = \{\mathbf{x} : \mathbf{x} \in \mathcal{H},\ D_{T^*}\mathbf{x} = \mathbf{0}\} = \{\mathbf{x} : \mathbf{x} \in \mathcal{H},\ \|T^*\mathbf{x}\| = \|\mathbf{x}\|\}. \tag{B.19}$$

Following the above observations we define:

Definition B.30 (*Defect operators, defect spaces, defect indices*) Let T be a contraction operator on $\mathcal{H}$. The operators D_T and D_{T^*} defined in Eq. (B.17) are called the *defect operators*, the subspaces

$$\mathcal{D}_T = \overline{D_T\mathcal{H}} = \mathcal{N}_{D_T}^{\perp}, \quad \mathcal{D}_{T^*} = \overline{D_{T^*}\mathcal{H}} = \mathcal{N}_{D_{T^*}}^{\perp} \tag{B.20}$$

are called the *defect subspaces* and

$$\sigma_T = \dim(\mathcal{D}_T), \qquad \sigma_{T^*} = \dim(\mathcal{D}_{T^*}) \tag{B.21}$$

are called the *defect indices* for the contraction T. □

Observe also that

$$\sigma_T = \dim(\mathcal{D}_T) = 0 \Rightarrow \mathcal{N}_{D_T} = \mathcal{H} \Rightarrow \forall \mathbf{x} \in \mathcal{H}, \ \|T\mathbf{x}\| = \|\mathbf{x}\|$$

so that in this case T is an isometry. Similarly,

$$\sigma_{T^*} = \dim(\mathcal{D}_{T^*}) = 0 \Rightarrow \mathcal{N}_{D_{T^*}} = \mathcal{H} \Rightarrow \forall \mathbf{x} \in \mathcal{H}, \ \|T^*\mathbf{x}\| = \|\mathbf{x}\|$$

and in this case T^* is an isometry. Hence, if $\sigma_T = \sigma_{T^*} = 0$ then T is a unitary operator.

Let A be an operator on a Hilbert space $\mathcal{K}$ and let $\mathcal{H}$ be a subspace of $\mathcal{K}$. Denote by $P_{\mathcal{H}}$ the projection of $\mathcal{K}$ onto $\mathcal{H}$. The projection (or compression) of A on $\mathcal{H}$ is defined to be $B = P_{\mathcal{H}} A P_{\mathcal{H}}$.

Definition B.31 (*Dilation*) Let $\mathcal{H}$ be a subspace of a Hilbert space $\mathcal{K}$. Let A be an operator on $\mathcal{K}$ and let B be an operator on $\mathcal{H}$. Then A is called a dilation of B if

$$B^n = P_{\mathcal{H}} A^n P_{\mathcal{H}}, \quad \forall n \in \mathbb{N}$$

□

Let B be an operator on a Hilbert space $\mathcal{H}$. Let A be a dilation of B on a Hilbert space $\mathcal{K}$ and let A' be a dilation of B on a Hilbert space $\mathcal{K}'$. The dilations A and A' are called *isomorphic* if there exists a unitary mapping $U : \mathcal{K} \mapsto \mathcal{K}'$ such that:

1. $U\mathbf{x} = \mathbf{x}, \ \forall \mathbf{x} \in \mathcal{H}$.
2. $A = U^* A' U$.

Our discussion below will focus mainly on contractions of class $C_{0\cdot}$ and one of our goals will be to construct functional models for this class of contractions. However, it is quite illuminating to prove a more general result, concerning the existence of an isometric dilation for a general contraction operator. We have the following theorem:

Theorem B.37 (Existence of isometric dilation) *Let T be a contraction on a Hilbert space $\mathcal{H}$. Then there exists a Hilbert space $\mathcal{K}_+$ with $\mathcal{H} \subseteq \mathcal{K}_+$ and an operator W on $\mathcal{K}_+$ such that W is an* isometric dilation *of T, i.e., W is an isometry on $\mathcal{K}_+$ and is a dilation of T. Moreover, the dilation Hilbert space $\mathcal{K}_+$ can be chosen minimal in the sense that*

$$\mathcal{K}_+ = \bigvee_{n=0}^{\infty} W^n \mathcal{H}. \tag{B.22}$$

The minimal isometric dilation of T is determined up to an isomorphism. Furthermore, $\mathcal{H}$ is invariant for W^* and

$$T P_{\mathcal{H}} = P_{\mathcal{H}} W, \qquad T^* = W^* | \mathcal{H}$$

□

Proof Define the Hilbert space

$$\tilde{\mathcal{K}}_+ := \bigoplus_{n=0}^{\infty} \mathcal{H}$$

The elements of $\tilde{\mathcal{K}}_+$ are infinite sequences

$$f = (\mathbf{x}_n)_{n=0}^{\infty}, \quad \mathbf{x}_n \in \mathcal{H}, \quad \|f\|^2 = \sum_{n=0}^{\infty} \|\mathbf{x}_n\|^2 < \infty$$

and the inner product of two arbitrary elements $f, g \in \tilde{\mathcal{K}}_+$, with $f = (\mathbf{x}_n)_{n=0}^{\infty}$, $g = (\mathbf{y}_n)_{n=0}^{\infty}$, is given by

$$\langle f, g \rangle = \sum_{n=0}^{\infty} \langle \mathbf{x}_n, \mathbf{y}_n \rangle_{\mathcal{H}}$$

We define a mapping $\hat{U} : \mathcal{H} \mapsto \tilde{\mathcal{K}}_+$ by

$$\mathbf{x} \in \mathcal{H}, \quad \hat{U}\mathbf{x} = (\mathbf{x}_n)_{n=0}^{\infty} = \begin{cases} \mathbf{x}, & n = 0 \\ \mathbf{0}, & n \geq 1 \end{cases} \tag{B.23}$$

The mapping $\hat{U}$ unitarily embeds $\mathcal{H}$ as a subspace of $\tilde{\mathcal{K}}_+$. The orthogonal projection of $\tilde{\mathcal{K}}_+$ onto $\mathcal{H}$ is then defined by

$$\forall f = (\mathbf{x}_n)_{n=0}^{\infty} \in \tilde{\mathcal{K}}_+, \quad (P_{\mathcal{H}} f)_n = \begin{cases} \mathbf{x}_0, & n = 0 \\ \mathbf{0}, & n \geq 1 \end{cases} \tag{B.24}$$

Given the contraction T we now define an operator $\hat{W} : \tilde{\mathcal{K}}_+ \mapsto \tilde{\mathcal{K}}_+$ by

$$f = (\mathbf{x}_n)_{n=0}^{\infty} \in \tilde{\mathcal{K}}_+, \quad (\hat{W} f)_n = \begin{cases} T\mathbf{x}_0, & n = 0 \\ D_T \mathbf{x}_0, & n = 1 \\ \mathbf{x}_{n-1}, & n \geq 2 \end{cases}, \quad D_T = (1 - T^*T)^{1/2} \tag{B.25}$$

Therefore

$$\|\hat{W} f\|^2 = \sum_{n=0}^{\infty} \|(\hat{W} f)_n\|_{\mathcal{H}}^2 = \|T\mathbf{x}_0\|_{\mathcal{H}}^2 + \|D_T \mathbf{x}_0\|^2 + \sum_{n=1}^{\infty} \|\mathbf{x}_n\|^2$$

and since

$$\|T\mathbf{x}_0\|^2_{\mathcal{H}} + \|D_T\mathbf{x}_0\|^2 = \|T\mathbf{x}_0\|^2_{\mathcal{H}} + \|(I - T^*T)^{1/2}\mathbf{x}_0\|^2_{\mathcal{H}} = \|T\mathbf{x}_0\|^2_{\mathcal{H}} + \langle \mathbf{x}_0, (I - T^*T)\mathbf{x}_0\rangle =$$
$$= \|T\mathbf{x}_0\|^2_{\mathcal{H}} + \langle \mathbf{x}_0, \mathbf{x}_0\rangle - \langle \mathbf{x}_0, T^*T\mathbf{x}_0\rangle = \|T\mathbf{x}_0\|^2_{\mathcal{H}} + \|\mathbf{x}_0\|^2_{\mathcal{H}} - \|T\mathbf{x}_0\|^2_{\mathcal{H}} = \|\mathbf{x}_0\|^2_{\mathcal{H}}$$

we get that

$$\|\hat{W} f\|^2 = \sum_{n=0}^{\infty} \|(\hat{W} f)_n\|^2_{\mathcal{H}} = \sum_{n=0}^{\infty} \|\mathbf{x}_n\|^2_{\mathcal{H}} = \|f\|^2$$

Hence $\hat{W}$ is an isometry on $\tilde{\mathcal{K}}_+$. Furthermore, using Eq. (B.25), it is easily shown by induction that

$$f = (\mathbf{x}_n)_{n=0}^{\infty} \in \tilde{\mathcal{K}}_+, \quad (\hat{W}^k f)_n = \begin{cases} T^k\mathbf{x}_0, & n = 0 \\ D_T T^{(k-n)}\mathbf{x}_0, & 1 \le n \le k \\ \mathbf{x}_{n-k}, & n \ge k+1 \end{cases}, \quad D_T = (1 - T^*T)^{1/2}$$

For simplicity we shall identify $\mathcal{H}$ and its embedding into $\tilde{\mathcal{K}}_+$ by $\hat{U}$. In this case, using the projection $P_{\mathcal{H}}$, we get that

$$\forall \mathbf{x} \in \mathcal{H}, \quad \Rightarrow \quad (\hat{W}^k\mathbf{x})_n = \begin{cases} T^k\mathbf{x}, & n = 0 \\ D_T T^{(k-n)}\mathbf{x}, & 1 \le n \le k \\ \mathbf{0}, & n \ge k+1 \end{cases}$$

$$\Rightarrow \quad (P_{\mathcal{H}}\hat{W}^k\mathbf{x})_n = \begin{cases} T^k\mathbf{n}, & n = 0 \\ \mathbf{0}, & n \ge 1 \end{cases} \quad \Rightarrow P_{\mathcal{H}}\hat{W}^k\mathbf{x} = T^k\mathbf{x}.$$

Hence W is an isometric dilation of T. As constructed, the isometric dilation W above is not minimal in the sense of Eq. (B.22). However, if we set

$$\mathcal{K}_+ := \bigvee_{n=0}^{\infty} \hat{W}^n\mathcal{H}$$

where, as above, $\mathcal{H}$ is considered to be a subspace of $\tilde{\mathcal{K}}_+$ (so that $\mathcal{K}_+ \subset \tilde{\mathcal{K}}_+$) and define

$$W := \hat{W}|\mathcal{K}_+$$

then W is a minimal isometric dilation of T on $\mathcal{K}_+$.

Assume that W is a minimal isometric dilation of T on a dilation Hilbert space $\mathcal{K}_+$. Then we have

$$\forall \mathbf{x} \in \mathcal{H}, \quad TP_{\mathcal{H}}W^n\mathbf{x} = TT^n\mathbf{x} = T^{(n+1)}\mathbf{x} = P_{\mathcal{H}}W^{(n+1)}\mathbf{x} = P_{\mathcal{H}}WW^n\mathbf{x}$$

and since W is minimal we get that on $\mathcal{K}_+$ we have $TP_{\mathcal{H}} = P_{\mathcal{H}}W$. Moreover, for any $\mathbf{x} \in \mathcal{H}$ and $\mathbf{y} \in \mathcal{K}_+$ we have

$$\langle T^*\mathbf{x}, \mathbf{y}\rangle_{\mathcal{K}_+} = \langle T^*\mathbf{x}, P_{\mathcal{H}}\mathbf{y}\rangle_{\mathcal{K}_+} = \langle \mathbf{x}, TP_{\mathcal{H}}\mathbf{y}\rangle_{\mathcal{K}_+} = \langle \mathbf{x}, P_{\mathcal{H}}W\mathbf{y}\rangle_{\mathcal{K}_+} = \langle W^*\mathbf{x}, \mathbf{y}\rangle_{\mathcal{K}_+}$$

and since $\mathbf{y} \in \mathcal{K}_+$ is arbitrary we get that $T^* = W^*|\mathcal{H}$.

We are left with the task of showing that all minimal isometric dilations of T are isomorphic. Assume that W_1 and W_2 are two isometric dilations of T, respectively on the Hilbert spaces $\mathcal{K}_+$ and $\mathcal{K}'_+$. Assuming that $n \geq m$ we then get that

$$\begin{aligned}\langle W_1^n\mathbf{x}, W_1^m\mathbf{x}'\rangle &= \langle (W_1^*)^m W_1^n\mathbf{x}, \mathbf{x}'\rangle = \langle W_1^{(n-m)}\mathbf{x}, \mathbf{x}'\rangle = \langle T^{(n-m)}\mathbf{x}, \mathbf{x}'\rangle = \\ &= \langle W_2^{(n-m)}\mathbf{x}, \mathbf{x}'\rangle = \langle (W_2^*)^m W_2^n\mathbf{x}, \mathbf{x}'\rangle = \langle W_2^n\mathbf{x}, W_2^m\mathbf{x}'\rangle \end{aligned} \tag{B.26}$$

with a similar argument valid for $m \geq n$. By Eq. (B.26) we get that for all finite sequences $(\mathbf{x}_n)_{n=0}^N$, $(\mathbf{y}_n)_{n=0}^N$, $\mathbf{x}_n, \mathbf{y}_n \in \mathcal{H}$

$$\Big\langle \sum_{n=0}^N W_1^n\mathbf{x}_n, \sum_{n=0}^N W_1^n\mathbf{y}_n\Big\rangle = \Big\langle \sum_{n=0}^N W_2^n\mathbf{x}_n, \sum_{n=0}^N W_2^n\mathbf{y}_n\Big\rangle$$

Thus, if we set $\mathcal{H}_1 = \text{span}\{\sum_{n=0}^N W_1^n\mathbf{x}_n, \ \forall N \in \mathbb{N}\cup\{0\}, \forall \mathbf{x}_n \in \mathcal{H}, \ 0 \leq n \leq N\}$ and $\mathcal{H}_2 = \text{span}\{\sum_{n=0}^N W_2^n\mathbf{x}_n, \ \forall N \in \mathbb{N}\cup\{0\}, \forall \mathbf{x}_n \in \mathcal{H}, \ 0 \leq n \leq N\}$ and define a mapping $\tilde{U} : \mathcal{H}_1 \mapsto \mathcal{H}_2$ by

$$\tilde{U}\Big(\sum_{n=0}^N W_1^n\mathbf{x}_n\Big) = \sum_{n=0}^N W_2^n\mathbf{x}_n$$

then $\tilde{U}$ is an isometric isomorphism of $\mathcal{H}_1$ onto $\mathcal{H}_2$. Finally, if W_1 is minimal on $\mathcal{K}_+$ and W_2 is minimal on $\mathcal{K}'_+$ the mapping $\tilde{U}$ can be extended to a unitary mapping of $\mathcal{K}_+$ onto $\mathcal{K}'_+$ which we shall denote also by $\tilde{U}$. By its definition we have $\tilde{U}\mathbf{x} = \mathbf{x}$, $\forall \mathbf{x} \in \mathcal{H}$ and

$$\begin{aligned}\tilde{U}W_1\Big(\sum_{n=0}^N W_1^n\mathbf{x}_n\Big) &= \tilde{U}\Big(\sum_{n=0}^N W_1^{(n+1)}\mathbf{x}_n\Big) = \tilde{U}\Big(\sum_{n=0}^N W_2^{(n+1)}\mathbf{x}_n\Big) = W_2\Big(\sum_{n=0}^N W_2^n\mathbf{x}_n\Big) = \\ &= W_2\tilde{U}\Big(\sum_{n=0}^N W_1^n\mathbf{x}_n\Big)\end{aligned}$$

which extends to the intertwining relation $\tilde{U}W_1 = W_2\tilde{U}$ between $\mathcal{K}_+$ and $\mathcal{K}'_+$. ■

Theorem B.38 (Existence of unitary dilation) *Let T be a contraction on a Hilbert space $\mathcal{H}$. Then there exists a Hilbert space $\mathcal{K}$ with $\mathcal{H} \subseteq \mathcal{K}$ and an operator U on $\mathcal{K}$ such that U is a* unitary dilation *of T, i.e., U is unitary on $\mathcal{K}$ and is a dilation of T. Moreover, the dilation Hilbert space $\mathcal{K}$ can be chosen minimal in the sense that*

$$\mathcal{K} = \bigvee_{n=-\infty}^{\infty} W^n \mathcal{H}. \tag{B.27}$$

The minimal unitary dilation of T is determined up to an isomorphism. □

Proof Given the contraction T on $\mathcal{H}$, by Theorem B.37 there exists an isometric dilation W of T on a Hilbert space $\mathcal{K}_+$ such that $\mathcal{H} \subset \mathcal{K}_+$. By Proposition B.7 the isometric dilation W can be extended to a unitary operator U on a Hilbert space $\tilde{\mathcal{K}}$ such that $\mathcal{K}_+ \subset \tilde{\mathcal{K}}$. Thus U is a unitary dilation of T. This unitary dilation is in general not minimal but if we now define

$$\mathcal{K} = \bigvee_{n=-\infty}^{\infty} U^n \mathcal{H}$$

we obtain a subspace $\mathcal{K} \subset \tilde{\mathcal{K}}$ such that the restriction of U to $\mathcal{K}$ is a minimal unitary dilation of T. Finally, the proof that all minimal unitary dilations of T are isomorphic is similar to the case of minimal isometric dilations in Theorem B.37. ■

It is evident that the proofs of Theorems B.37 and B.38 do not offer a method for the construction of functional models for contractions. In fact, the construction of functional models for general contractions is quite involved (see Sz.-Nagy and Foias 1970) and is beyond the scope of our present discussion. Instead, we shall construct functional models directly for contractions of class $C_{0\cdot}$ which, as is mentioned above, as the main objects of interest for our current discussion.

B.5 Functional Models for Contractions of Class $C_{0\cdot}$

Our goal is to find a universal functional representation for contractions of class $C_{0\cdot}$, i.e., a representation of a class $C_{0\cdot}$ contraction in terms of a corresponding operator operating on a well defined (vector valued) function space. Indeed, we shall find that we can obtain a universal representation of a $C_{0\cdot}$ contraction in terms of the compression of the adjoint of the canonical shift operator on Hardy space onto a well defined subspace. Our first step towards this functional representation is the following theorem:

Theorem B.39 (Model operators for $C_{0\cdot}$ contractions) *Let T be a contraction operator on a Hilbert space $\mathcal{H}$ and assume that $T \in C_{0\cdot}$ (i.e., $\lim_{n\to\infty} \|T^n \mathbf{x}\| = 0, \forall \mathbf{x} \in \mathcal{H}$). Let $D_T = (I - T^*T)^{1/2}$ be the defect operator for T and let $\mathcal{D}_T = \overline{D_T \mathcal{H}}$ be the corresponding defect subspace. Then there exists a Hilbert space $\mathcal{K}$, a unilateral shift S on $\mathcal{K}$ with multiplicity equal to* $\dim \mathcal{D}_T$ *and a subspace $\mathcal{M}$ of $\mathcal{K}$, such that T is unitarily equivalent to $S^*|\mathcal{M}$.* □

Proof Recall the definition of the defect subspace $\mathcal{D}_T$,

$$\mathcal{D}_T = \overline{D_T \mathcal{H}}, \quad D_T = (I - T^*T)^{1/2}$$

where D_T is the defect operator. Consider the Hilbert space $\ell_+^2(\mathcal{D}_T)$ of square summable, $\mathcal{D}_T$ valued, sequences

$$\ell_+^2(\mathcal{D}_T) = \{(\mathbf{x}_n)_{n=0}^{\infty}, \mathbf{x}_n \in \mathcal{D}_T, \sum_{n=0}^{\infty} \|\mathbf{x}_n\|^2 < \infty\}$$

On $\ell_+^2(\mathcal{D}_T)$ we define the canonical unilateral shift operator S_1,

$$\mathbf{x} = (\mathbf{x}_n)_{n=0}^{\infty} \in \ell_+^2(\mathcal{D}_T), \quad (S_1\mathbf{x})_n = \begin{cases} 0, & n = 0 \\ \mathbf{x}_{n-1}, & n = 1, 2 \end{cases} .$$

Now define a mapping $W : \mathcal{H} \mapsto \ell_+^2(\mathcal{D}_T)$ by

$$(W\mathbf{x})_n := D_T T^n \mathbf{x} = (I - T^*T)^{1/2} T^n \mathbf{x} .$$

In fact, for any element $\mathbf{x} \in \mathcal{H}$ we have

$$\begin{aligned} \|(W\mathbf{x})_n\|^2 &= \|D_T T^n \mathbf{x}\|^2 = \|(I - T^*T)^{1/2} T^n \mathbf{x}\|^2 = \\ &= \langle (I - T^*T)^{1/2} T^n \mathbf{x}, (I - T^*T)^{1/2} T^n \mathbf{x} \rangle = \langle \mathbf{x}, (T^*)^n (I - T^*T) T^n \mathbf{x} \rangle = \\ &= \langle \mathbf{x}, (T^*)^n T^n \mathbf{x} \rangle - \langle \mathbf{x}, (T^*)^{(n+1)} T^{(n+1)} \mathbf{x} \rangle = \|T^n \mathbf{x}\|^2 - \|T^{(n+1)} \mathbf{x}\|^2, \end{aligned}$$

and hence

$$\begin{aligned} \|W\mathbf{x}\|^2 &= \sum_{n=0}^{\infty} \|(W\mathbf{x})_n\|^2 = \lim_{n\to\infty} \sum_{k=0}^{n} \|(W\mathbf{x})_n\|^2 = \lim_{n\to\infty} \sum_{k=0}^{n} \left(\|T^k \mathbf{x}\|^2 - \|T^{(k+1)} \mathbf{x}\|^2 \right) = \\ &= \|\mathbf{x}\|^2 - \lim_{n\to\infty} \|T^{(n+1)} \mathbf{x}\|^2 = \|\mathbf{x}\|^2 . \end{aligned}$$

Thus W is an isometric isomorphism of $\mathcal{H}$ onto $\mathcal{M} = W\mathcal{H}$. In addition, for any $\mathbf{x} \in \mathcal{H}$,

$$\begin{aligned} (S_1^* W\mathbf{x})_n &= (W\mathbf{x})_{n+1} = D_T T^{(n+1)} \mathbf{x} = D_T T^n (T\mathbf{x}) = (WT\mathbf{x})_n, \; n = 0, 1, \ldots \\ &\Rightarrow S_1^* W\mathbf{x} = WT\mathbf{x} . \end{aligned}$$

■

Let $W_1 : \mathcal{H} \mapsto \ell_+^2(\mathcal{D}_T)$ be the map constructed in the proof of Theorem B.39 and define $\tilde{T} = W_1 T W_1^*$. In this case $\tilde{T}$ is unitarily equivalent to T and we have $\tilde{T} = S_1^* | \mathcal{M}$ with $\mathcal{M} = W\mathcal{H}$ an invariant subspace for S_1^*. In this case for any $\mathbf{x}, \mathbf{y} \in \mathcal{M}$ we have

$$\langle \tilde{T}\mathbf{x}, \mathbf{y} \rangle = \langle S_1^* \mathbf{x}, \mathbf{y} \rangle = \langle \mathbf{x}, S_1 \mathbf{y} \rangle = \langle P_{\mathcal{M}} \mathbf{x}, S_1 \mathbf{y} \rangle = \langle \mathbf{x}, P_{\mathcal{M}} S_1 \mathbf{y} \rangle = \langle \mathbf{x}, \tilde{T}^* \mathbf{y} \rangle \Rightarrow \tilde{T}^* = P_{\mathcal{M}} S_1 | \mathcal{M} .$$

Moreover, we have

$$\langle \tilde{T}^n \mathbf{x}, \mathbf{y} \rangle = \langle (S_1^*)^n \mathbf{x}, \mathbf{y} \rangle = \langle \mathbf{x}, S_1^n \mathbf{y} \rangle = \langle P_{\mathcal{M}} \mathbf{x}, S_1^n \mathbf{y} \rangle = \langle \mathbf{x}, P_{\mathcal{M}} S_1^n \mathbf{y} \rangle =$$
$$= \langle \mathbf{x}, (\tilde{T}^n)^* \mathbf{y} \rangle = \langle \mathbf{x}, (\tilde{T}^*)^n \mathbf{y} \rangle \Rightarrow (\tilde{T}^*)^n = P_{\mathcal{M}} S_1^n | \mathcal{M}.$$

We conclude that S_1 is an isometric dilation of $\tilde{T}^*$ on $\ell_+^2(\mathcal{D}_T)$. Furthermore, since $\mathcal{M}$ is invariant for S_1^*, we have

$$\forall \mathbf{x} \in \mathcal{M},\ \mathbf{y} \in \mathcal{M}^\perp, \quad 0 = \langle S_1^* \mathbf{x}, \mathbf{y} \rangle = \langle \mathbf{x}, S_1 \mathbf{y} \rangle \Rightarrow S_1 \mathcal{M}^\perp \subseteq \mathcal{M}^\perp,$$

hence $\mathcal{M}^\perp$ is an invariant subspace for S_1. Thus, the representation of T in Theorem B.39, is characterized by the characterization of invariant subspaces for the unilateral shift S_1. First we shall need a definition:

Definition B.32 Let $S \in \mathcal{B}(\mathcal{H})$ be a unilateral shift operator. An operator $A \in \mathcal{B}(\mathcal{H})$ is called:

1. *S-analytic* if $AS = SA$,
2. *S-inner* if A is S-analytic and a partial isometry on $\mathcal{H}$,
3. *S-outer* if A is S-analytic and $\overline{A\mathcal{H}}$ is a reducing subspace for S.

The characterization of invariant subspaces for unilateral shifts is achieved by the Beurling–Lax theorem:

Theorem B.40 (Beurling–Lax theorem) *Let S be a unilateral shift operator on a Hilbert space $\mathcal{H}$. A subspace $\mathcal{M} \subseteq \mathcal{H}$ is invariant under S iff there exists an S-inner operator A on $\mathcal{H}$ such that $\mathcal{M} = A\mathcal{H}$.* □

Proof First assume that $\mathcal{M} = A\mathcal{H}$ with A an S-inner operator. In this case we have

$$S\mathcal{M} = SA\mathcal{H} = AS\mathcal{H} \subseteq A\mathcal{H} = \mathcal{M}$$

so that $\mathcal{M}$ is invariant for S. To prove the converse statement we use the counting lemma:

Lemma B.3 (Counting lemma) *Let P be an orthogonal projection on a separable Hilbert space $\mathcal{H}$ and let $\{\mathbf{e}_j\}$ be any orthonormal basis for $\mathcal{H}$. Then we have*

$$\dim(P\mathcal{H}) = \sum_j \|P\mathbf{e}_j\|^2.$$

□

Proof of Lemma Let $\{\mathbf{h}_k\}$ be an orthonormal basis for $P\mathcal{H}$. Then

$$\sum_j \|P\mathbf{e}_j\|^2 = \sum_j \langle P\mathbf{e}_j, P\mathbf{e}_j \rangle = \sum_j \sum_k |\langle P\mathbf{e}_j, \mathbf{h}_k \rangle|^2 = \sum_j \sum_k |\langle \mathbf{e}_j, \mathbf{h}_k \rangle|^2 =$$
$$= \sum_k \sum_j |\langle \mathbf{e}_j, \mathbf{h}_k \rangle|^2 = \sum_k \|\mathbf{h}_k\|^2 = \dim(P\mathcal{H}).$$

■

We can now continue with the proof of the theorem. Assume that $\mathcal{M} \subset \mathcal{H}$ is invariant for S. Denote by $P_{\mathcal{M}}$ the orthogonal projection of $\mathcal{H}$ on $\mathcal{M}$. In this case $SP_{\mathcal{M}}S^*$ is the orthogonal projection on $S\mathcal{M}$ and hence $\hat{P} = P_{\mathcal{M}} - SP_{\mathcal{M}}S^*$ is the orthogonal projection on $\mathcal{M} \ominus S\mathcal{M}$. Note that if $Ker(S^*)$ is infinite dimensional then $\dim(Ker(S^*)) = \dim(\mathcal{H}) \geq \dim(\hat{P}\mathcal{H})$. Moreover, if $\dim(Ker(S^*)) = m$ and $\{\mathbf{e}_i\}_{i=1}^m$ is an orthonormal basis for $Ker(S^*)$ then $\{S^n\mathbf{e}_i : 1 \leq i \leq m,\ n = 0, 1, 2, \ldots\}$ is an orthonormal basis for $\mathcal{H}$. In this case by the counting lemma we obtain

$$\dim(\hat{P}\mathcal{H}) = \sum_{i=1}^{m}\sum_{n=0}^{\infty}\langle \hat{P}S^n\mathbf{e}_i, S^n\mathbf{e}_i\rangle = \lim_{k\to\infty}\sum_{i=1}^{m}\sum_{n=0}^{k}\langle (P_{\mathcal{M}} - SP_{\mathcal{M}}S^*)S^n\mathbf{e}_i, S^n\mathbf{e}_i\rangle =$$

$$= \lim_{k\to\infty}\sum_{i=1}^{m}\left[\sum_{n=2}^{k}\left(\langle (S^*)^n P_{\mathcal{M}}S^n\mathbf{e}_i, \mathbf{e}_i\rangle - \langle (S^*)^{(n-1)}P_{\mathcal{M}}S^{(n-1)}\mathbf{e}_i, \mathbf{e}_i\rangle\right) + \langle P_{\mathcal{M}}S\mathbf{e}_i, S\mathbf{e}_i\rangle\right] =$$

$$= \lim_{k\to\infty}\sum_{i=1}^{m}\langle (S^*)^k P_{\mathcal{M}}S^k\mathbf{e}_i, \mathbf{e}_i\rangle = \lim_{k\to\infty}\sum_{i=1}^{m}\langle P_{\mathcal{M}}S^k\mathbf{e}_i, S^k\mathbf{e}_i\rangle \leq \sum_{i=1}^{m}\|\mathbf{e}_i\|^2 = \dim(Ker(S^*))\,.$$

Since $\dim(\hat{P}\mathcal{H}) \leq \dim(Ker(S^*))$ there exists an isometric isomorphism W mapping $\hat{P}\mathcal{H}$ onto a subspace of $Ker(S^*)$. If we set $J := W\hat{P}$ we then have $J^*J = (W\hat{P})^*W\hat{P} = \hat{P}W^*W\hat{P} = \hat{P}$, i.e.,

$$J^*J = P_{\mathcal{M}} - SP_{\mathcal{M}}S^*\,,$$

and hence

$$S^kJ^*J(S^*)^k = S^kP_{\mathcal{M}}(S^*)^k - S^{(k+1)}P_{\mathcal{M}}(S^*)^{(k+1)}, \quad k = 0, 1, 2, \ldots\,.$$

Therefore,

$$\sum_{k=0}^{n} S^kJ^*J(S^*)^k = P_{\mathcal{M}} - S^{(n+1)}P_{\mathcal{M}}(S^*)^{(n+1)}\,,$$

and for every $\mathbf{x}, \mathbf{y} \in \mathcal{H}$,

$$\langle P_{\mathcal{M}}\mathbf{x}, \mathbf{y}\rangle - \langle S^{(n+1)}P_{\mathcal{M}}(S^*)^{(n+1)}\mathbf{x}, \mathbf{y}\rangle =$$

$$= \sum_{k=0}^{n}\langle J(S^*)^k\mathbf{x}, J(S^*)^k\mathbf{y}\rangle = \Big\langle \sum_{k=0}^{n} S^kJ(S^*)^k\mathbf{x}, \sum_{k=0}^{n} S^kJ(S^*)^k\mathbf{y}\Big\rangle\,. \tag{B.28}$$

By Eq. (B.28) we obtain

$$\|P_{\mathcal{M}}\mathbf{x}\|^2 - \|P_{\mathcal{M}}(S^*)^{(n+1)}\mathbf{x}\|^2 = \Big\|\sum_{k=0}^{n} S^k J(S^*)^k\mathbf{x}\Big\|^2$$

$$\Rightarrow \ \|\mathbf{x}\|^2 \geq \Big\|\sum_{k=0}^{n} S^k J(S^*)^k\mathbf{x}\Big\|^2 = \sum_{k=0}^{n} \big\|J(S^*)^k\mathbf{x}\big\|^2 \qquad \text{(B.29)}$$

This inequality shows that, for each $\mathbf{x} \in \mathcal{H}$ the limit $\lim_{n\to\infty}\sum_{k=0}^{n} S^k J(S^*)^k\mathbf{x}$ exists in $\mathcal{H}$. Therefore, we may define an operator $A \in \mathcal{B}(\mathcal{H})$ by

$$A^* = s - \lim_{n\to\infty}\sum_{k=0}^{n} S^k J(S^*)^k$$

Note that by Eq. (B.29) A^* is a contraction. Furthermore, since

$$S^*\left(\sum_{k=0}^{n} S^k J(S^*)^k\right) = \sum_{k=1}^{n} S^{(k-1)} J(S^*)^k = \sum_{k=0}^{n} S^k J(S^*)^{(k+1)} = \left(\sum_{k=0}^{n} S^k J(S^*)^k\right) S^*$$

we have, for all $\mathbf{x} \in \mathcal{H}$,

$$\|(S^*A^* - A^*S^*)\mathbf{x}\| =$$

$$= \lim_{n\to\infty}\left\|\left(S^*A^* - S^*\left(\sum_{k=0}^{n} S^k J(S^*)^k\right) + \left(\sum_{k=0}^{n} S^k J(S^*)^k\right)S^* - A^*S^*\right)\mathbf{x}\right\|$$

$$\leq \lim_{n\to\infty}\left\|S^*\left(A^* - \sum_{k=0}^{n} S^k J(S^*)^k\right)\mathbf{x}\right\| + \lim_{n\to\infty}\left\|\left(\sum_{k=0}^{n} S^k J(S^*)^k - A^*\right)S^*\mathbf{x}\right\| = 0$$

$$\Rightarrow \ S^*A^* = A^*S^*,$$

and we conclude that A is S-analytic. Going back to Eq. (B.28) and taking the limit $n \to \infty$ we get that

$$\forall \mathbf{x}, \mathbf{y} \in \mathcal{H}, \ \langle P_{\mathcal{M}}\mathbf{x}, \mathbf{y}\rangle - \lim_{n\to\infty}\langle S^{(n+1)} P_{\mathcal{M}}(S^*)^{(n+1)}\mathbf{x}, \mathbf{y}\rangle = \langle A^*\mathbf{x}, A^*\mathbf{y}\rangle$$

$$\Rightarrow \ \forall \mathbf{x}, \mathbf{y} \in \mathcal{H}, \ \langle P_{\mathcal{M}}\mathbf{x}, \mathbf{y}\rangle = \langle AA^*\mathbf{x}, \mathbf{y}\rangle \ \Rightarrow \ P_{\mathcal{M}} = AA^*.$$

Since $P_{\mathcal{M}}$ is a projection we get that A is a partial isometry and $P_{\mathcal{M}} = AA^*$ is a projection onto the final space of A. Since A is S-analytic and a partial isometry then A is S-inner. We conclude the proof by noting that $\mathcal{M} = P_{\mathcal{M}}\mathcal{H} = A\mathcal{H}$. ■

To Summarize our conclusions from Theorems B.39 and B.40, we have shown that if T is a class $C_{0\cdot}$ contraction then T is unitarily equivalent to the operator $\tilde{T} = (S_1)^*|\mathcal{M}$ where S_1 is the canonical unilateral shift operator on $\ell_+^2(\mathcal{D}_T)$, $\mathcal{M}^\perp$ is a closed subspace of $\ell_+^2(\mathcal{D}_T)$ invariant for S_1 and there exists an S_1-inner operator A such that $\mathcal{M}^\perp = A\ell_+^2(\mathcal{D}_T)$. In addition S_1 is an isometric dilation of $\tilde{T}^*$, with $\tilde{T}^* = P_{\mathcal{M}}S_1|\mathcal{M}$, i.e., $\tilde{T}^*$ is the compression of S_1 into $\mathcal{M}$.

A particular case of Theorem B.40 of special interest to us is when the Hilbert space is $\mathcal{H}^2_{\mathcal{N}}(D)$ and the unilateral shift operator is S_2, i.e., the canonical shift operator on $\mathcal{H}^2_{\mathcal{N}}(D)$. In this case we have:

Theorem B.41 (Beurling–Lax theorem for the unit disk) *Let S_2 be the canonical unilateral shift on $\mathcal{H}^2_{\mathcal{N}}(D)$. A subspace $\mathcal{M} \subseteq \mathcal{H}^2_{\mathcal{N}}(D)$ is invariant under S_2 iff there exists an operator valued inner function Θ such that $\mathcal{M} = \Theta\mathcal{H}^2_{\mathcal{N}}(D)$.* □

An operator valued inner function on $\mathcal{H}^2_{\mathcal{N}}(D)$ is an operator valued function defined in $B_1(0)$ with values in $\mathcal{B}(\mathcal{N})$ such that:

1. $\Theta(z)$ is an analytic operator valued function on $B_1(0)$
2. For each $z \in B_1(0)$ we have $\|\Theta(z)\| \leq 1$
3. The strong limit $s - \lim_{r\to 1^-} \Theta(re^{it})$ exists a.e. for $t \in [0, 2\pi]$ and the boundary value operators $\Theta(e^{it})$ are partial isometries on $\mathcal{N}$ with fixed initial space for all t.

Applying the isometric isomorphism W_2 of Eq. (B.6) to the dilation space $\ell^2_+(\mathcal{D}_T)$ and using Theorem B.41, we obtain a functional model for the $C_{0\cdot}$ contraction T. Thus we have:

Theorem B.42 *Let T be a contraction of class $C_{0\cdot}$ on a Hilbert space $\mathcal{H}$. Let $D_T = (I - T^*T))^{1/2}$ be the defect operator for T and let $\mathcal{D}_T = \overline{D_T\mathcal{H}}$ be the corresponding defect subspace. Then T is unitarily equivalent to an operator $\hat{T} = (S_2)^*|\hat{\mathcal{M}}$ where S_2 is the canonical shift operator on $\mathcal{H}^2_{\mathcal{D}_T}(B_1(0))$ and $\hat{\mathcal{M}}$ is an invariant subspace for $(S_2)^*$ given by $\hat{\mathcal{M}} = \mathcal{H}^2_{\mathcal{D}_T}(B_1(0)) \ominus \Theta\mathcal{H}^2_{\mathcal{D}_T}(B_1(0))$, where Θ is an operator valued inner function. In addition S_2 is an isometric dilation of $\hat{T}^*$ with $\hat{T}^* = P_{\hat{\mathcal{M}}}S_2|\hat{\mathcal{M}}$ (i.e., $\hat{T}$ is the compression of S_2 into $\hat{\mathcal{M}}$).* □

Our last step is to apply the isometric isomorphism W_3 defined in Eq. (B.8) to transform the whole structure described in Theorem B.42 to the upper half-plane. We obtain:

Theorem B.43 *Let T be a contraction of class $C_{0\cdot}$ on a Hilbert space $\mathcal{H}$. Let $D_T = (I - T^*T))^{1/2}$ be the defect operator for T and let $\mathcal{D}_T = \overline{D_T\mathcal{H}}$ be the corresponding defect subspace. Then T is unitarily equivalent to an operator $\hat{T} = S_3^*|\hat{\mathcal{M}}$ where S_3 is the canonical shift operator on $\mathcal{H}^2(\mathbb{C}^+; \mathcal{D}_T)$ and $\hat{\mathcal{M}}$ is an invariant subspace for S_3^* given by $\hat{\mathcal{M}} = \mathcal{H}^2(\mathbb{C}^+; \mathcal{D}_T) \ominus \Theta\mathcal{H}^2(\mathbb{C}^+; \mathcal{D}_T)$, where Θ is an operator valued inner function. In addition S_3 is an isometric dilation of $\hat{T}^*$ with $\hat{T}^* = P_{\hat{\mathcal{M}}}S_3|\hat{\mathcal{M}}$ (i.e., $\hat{T}$ is the compression of S_3 into $\hat{\mathcal{M}}$).* □

With Theorems B.42 and B.43 we obtain Hardy space functional representations for contraction operators of class $C_{0\cdot}$ (on the unit disk and the upper half-plane, respectively). However, our goal is to find functional models for continuous, contractive semigroups. As a first step towards this goal we present in the next section a short discussion of the functional calculus of c.u.n. contractions and then continue to continuous semigroups in the subsequent section.

B.6 Functional Calculus for c.n.u. Contractions

Consider the set of functions

$$A = \left\{ f(\cdot) \; : \; f(z) = \sum_{k=0}^{\infty} a_k z^k, \quad \sum_{k=0}^{\infty} |a_k| < \infty \right\}.$$

The functions $f \in A$ are holomorphic on $B_1(0)$ and continuous on $\overline{B_1(0)}$. If, for a function f defined on $B_1(0)$, we define

$$\tilde{f}(z) = \overline{f(\overline{z})}$$

then the set A is, in fact, an algebra with respect to standard pointwise addition and multiplication of functions and involution defined by $f \to \tilde{f}$. Note that

$$f(z) = \sum_{k=0}^{\infty} a_k z^k \; \Rightarrow \; \tilde{f}(z) = \sum_{k=0}^{\infty} \overline{a_k} z^k .$$

For a contraction T on a Hilbert space $\mathcal{H}$ we define a correspondence $A \mapsto \mathcal{B}(\mathcal{H})$ by

$$f \in A \longrightarrow f(T) = \sum_{k=0}^{\infty} a_k T^k . \tag{B.30}$$

The operator series on the right hand side above is converging. Indeed, if T is a contraction and $f \in A$ we define the partial sums $S_n(T) = \sum_{k=0}^{n} a_k T^k$ and then

$$\| S_{n'} - S_n \| = \left\| \sum_{k=n+1}^{n'} a_k T^K \right\| \leq \sum_{k=n+1}^{n'} |a_k| \| T \|^k \leq \sum_{k=n+1}^{n'} |a_k|, \quad n' > n$$

so that $\{S_n\}_{n=0}^{\infty}$ is a Cauchy sequence and its limit in operator norm is a well defined bounded operator. In addition we have

$$(a(T))^* = \big(\sum_{k=0}^{\infty} a_k T^k\big)^* = \sum_{k=0}^{\infty} \overline{a_k} \, (T^*)^k = \tilde{f}(T^*) .$$

Let U be the minimal unitary dilation of T. In this case, given a function $f(z) = \sum_{k=0}^{\infty} a_k z^k \in A$, for every partial sum $S_n(T) = \sum_{k=0}^{n} a_k T^k$ we have

$$S_n(T) = \sum_{k=0}^{n} a_k T^k = \sum_{k=0}^{n} a_k (P_{\mathcal{H}} U^k | \mathcal{H}) = P_{\mathcal{H}} \big(\sum_{k=0}^{n} a_k U^k \big) | \mathcal{H} = P_{\mathcal{H}} S_n(U) | \mathcal{H}$$

and the operator norm convergence of $S_n(U)$ to $f(U)$ implies that

$$f(T) = P_{\mathcal{H}} f(U)|\mathcal{H}, \quad \forall f \in A$$

and also the von Neumann inequality

$$\|f(T)\| \le \sup_{z \in B_1(0)} |f(z)|, \quad \forall f \in A\,.$$

For a function $f(z)$ holomorphic on $B_1(0)$ we set

$$f_r(z) := f(rz), \quad 0 < r < 1,\; z \in B_1(0)$$

and then $f_r \in A$, since $\sum_{k=0}^{\infty} |a_k| r^k < \infty$. Furthermore, for $f \in H^{\infty}(B_1(0))$ we have

$$|f_r(z)| \le \|f\|_{\infty}, \quad 0 < r < 1,\; z \in B_1(0)\,.$$

Since any $f \in H^{\infty}(B_1(0))$ is holomorphic in $B_1(0)$ then $f_r(z) \in A$ for $0 < r < 1$ and the operators $f_r(T)$ are well defined. We now define:

Definition B.33 (*The space $\mathcal{H}^{\infty}_T$*) Let T be a contraction on a Hilbert space $\mathcal{H}$. Denote by $\mathcal{H}^{\infty}_T$ the subset of $\mathcal{H}^{\infty}(B_1(0))$ such that the strong limit $s - \lim_{r \to 1^-} f_r(T)$ exists and set

$$f(T) := s - \lim_{r \to 1^-} f_r(T), \quad f \in H^{\infty}_T\,.$$

□

This definition is consistent for functions in the algebra A. It can be shown that if $f(z) \in A$ then, with this definition, $f_r(T)$ converges to $f(T)$ not only in the strong sense but also in operator norm. The space $\mathcal{H}^{\infty}_T$ is a sub algebra of $\mathcal{H}^{\infty}(B_1(0))$ and the mapping of $\mathcal{H}^{\infty}_T$ into $\mathcal{B}(\mathcal{H})$ by Definition B.33 is an algebra homomorphism. In the context of the present discussion we are interested only in c.n.u. contractions. For this class of contractions the space $\mathcal{H}^{\infty}_T$ has a simple characterization:

Theorem B.44 (Functional calculus for c.n.u. contractions) *Let T be a c.n.u. (completely non unitary) contraction on a Hilbert space $\mathcal{H}$ and let U be its minimal unitary dilation defined on the dilation Hilbert space $\mathcal{K}$ with $\mathcal{H} \subset \mathcal{K}$. Then we have $\mathcal{H}^{\infty}_T = \mathcal{H}^{\infty}_U = \mathcal{H}^{\infty}(B_1(0))$. The mapping of $\mathcal{H}^{\infty}(B_1(0))$ to $\mathcal{B}(\mathcal{H})$ defined by*

$$f(z) = \sum_{k=0}^{\infty} a_k z^k \longrightarrow f(T) = s - \lim_{r \to 1^-} \sum_{k=0}^{\infty} a_k r^k T^k$$

is an algebra homomorphism of $\mathcal{H}^{\infty}(B_1(0|))$ into $\mathcal{B}(\mathcal{H})$ such that:

1. $f(T) = \begin{cases} I, & f(z) \equiv 1 \\ T, & f(z) = z \end{cases}.$

2. $\|f(T)\| \le \|f\|_\infty$.
3. $(f(T))^* = \tilde{f}(T^*)$.
4. $f(T) = P_{\mathcal{H}} f(U)|\mathcal{H}$.

□

The identification of the functional calculus for a c.n.u. contraction with an algebra homomorphism of $\mathcal{H}^\infty(B_1(0))$ into $\mathcal{B}(\mathcal{H})$ makes it possible to apply powerful tools from the structural theory of Hardy space functions to the analysis of T, as the next three propositions demonstrate. The proof of the first of these propositions is directly related to the functional models of Sect. B.5 and, hence, we include it along with its proof.

Proposition B.10 *For every non-outer function $f \in \mathcal{H}^\infty(B_1(0))$ there exists a c.n.u. contraction T on a Hilbert space $\mathcal{H}$ (with $\mathcal{H} \neq \{\mathbf{0}\}$) such that $f(T) = 0$.* □

Proof Let $f \in \mathcal{H}^\infty(B_1(0))$ and let $f(z) = f_{in}(z) f_{out}(z)$ be the canonical factorization of f into the product of an inner function and an outer function. Then, if T is a c.n.u contraction, we have $f(T) = f_{in}(T) f_{out}(T)$. Thus, if we find a contraction T such that $f_{in}(T) = 0$ then we would have also $f(T) = 0$. Now, given the inner function f_{in} define

$$\hat{\mathcal{M}} = \mathcal{H}^2(D) \ominus f_{in}\mathcal{H}^2(D)\,.$$

If S_2 is the canonical shift operator on $\mathcal{H}^2(D)$ then $T = P_{\hat{\mathcal{M}}} S_2|\hat{\mathcal{M}}$ is a contraction such that $T^* = S_2^*|\hat{\mathcal{M}}$ and $\lim_{n\to\infty} \|(T^*)^n f\| = 0$, $\forall f \in \hat{\mathcal{M}}$, i.e., T is c.n.u. (see Sect. B.5). Moreover, S_2 is a dilation of T, i.e.,

$$T^n = P_{\hat{\mathcal{M}}} S_2^n|\hat{\mathcal{M}}\,.$$

By the functional calculus for c.n.u. contractions we have

$$f_{in}(T) = P_{\hat{\mathcal{M}}} f_{in}(S_2)|\hat{\mathcal{M}} = P_{\hat{\mathcal{M}}} f_{in}(z)|\hat{\mathcal{M}}\,.$$

However $f_{in}(z)\hat{\mathcal{M}} \subset f_{in}(z)\mathcal{H}^2(D) = \hat{\mathcal{M}}^\perp$ and hence $f_{in}(T) = 0$. ■

Proposition B.11 *For every c.n.u. contraction T on a Hilbert space $\mathcal{H}$, and for every outer function $f(z) \in \mathcal{H}^\infty(B_1(0))$, the operator $f(T)$ is invertible with a dense domain in $\mathcal{H}$.* □

From the above two propositions we immediately obtain:

Proposition B.12 *Let $f \in H^\infty(B_1(0))$. Then $f(T)$ is invertible for all c.n.u. contractions iff f is an outer function. If f is outer then for every contraction T on a Hilbert space $\mathcal{H}$ the inverse $(f(T))^{-1}$ exists and has dense domain in $\mathcal{H}$.* □

B.7 One Parameter Contractive Semigroups and Their Cogenerators

In this section we consider continuous one parameter semigroups of contractions $\{T(s)\}_{s\geq 0}$ on a Hilbert space $\mathcal{H}$. The generator of such a semigroup is defined to be

$$\tilde{A}\mathbf{x} = i \lim_{s\to 0^+} [s^{-1}(T(s) - I)\mathbf{x}],$$

and $T(s) = \exp(-i\tilde{A}s)$. The generator $\tilde{A}$ is a closed operator on $\mathcal{H}$ with a dense domain in $\mathcal{H}$ and $i \in \rho(\tilde{A})$. Note that if $T(s)$ is a contraction we have

$$\mathrm{Re}\,\langle -i\tilde{A}\mathbf{x}, \mathbf{x}\rangle_{\mathcal{H}} = \lim_{s\to 0^+} [s^{-1}\mathrm{Re}\,\langle (T(s) - I)\mathbf{x}, \mathbf{x}\rangle_{\mathcal{H}} = \lim_{s\to 0^+} [s^{-1}(\mathrm{Re}\,\langle (T(s))\mathbf{x}, \mathbf{x}\rangle_{\mathcal{H}} - \|\mathbf{x}\|^2_{\mathcal{H}})]$$

and since

$$\mathrm{Re}\,\langle (T(s))\mathbf{x}, \mathbf{x}\rangle_{\mathcal{H}} - \|\mathbf{x}\|^2_{\mathcal{H}} \leq 0$$

we have

$$\mathrm{Im}\,\langle \mathbf{x}, \tilde{A}\mathbf{x}\rangle_{\mathcal{H}} \leq 0, \quad \forall \mathbf{x} \in D(\tilde{A}).$$

This implies also that

$$\|(\tilde{A} + i)\mathbf{x}\|^2 - \|(\tilde{A} - i)\mathbf{x}\|^2 = 4\mathrm{Im}\,\langle \mathbf{x}, \tilde{A}\mathbf{x}\rangle_{\mathcal{H}} \leq 0, \quad \forall \mathbf{x} \in D(\tilde{A})$$

Since $i \in \rho(\tilde{A})$ then $(\tilde{A} - i)$ is boundedly invertible. Therefore, given any $\mathbf{y} \in \mathcal{H}$ and setting $\mathbf{x} = (\tilde{A} - i)^{-1}\mathbf{y}$ we have

$$\|(\tilde{A} + i)(\tilde{A} - i)^{-1}\mathbf{y}\|^2 - \|\mathbf{y}\|^2 \leq 0, \quad \forall \mathbf{y} \in \mathcal{H}.$$

Hence, the operator

$$T = (\tilde{A} + i)(\tilde{A} - i)^{-1}$$

is a contraction on $\mathcal{H}$. We define

Definition B.34 (*Cogenerator*) Let $\{T(s)\}_{s\geq 0}$ be a continuous, one parameter semigroup of contractions on a Hilbert space $\mathcal{H}$. Let $\tilde{A}$ be the generator of this semigroup. We call the operator

$$T = (\tilde{A} + i)(\tilde{A} - i)^{-1}$$

the *cogenerator* of the semigroup $\{T(s)\}_{s\geq 0}$. □

Note that

$$T - I = (\tilde{A} + i)(\tilde{A} - i)^{-1} - I = [(\tilde{A} + i) - (\tilde{A} - i)](\tilde{A} - i)^{-1} = 2i(\tilde{A} - i)^{-1}$$

$$\Rightarrow (T - I)^{-1} = -\frac{i}{2}(\tilde{A} - i), \quad T + I = 2i(\tilde{A} - i)^{-1} + 2I = 2\tilde{A}(\tilde{A} - i)^{-1}$$

so that

$$\tilde{A} = i(T + I)(T - I)^{-1}$$

The relation between $\tilde{A}$ and T implies the T can replace $\tilde{A}$ in the study of the semigroup $\{T(s)\}_{s\geq 0}$. The advantage of T in this respect is apparent, since T is a contraction whereas $\tilde{A}$ may not be bounded. The following theorem is proved by Sz-Nagy and Foias (1970):

Theorem B.45 *Let T be a contraction on a Hilbert space $\mathcal{H}$, There exists a continuous, contractive one parameter semigroup $\{T(s)\}_{s\geq 0}$ whose cogenerator is T iff 1 is not an eigenvalue of T. If T is the cogenerator of the semigroup $\{T(s)\}_{s\geq 0}$ then we have*

$$T(s) = e_s(T) = \exp\left(s\,\frac{T+I}{T-I}\right), \quad s \geq 0, \tag{B.31}$$

and

$$T = \lim_{s\to 0^+} \varphi_s(T(s)),$$

where $\varphi_s(x) \in A$ is defined by

$$\varphi_s(x) = \frac{x-1+s}{x-1-s} = \frac{1-s}{1+s} - \frac{2s}{1+s}\sum_{n=1}^{\infty}\left(\frac{x}{1+s}\right)^n$$

□

As mentioned above a continuous, contractive, one parameter semigroup $\{T(s)\}_{s\geq 0}$ can be studied by studying its cogenerator T. In this respect we have, for example:

Proposition B.13 *A continuous, contractive one parameter semigroup $\{T(s)\}_{s\geq 0}$ consists of normal, self-adjoint or unitary operators iff its congenerator T is, respectively, normal, self-adjoint or unitary.* □

Assume, for example, that the cogenerator T is normal with a spectral projection valued measure μ. In this case we have the spectral representations

$$f(T) = \int_{\sigma(T)} f(\lambda)d\mu(\lambda), \quad f \in H_T^{\infty}$$

and in particular,

$$T(s) = e_s(T) = \int_{\sigma(T)} e_s(\lambda)d\mu(\lambda) = \int_{\sigma(T)} \exp\left(s\,\frac{\lambda+1}{\lambda-1}\right)d\mu(\lambda)\,.$$

Assume that $\{T(s)\}_{s\geq 0}$, and therefore the congenerator T, are unitary. We then have

$$T(s) = \int_{|\lambda|=1} \exp\left(s\frac{\lambda+1}{\lambda-1}\right) d\mu(\lambda) = \int_{-\infty}^{\infty} e^{ist} d\tilde{\mu}(t),$$

where the last equality is obtained by the transformation

$$t = -i\frac{\lambda+1}{\lambda-1}, \quad \lambda \neq 1, \ |\lambda| = 1.$$

B.8 Dilations of C_0. Class Semigroups

Let $\{T(s)\}_{s\geq 0}$ be a continuous, contractive, one parameter semigroup on a Hilbert space $\mathcal{H}$ and let T be its cogenerator. Let U be the minimal unitary dilation of T, defined on a Hilbert space $\mathcal{K} = \bigvee_{n=-\infty}^{\infty} U^n\mathcal{H}$. Since 1 is not an eigenvalue of T then i is not an eigenvalue of U and hence U is the cogenerator of a semigroup $\{U(s)\}_{s\geq 0}$ of unitary operators. By Theorem B.45 we get that

$$T(s) = e_s(T), \quad U(s) = e_s(U)$$

and

$$T = \lim_{s\to 0^+} \varphi_s(T(s)), \quad U = \lim_{s\to 0^+} \varphi_s(U(s))$$

By part 4 of Theorem B.44 we get that

$$T(s) = P_{\mathcal{H}}U(s)|\mathcal{H}, \ s \geq 0,$$

so that the semigroup $\{U(S)\}_{s\geq 0}$ is a dilation of the semigroup $\{T(s)\}_{s\geq 0}$. Moreover, the dilation semigroup $\{U(s)\}_{s\geq 0}$ can be extended into a one parameter, continuous unitary group by setting

$$U(s) = \begin{cases} U(s), & s \geq 0 \\ (U(-s))^*, & s < 0 \end{cases}.$$

Furthermore, it can be shown that (Sz.-Nagy and Foias 1970)

$$\mathcal{K} = \bigvee_{n=-\infty}^{\infty} U^n\mathcal{H} = \bigvee_{s\in\mathbb{R}} U(s)\mathcal{H}.$$

Recall the standard classification of contraction operators in Definition B.28. The following proposition see Sz.-Nagy and Foias (1970), leads to a corresponding classification of continuous, contractive one parameter semigroups:

Proposition B.14 *Let $\{T(s)\}_{s\geq 0}$ be a one parameter, continuous, contractive semigroup and let T be its cogenerator. Let $\{U(s)\}_{s\in\mathbb{R}}$ be the minimal unitary dilation of $\{T(s)\}_{s\geq 0}$ and let U be the minimal unitary dilation of T (so that $U(s) = e_s(U)$). Then we have:*

$$s - \lim_{n\to\infty}((U^*)^n T^n \mathbf{x}) = s - \lim_{s\to\infty}(U(s))^* T(s)\mathbf{x}), \ \forall \mathbf{h} \in \mathcal{H}$$

$$s - \lim_{n\to\infty}(U^n (T^*)^n \mathbf{x}) = s - \lim_{s\to\infty}(U(s)(T(s))^* \mathbf{x}), \ \forall \mathbf{h} \in \mathcal{H}$$

and hence also

$$\lim_{n\to\infty} \|T^n \mathbf{x}\| = \lim_{s\to\infty} \|T(s)\mathbf{x}\| \ \forall \mathbf{x} \in \mathcal{H}, \quad \lim_{n\to\infty} \|(T^*)^n \mathbf{x}\| = \lim_{s\to\infty} \|(T(s))^* \mathbf{x}\|, \ \forall \mathbf{x} \in \mathcal{H}. \quad \text{(B.32)}$$

□

Following Proposition B.14 we may define various classes of contractive semigroups in analogy to the classes we defined for single contractions, i.e.,

Definition B.35 Let $\{T(s)\}_{s\geq 0}$ be a one-parameter, continuous contractive semigroup on a Hilbert space $\mathcal{H}$. We define the following classes:

1. $\{T(s\}_{s\geq 0}$ is of class $C_{0\cdot}$ if $s - \lim_{s\to\infty} T(s)\mathbf{x} = \mathbf{0}, \forall \mathbf{x} \in \mathcal{H}$.
2. $\{T(s)\}_{s\geq 0}$ is of class $C_{\cdot 0}$ if $s - \lim_{s\to\infty}(T(s))^* \mathbf{x} = \mathbf{0}, \forall \mathbf{x} \in \mathcal{H}$
3. $\{T(s\}_{s\geq 0}$ is of class $C_{1\cdot}$ if $s - \lim_{s\to\infty} T(s)\mathbf{x} \neq \mathbf{0}, \forall \mathbf{x} \in \mathcal{H}, \mathbf{x} \neq \mathbf{0}$.
4. $\{T(s)\}_{s\geq 0}$ is of class $C_{\cdot 1}$ if $s - \lim_{s\to\infty}(T(s))^* \mathbf{x} \neq \mathbf{0}, \forall \mathbf{x} \in \mathcal{H}, \mathbf{x} \neq \mathbf{0}$.

In addition, we define the classes $C_{\alpha\beta} = C_{\alpha\cdot} \cap C_{\cdot\beta}$, $\alpha, \beta \in \{0, 1\}$. □

As an immediate consequence of Proposition B.14 we have

Corollary B.9 *A contractive semigroup $\{T(s)\}_{s\geq 0}$ is of class $C_{0\cdot}$, $C_{\cdot 0}$, $C_{1\cdot}$, $C_{\cdot 1}$, etc. if and only if its cogenerator T belongs to the corresponding class of single contractions.* □

In Sect. B.5 we constructed a functional model for contractions of class $C_{0\cdot}$. As the congenerator of a class $C_{0\cdot}$ continuous contractive semigroup $\{T(s)\}_{s\geq 0}$ is a class $C_{0\cdot}$ contraction T, this implies the existence of a universal functional model for class $C_{0\cdot}$ continuous contractive semigroups. Our next task is to present this functional model.

Let $\{T(s)\}_{s\geq 0}$ be a continuous, contractive, semigroup of class $C_{0\cdot}$ on a Hilbert space $\mathcal{H}$ and let T be its cogenerator. Let $D_T = (I - T^*T)^{1/2}$ and $\mathcal{D}_T$ be, respectively, the defect operator and defect subspace of T. Let $\mathcal{H}^2(\mathbb{C}^+; \mathcal{D}_T)$ be the Hardy space of $\mathcal{D}_T$ valued functions defined on the upper half-plane and let S_3 be the canonical unilateral shift operator on $\mathcal{H}^2(\mathbb{C}^+; \mathcal{D}_T)$. By Theorem B.43 there exists a map $\hat{W}$: $\mathcal{H} \mapsto \mathcal{H}^2(\mathbb{C}^+; \mathcal{D}_T)$ mapping $\mathcal{H}$ isometrically onto a subspace $\hat{\mathcal{M}} \subset \mathcal{H}^2(\mathbb{C}^+; \mathcal{D}_T)$, with $\hat{\mathcal{M}}$ an invariant subspace for S_3^* and $\hat{\mathcal{M}} = \mathcal{H}^2(\mathbb{C}^+; \mathcal{D}_T) \ominus \Theta\mathcal{H}^2(\mathbb{C}^+; \mathcal{D}_T)$, where Θ is an operator valued inner function. The cogenerator T is then unitarily equivalent to the operator $\hat{T} = S_3^*|\hat{\mathcal{M}}$, T^* is unitarily equivalent to $\hat{T}^* = P_{\hat{\mathcal{M}}} S_3|\hat{\mathcal{M}}$ and S_3 is an isometric dilation of $\hat{T}^*$, i.e., $(\hat{T}^*)^n = P_{\hat{\mathcal{M}}} S_3^n|\hat{\mathcal{M}}$.

Turning to the continuous semigroup $\{T(s)\}_{s\geq 0}$ we may apply the mapping $\hat{W}$ and define a unitarily equivalent semigroup $\{\hat{T}(s)\}_{s\geq 0}$ on $\hat{\mathcal{M}}$ by $\hat{T}(s) := \hat{W}T(s)\hat{W}^*$. By Eq. (B.31) we then have

$$\hat{T}(s) = e_s(\hat{T}) = \exp\left(s\,\frac{\hat{T}+I}{\hat{T}-I}\right), \quad , \hat{T}^*(s) = e_s(\hat{T}^*) = \exp\left(s\,\frac{\hat{T}^*+I}{\hat{T}^*-I}\right) \quad s \geq 0\,,$$

and by applying the functional calculus to $\hat{T}^*$ and the function $e_s \in \mathcal{H}^\infty(B_1(0))$ we get that

$$\hat{T}^*(s) = e_s(\hat{T}^*) = P_{\hat{\mathcal{M}}} e_s(U_3)|\hat{\mathcal{M}} = P_{\hat{\mathcal{M}}} e_s(S_3)|\hat{\mathcal{M}}\,, \tag{B.33}$$

where U_3 is the extension of S_3 to a bilateral shift on $L^2(\mathbb{R}; \mathcal{D}_T)$. Now, since for every $f \in \mathcal{H}^2(\mathbb{C}^+; \mathcal{D}_T)$ we have

$$(S_3 f)(\sigma) = \frac{\sigma - i}{\sigma + i} f(\sigma)\,,$$

then

$$(e_s(S_3)f)(\sigma) = \exp\left(s\,\frac{\frac{\sigma-i}{\sigma+i}+1}{\frac{\sigma-i}{\sigma+i}-1}\right) f(\sigma) = e^{is\sigma} f(\sigma)\,.$$

Let $\hat{E}$ be the operator of multiplication by the independent variable on $L^2(\mathbb{R}; \mathcal{D}_T)$. Let $\{u(t)\}_{t\in\mathbb{R}}$ be the continuous, one parameter, unitary evolution group on $L^2(\mathbb{R}; \mathcal{D}_T)$ generated by $\hat{E}$, i.e. (see Eq. (4.1.1) of Chap. 4)

$$[u(t)f](E) = [e^{-i\hat{E}t} f](E) = e^{-i\sigma t} f(\sigma), \quad f \in L^2(\mathbb{R}; \mathcal{D}_T), \quad \sigma \in \mathbb{R}. \tag{B.34}$$

If in addition we denote by $\hat{P}_+$ the projection of $L^2(\mathbb{R}; \mathcal{D}_T)$ on $\mathcal{H}^2(\mathbb{C}^+; \mathcal{D}_T)$ we conclude from Eqs. (B.33) and (B.34) that

$$\hat{T}(s) = \hat{P}_+ u(s)|\hat{\mathcal{M}}, \quad s \geq 0\,. \tag{B.35}$$

Functional models for contractions of the type appearing in Eq. (B.35) stand at the heart of the Lax–Phillips scattering theory in Chap. 2 and the modified Lax–Phillips theory in Chap. 4 and appear also in Chap. 5. Note that, since the definition of the group $\{u(s)\}_{s\in\mathbb{R}}$ and the projection $\hat{P}_+$ are essentially independent of the semigroup $\{T(s)\}_{s\geq 0}$, then the functional model in Eq. (B.35) is determined by the subspace $\hat{\mathcal{M}}$ which, in turn, is defined by the inner function Θ. Thus, it is natural to investigate the relation between the inner function and the spectrum of the corresponding contraction. This is taken up (in a relatively restricted context) in the next section.

B.9 Contractions of Class C_0

We would like to study the relation between a given contraction of class $C_{0\cdot}$ and the inner function determining its functional model. A full description of this relation is beyond the scope of the present discussion (let alone the analogous discussion for contractions not of class $C_{0\cdot}$; see Sz.-Nagy and Foias (1970) and we shall restrict ourselves to the study of the particular case of contractions of class C_0 (to be defined below), which is nevertheless sufficient in many cases and effectively demonstrates the correspondence between properties of the contraction and the associated inner function.

The class C_0 of contraction operators is a particular subclass of the class C_{00}. Recall that the functional model for contractions obtain in Sect. B.8 is valid for contractions of class $C_{0\cdot}$ and hence is valid for contractions of class C_0. The definition of the class C_0 will be given below, but first we motivate the introduction of this class by the following two propositions:

but first we motivate the introduction of this class by the following two propositions:

Proposition B.15 *Let T be a c.n.u. contraction on a Hilbert space $\mathcal{H}$. Let $f \in H^\infty(B_1(0))$ be a non-zero function and let $\mathbf{x} \in \mathcal{H}$ be such that $f(T)\mathbf{x} = \mathbf{0}$. Then we have* $\lim_{n\to\infty} \|T^n\mathbf{x}\| = 0$. □

Proof Let T be a contraction on a Hilbert space $\mathcal{H}$ and let U be its minimal unitary dilation defined on a Hilbert space $\mathcal{K}$ such that $\mathcal{H} \subset \mathcal{K}$. It is possible to prove that the limit

$$\lim_{n\to\infty} U^{-n}T^n\mathbf{x} = L\mathbf{x}$$

exists for all $\mathbf{x} \in \mathcal{H}$. For a fixed $m \in \mathbb{N} \cup \{0\}$ we then have

$$L\mathbf{x} = \lim_{n\to\infty} U^{-(n+m)}T^{(n+m)}\mathbf{x} = \lim_{n\to\infty} U^{-m}U^{-n}T^nT^m\mathbf{x} = U^{-m}LT^m\mathbf{x}$$

and hence

$$U^mL\mathbf{x} = LT^m\mathbf{x}. \tag{B.36}$$

The fact that Eq. (B.36) is valid for all non-negative powers of the contraction T (and its minimal unitary dilation U) implies that if T is c.n.u., so that $H^\infty_T = H^\infty_U = H^\infty(B_1(0))$, we can use the functional calculus for T to obtain

$$f(U)L\mathbf{x} = Lf(T)\mathbf{x},\ \forall f \in H^\infty(B_1(0)).$$

Assume that $f \in H^\infty(B_1(0))$ and $\mathbf{x} \in \mathcal{H}$ are such that $f(T)\mathbf{x} = \mathbf{0}$. In this case we have $f(U)L\mathbf{h} = \mathbf{0}$. We observe that for $f \in H^\infty(B_1(0))$ the boundary value function $f(e^{it})$, $t \in [0, 2\pi]$, exists and $f(e^{it}) \neq 0$ a.e. with respect to Lebesgue measure.

Thus, $f(U)$ is an invertible operator and we conclude that $L\mathbf{x} = \lim_{n\to\infty} U^{-n}T^n\mathbf{x} = \mathbf{0}$. Hence we get that

$$\lim_{n\to\infty} \|T^n\mathbf{x}\| = \lim_{n\to\infty} \|U^{-n}T^n\mathbf{x}\| = \|L\mathbf{x}\| = 0\,.$$

■

Proposition B.16 *Let T be a c.n.u. contraction and assume that there exists a function $f \in H^\infty(B_1(0))$, not identically zero, such that $f(T) = 0$. Then we have $s - \lim_{n\to\infty} T^n = 0$ and $s - \lim_{n\to\infty}(T^*)^n = 0$, i.e., $T \in C_{00}$.* □

Proof Assume that $f \in H^\infty(B_1(0))$ is such that $f(T) = 0$. According to the proposition above we have in this case $\lim_{n\to\infty} \|T^n\mathbf{x}\| = 0$, $\forall \mathbf{x} \in \mathcal{H}$. In addition we have $\hat{f}(T^*) = (f(T))^* = 0$ so that, again by the proposition above, we get that $\lim_{n\to\infty} \|(T^*)^n\mathbf{x}\| = 0$, $\forall \mathbf{x} \in \mathcal{H}$. ■

The two propositions above motivate the following definition:

Definition B.36 (*Class C_0 of contractions*) The collection of all c.n.u contractions T such that there exists a function $f \in H^\infty(B_1(0))$ with $f(T) = 0$ is called class C_0. □

- By the above propositions we have, of course, $C_0 \subset C_{00}$.

By definition, for a contraction $T \in C_0$ there exists a function $f \in H^\infty(B_1(0))$ such that $f(T) = 0$. Let f be such a function and consider the canonical factorization $f(z) = f_{in}(z) f_{out}(z)$ where $f_{in}(z)$ and $f_{out}(z)$ are, respectively, the inner and outer parts of f. By the functional calculus for c.n.u. contractions (recall that $f \in C_0$ is c.n.u.) we have then $0 = f(T) = f_{in}(T) f_{out}(T)$. However, since $f_{out}(z)$ is outer then $f_{out}(T)$ is invertible (Proposition B.11) and we must have $f_{in}(T) = 0$. Therefore, for $T \in C_0$ we can always take the function f satisfying $f(T) = 0$ to be an *inner function*. For $T \in C_0$ let

$$A_T = \{f \in H^\infty(B_1(0)) \ : \ f(T)\} \tag{B.37}$$

be set of all functions in $H^\infty(B_1(0))$ annihilating T. By definition of C_0 this set is not empty. We make then the following definition:

Definition B.37 (*minimal function*) Let $T \in C_0$. An inner function m_T such that every other function in A_T is a multiple of m_T, if it exists, is called a *minimal function* for T. □

It is clear that the minimal function m_T, if it exists, is unique up to multiplication be a constant of modulus 1. This is because of the fact that if two inner functions are divisors of each other then they are identical up to multiplication by a constant with unit modulus.

Proposition B.17 *For every contraction $T \in C_0$ there exists a minimal function.* □

Proof Let $T \in C_0$. Let I be an index set and let $[f_\alpha(z)]_{\alpha\in I}$ be a class of inner functions such that $f_\alpha(T) = 0, \forall\alpha \in I$ (of course, in this case $f_\alpha \in A_T$). Let U be the minimal unitary dilation of T and let $E(t), t \in [0, 2\pi]$ be the spectral projection valued measure of U. Then, for every $\mathbf{h}, \mathbf{g} \in \mathcal{H}$ the function $\langle \mathbf{h}, E(t)\mathbf{g}\rangle_{\mathcal{H}}$ is absolutely continuous with respect to Lebesgue measure. We therefore have

$$0 = \langle T^n f_\alpha(T)\mathbf{h}, \mathbf{g}\rangle_{\mathcal{H}} = \langle P_{\mathcal{H}} U^n f_\alpha(U)\mathbf{h}, \mathbf{g}\rangle_{\mathcal{H}} = \langle U^n f_\alpha(U)\mathbf{h}, \mathbf{g}\rangle_{\mathcal{H}} =$$

$$= \int_{\sigma(U)} e^{int} f_\alpha(e^{it}) d\langle \mathbf{h}, E(t)\mathbf{g}\rangle_{\mathcal{H}} = \int_0^{2\pi} e^{int} f_\alpha(e^{it})\varphi_{\mathbf{h},\mathbf{g}}(t)dt, \quad n = 0, 1, 2, \ldots$$

where

$$\varphi_{\mathbf{h},\mathbf{g}}(t) = \frac{d\langle \mathbf{h}, E(t)\mathbf{g}\rangle_{\mathcal{H}}}{dt}.$$

Observe that $\varphi_{\mathbf{h},\mathbf{g}}(t) \in L^1([0, 2\pi])$. By Proposition (B.17), if $\tilde{f}_{in}(z)$ is the largest common inner divisor of all of the functions f_α then we have also

$$\langle \tilde{f}_{in}(T)\mathbf{h}, \mathbf{g}\rangle_{\mathcal{H}} = \int_0^{2\pi} \tilde{f}_{in}(e^{it})\varphi_{\mathbf{h},\mathbf{g}}(t)dt = 0$$

and since $\mathbf{h}$ and $\mathbf{g}$ are arbitrary, we conclude that $\tilde{f}_{in}(T) = 0$. Now, given a contraction $T \in C_0$ we take m_T to be the largest common inner divisor of all functions in the class of functions A_T defined in Eq. (B.37). Then $m_T(T) = 0$ and m_T is necessarily the minimal function for T. ∎

The role played by the minimal function m_T of a contraction T of class C_0 is analogous to the role of the minimal polynomial of a finite matrix in linear algebra. The next theorem is describes the relation between the minimal function of a contraction $T \in C_0$ and the spectrum of T:

Theorem B.46 *Let T be a contraction of class C_0 and let m_T be its minimal function. Let $\sigma(T)$ be the spectrum of T. Denote by S_T the set of zeros of m_T in $B_1(0)$ and the complement in the unit circle C of the union of arcs on which m_T is holomorphic. Then $\sigma(T) = S_T$.* □

Proof Let $z_0 \in \overline{B_1(0)}\backslash S_T$. If $z_0 \in B_1(0)$ then, by the definition of S_T, it is immediate that $m_T(z_0) \neq 0$. If $z_0 \in C\backslash(C \cap S_T)$ then by the analyticity property of m_T in a neighborhood of z_0 and by the fact that m_T is an inner function, so that $|m_T(e^{it})| = 1$ a.e., we also get that $m_T(z_0) \neq 0$. Now set

$$f(z) = \frac{m_T(z) - m_T(z_0)}{z - z_0}.$$

Since m_T is analytic in a neighborhood of z_0 then $f \in H^\infty(B_1(0))$ and, using the functional calculus for c.n.u. contractions from Theorem (B.46) and the fact that m_T

is the minimal function for T, we get that

$$(z_0 I - T)f(T) = f(T)(z_0 I - T) = m_T(z_0)I - m_T(T) = m_T(z_0)I$$

By the last equation $(z_0 I - T)$ is boundedly invertible and we have

$$(z_0 I - T)^{-1} = \frac{1}{m_T(z_0)} f(T) \in \mathcal{B}(\mathcal{H})$$

Hence $z_0 \in \rho(T)$, where $\rho(T)$ is the resolvent set of T. To summarize, we have found that if $z_0 \in \overline{B_1(0))} \backslash S_T$ then $z_0 \in \rho(T)$ and hence $\sigma(T) \subseteq S_T$. Therefore, we are left with the task of showing that $S_T \subseteq \sigma(T)$. Let $z_0 \in B_1(0)$ be a point such that $m_T(z_0) = 0$. By the standard multiplicative decomposition of an inner function we have then

$$m_T(z) = \frac{z - z_0}{1 - \overline{z_0} z} \tilde{m}_{z_0}(z)$$

where $\tilde{m}_{z_0}(z)$ is an inner function. Using the functional calculus for c.n.u. contractions and the fact that m_T is the minimal function for T we then get that

$$0 = m_T(T) = (1 - \overline{z_0}\, T)^{-1}(T - z_0 I)\, \tilde{m}_{z_0}(T) \;\Rightarrow\; 0 = (T - z_0 I)\, \tilde{m}_{z_0}(T)$$

Since $\tilde{m}_{z_0}$ is not a multiple of m_T we have $\tilde{m}_{z_0}(T) \neq 0$ and hence the operator $(T - z_0 I)$ is not invertible and $z_0 \in \sigma(T)$. Thus, we have $S_T \cap B_1(0) \subseteq \sigma(T)$.

It remains to be shown that $S_T \cap C \subset \sigma(T)$. For this, we first note that,

$$S_T \cap C \subseteq \sigma(T) \;\Leftrightarrow\; C \backslash (S_T \cap C) \supseteq C \backslash (\sigma(T) \cap C) = C \cap \rho(T)\,.$$

Observe that, by the definition of S_T, the set $C \backslash (S_T \cap C)$ consists of open arcs on C on which m_T is analytic. Hence, our goal is to show that for any open arc $C_o \subset \rho(T) \cap C$ the minimal function m_T is analytic on C_0. Consider the canonical factorization of m_T in the form

$$m_T(z) = C\, B(z) S(z)$$

where C is a constant of unit modulus, $B(z)$ is a Blaschke product and $S(z)$ is an inner function of the form

$$S(z) = \exp\left(- \int_0^{2\pi} \frac{e^{it} + z}{e^{it} - z} d\mu(t) \right)$$

with $\mu(t)$ being a finite measure on $[0, 2\pi]$, singular with respect to Lebesgue measure. Since $\rho(T)$ is an open set, every point in C_0 has a positive distance from $\sigma(T)$ and hence the zeros of the Blaschke product $B(z)$ cannot accumulate at any point in

C_o. Hence $B(z)$ is analytic on C_o and the only possibility for m_T to be not analytic on C_o is if $S(z)$ is not analytic on C_o. In fact, we shall show that $\mu(C_o) = 0$, in which case $S(z)$ is immediately analytic on C_o.

Assume that $\mu(C_0) > 0$. In this case there is an arc $C_1 \subset C_o$ such that $\mu(C_1) > 0$. Setting

$$m_1(z) = \exp\left(- \int_{C_1} \frac{e^{it} + z}{e^{it} - z} d\mu(t) \right)$$

we have that $m_1(z)$ is an inner divisor of $m_T(z)$. Moreover, since $|m_1(0)| = \exp(-\int_{C_1} d\mu(t)) = \exp(-\mu(C_1)) < 1$, the function $m_1(z)$ is not a constant. We can then express m_T in the form $m_T(z) = m_1(z)m_2(z)$ where $m_2(z)$ is also an inner function and in this case we would have

$$0 = m_T(T) = m_1(T)m_2(T).$$

Now, if $m_1(T)$ is invertible then $m_2(T) = 0$ and in this case, since m_T is minimal, we find that m_T is a divisor of m_2, which is impossible. We conclude that $m_1(T)$ is not invertible and

$$\mathcal{H}_1 = \{\mathbf{h} \in \mathcal{H} \, : \, m_1(T)\mathbf{h} = \mathbf{0}\} \neq \{\mathbf{0}\}$$

By its very definition, the subspace $\mathcal{H}_1$ is invariant for T. Setting $T_1 := T|\mathcal{H}_1$, the minimal function m_{T_1} of T_1 is a divisor of m_T (since $m_1(T_1) = m_1(T)|\mathcal{H}_1$) and has the form

$$m_{T_1}(z) = \exp\left(- \int_0^{2\pi} \frac{e^{it} + z}{e^{it} - z} d\mu_1(t) \right)$$

where mu_1 is a nonnegative, singular measure on $[0, 2\pi]$ vanishing on $C \backslash C_1$. The subset $S_{T_1} \subset C$ defined by m_{T_1} is then contained in C_1 and, furthermore, $\sigma(T_1) \subseteq S_{T_1} \subseteq C_1$.

Let $z \in \mathbb{C}$ be a point such that $|z| > 0$. Then, if $\mathcal{H}_1 \subseteq \mathcal{H}$ is invariant for T the, linear transformation $(zI - T)$ maps $\mathcal{H}_1$ onto itself (this is a consequence of the relations $\mathbf{g} = (zI - T)\mathbf{h} \Leftrightarrow \mathbf{h} = \sum_{n=0}^{\infty} z^{-(n+1)} T^n \mathbf{g}$). Hence, if $\mathcal{H}_1$ is the subspace defined above, we have

$$(zI_{\mathcal{H}_1} - T_1)^{-1} = (zI - T)^{-1}|\mathcal{H}_1, \quad \forall |z| > 1$$

If the point z then tends to a point $z_0 \in C \cap \rho(T)$ the operator $(zI_{\mathcal{H}_1} - T_1)^{-1}$ remains bounded, so that $z_0 \in \rho(T_1)$. We conclude that $C_1 \subset \rho(T_1)$. However, we already concluded that $\sigma(T_1) \subseteq C_1$ so that $\sigma(T_1) \subseteq \rho(T_1)$. This is a contradiction and hence $\mu(C_1) = 0$ and, moreover, $\mu(C_0) = 0$ implying the analyticity of $S(z)$ on C_0. Thus m_T is analytic on C_0 and the theorem is proved. ∎

The following theorem, which we state without proof, supplements Theorem B.46 above:

Theorem B.47 *For a contraction $T \in C_0$ on $\mathcal{H}$, the points $\alpha \in \sigma(T) \cap B_1(0)$ are eigenvalues of T. As a characteristic value of T such a point α has a finite index equal to tits multiplicity as a zero of the minimal function m_T. Moreover, for such an α $\mathcal{H}$ decomposes into a direct sum of two subspaces, hyperinvariant for T (i.e., invariant for T and for any operator commuting with T), say $\mathcal{H}_\alpha$ and $\mathcal{H}'_\alpha$. The space $\mathcal{H}_\alpha$ consists of characteristic vectors of T associated with α (and the zero vector) and $T'_\alpha = T|\mathcal{H}'_\alpha$ has α in its resolvent set.* □

References

Adler, S.L.: Phys. Rev. **140**, 736 (1965)
Aguilar, J., Combes, J.M.: Commun. Math. Phys. **22**, 269 (1971)
Aharonov, Y., Oppenheim, J., Popescu, S., Reznik, B., Unruh, W.G.: Lecture Notes in Physics, vol. 517, p. 204. Springer, Berlin (1999)
Aharonovich, I., Horwitz, L.P.: J. Math. Phys. **47**, 122902 (2006)
Aharonovich, I., Horwitz, L.P.: J. Math. Phys. **51**, 052903 (2010)
Aharonovich, I., Horwitz, L.P.: J. Math. Phys. **52**, 082901 (2011)
Aharonovich, I., Horwitz, L.P.: J. Math. Phys. **53**, 032902 (2012)
Appell, P.: Dynamique des Systemes Mecaniques Analytique. Gautier-Villars, Paris (1953)
Ar'nold, V.I.: Mathematical Methods of Classical Mechanics. Springer, Berlin (1978)
Bailey, T., Schieve, W.C.: Nuovo Cimento Soc. Ital. Fis. A **47A**, 231–250 (1978)
Balslev, E., Combes, J.M.: Commun. Math. Phys. **22**, 280 (1971)
Baumgartel, H.: Math. Nachr. **75**, 137 (1976); Ann. Henri Poincare **10**, 123 (2009)
Baumgartel, W.: Math. Nachr. **69**, 107–121 (1975)
Bekenstein, J.D.: Phys. Rev. D **70**, 083509 (2004)
Bekenstein, J.D., Sanders, R.H.: A Primer to Relativistic MOND Theory. EAS Pub. Ser. **20**, 225 (2006)
Ben Ari, T., Horwitz, L.P.: Phys. Lett. A **332**, 168 (2004)
Ben Zion, Y., Horwitz, L.P.: Phys. Rev. E **76**, 046220 (2007). See also (Ben Zion 2008), (Ben Zion 2010)
Ben Zion, Y., Horwitz, L.P.: Phys. Rev. E **78**, 036209 (2008)
Ben, Zion Y., Horwitz, L.P.: Phys. Rev. E **76**, 046220 (2007); Phys. Rev. E **78**, 036209 (2008); Phys. Rev. E **81**, 046217 (2010)
Bohigas, O., Giannoni, M.O., Schmidt, C.: Phys. Rev. Lett. **52**, 1 (1884)
Bohm, A., Gadella, M.: Dirac Kets, Gamow Vectors and Gel'fand Triplets: The rigged Hilbert space Formulation of Quantum mechanics. Lecture notes in Physics, vol. 348. Springer, Berlin (1989)
Born, M.: Einstein's Theory of Relativity. Dover, New York (1962)
Caianai, L., Casetti, L., Clementi, C., Pettini, M.: Phys. Rev. Lett **79**, 4361 (1997)
Calderon, E., Horwitz, L., Kupferman, R., Shnider, S.: Chaos **23**, 013120 (2013)
Cartan, H.: Calcul Differential et Formes Differentielle. Hemna, Paris (1967)
Casetti, L., Pettini, M.: Phys. Rev. E **48**, 4320 (1993)
Casetti, L., Clementi, C., Pettini, M.: Phys. Rev. E **54**, 5669 (1996)
Cohen, E., Horwitz, L.P.: Hadronic J. **24**, 593 (2001)
Cohen, I.B., Whitman, A.: The Principia: Mathematical Principles of Natural Philosophy: A New Translation. University of California Press, Berkeley (1999)
Cornfeld, I.P., Fomin, S.V., Sinai, Y.G.: Ergodic Theory. Springer, New York (1982)

L. Horwitz and Y. Strauss, *Unstable Systems*, Mathematical Physics Studies,
https://doi.org/10.1007/978-3-030-31570-2

Curtiss, W.D., Miller, F.R.: Differentiable Manifolds and Theoretical Physics. Academic, New York (1985)
da Silva, A.: Lectures on Symplectic Geometry. Lecture Notes in Mathematics 1764, Springer, New York (2006)
de la Madrid, R., Gadella, M.: Amer. J. Phys. **70**, 626 (2002)
Duffing G., *Erzwungene Schwingung, sei Varanderlicher Eigenfrequenz und ihre Technische Bedeutung*, Verlag Braunschweig (1918)
Duren, P.L.: Theory of H^p Spaces. Academic, New York (1970)
Eisenberg, E., Horwitz, L.P.: Time, irreversibility and unstable systems in quantum Physics. In: Prigogine, I., Rice, S. (eds.) Advances in Chemical Physics, vol. XCIX. Wiley, New York (1997)
Eisenhardt, L.P.: A Treatise on the Differential Geometry of Curves and Surfaces. Ginn, Boston (1909) [Dover, New York (2004)]
Fanchi, J.R.: Parametrized Relativistic Quantum Theory. Kluwer, Dordrecht (1993)
Flesia, C., Piron, C.: Helv. Phys. Acta **57**, 697 (1984)
Floquet, G.: Annales de l'Ecolé Normale Superieure **12**, 47 (1883)
Frankel, T.: The Geometry of Physics. Cambridge University Press, Cambridge (1997)
Friedrichs, K.O.: Commun. Pure Appl. Phys. Math. **1**, 361 (1950)
Galapon, E.A.: Proc. Roy. Soc. A Lond. **458**, 451 (2002)
Gamow, G.: Zeits. f. Physik **51**, 204 (1928)
Gershon, A., Horwitz, L.P.: J. Math. Phys. **50**, 102704 (2009)
Guckenheimer, J., Holmes, P.: Nonlinear Oscillations, Dynamical Systems, and Bifurcations of Vector Fields. Springer, New York (1983)
Gutzwiller, M.C.: Chaos in Classical and Quantum (Mechanics). Springer, New York (1990)
Hadamard, J.S.: J. Math. Pures et Appl. **4**, 27 (1898)
Henon, M., Heiles, C.: Astrophys. J. **69**, 73 (1964)
Hille, E.: Ordinary Differential Equations in the Complex Plane. Wiley, New York (1976)
Hislop, P.D., Sigal, I.M.: Itroduction to Spectral Theory: With Application to Schrödinger Operators. Springer, New York (1996)
Hoffman, K.: Banach Spaces of Analytic Functions. Prentice-Hall, Englewood cliffs (1962)
Holevo, A.S.: Probabilistic and Statistical Aspects of Quantum Theory. North Holland, Amsterdam (1982)
Horwitz, L.P.: Found. Phys. **25**, 1335 (1995)
Horwitz, L.P.: Relativistic Quantum Mechanics, Fundamental Theories of Physics 180. Springer, Dordrecht (2015)
Horwitz, L.P., Ben, Zion Y.: Phys. Rev. E **76**, 046220 (2007)
Horwitz, L.P., Ben, Zion Y.: Phys. Rev. E **78**, 06209 (2008)
Horwitz, L.P., Katznelson, E.: Phys. Rev. Lett. **50**, 1184 (1983)
Horwitz, L.P., Marchand, J.-P.: Rocky Mt. J. Math. **1**, 225 (1971)
Horwitz, L.P., Piron, C.: Helv. Phys. Acta **46**, 315 (1973)
Horwitz, L.P., Piron, C.: Helv. Phys. Acta **66**(7–8), 693 (1993)
Horwitz, L.P., Sigal, I.M.: Helv. Phys. Acta **51**, 685–715 (1978)
Horwitz, L.P., LaVita, J.A., Marchand, J.-P.: J. Math. Phys. **12**, 2537 (1971a)
Horwitz, L.P., Ben, Zion Y., Lewkowicz, M., Schiffer, M., Levitan, J.: Phys. Rev. Lett. **98**, 234301 (2007a)
Horwitz, L.P., Gershon, A., Schiffer, M.: Found. Phys. **41**, 141 (2010)
Horwitz, L.P., Yahalom, A., Lewkowicz, M., Levitan, J.: Int. J. Modern Phys. D **20**, 2787 (2011d)
Horwitz, L.P., Yahalom, A., Levityan, J., Lewkowicz, M.: Front. Phys. **12**, 124501 (2017)
Howland, J.S.: Math. Ann. **207**, 315 (1974)
Howland, J.S.: Indiana Math. J. **28**, 471 (1979)
Hunziker, W.: Ann. I.H.P. Phys. Theor. **45**, 339 (1986)
Itzykson, C., Zuber, J.B.: Quantum Field Theory. McGraw Hill, New York (1980)
Jacobi, C.G.J.: Vorlesungen über Dynamik. Verlag Reimer, Berlin (1884)
Jacobi, C.G.J.: Vorlesungen über Dynamik. Verlag G. Reiner, Berlin (1844)

Kandrup, H.E., Sideris, I., Bohn, C.L.: Phys. Rev. E **65**, 016214 (2001)
Kruglov, V., Makarov, K.A., Pavlov, B., Yafyasov, A.: Exponential decay in quantum mechanics. In: Computation, Physics and Beyond. Lecture Notes in Computer Science, vol. 7160. Springer, Heidelberg (2012)
Landau, L.D.: Mechanics. Mir, Moscow (1969)
Lax, P.D., Phillips, R.S.: Scattering Theory. Elsevier, Academic, New York (1967)
Lee, T.D.: Phys. Rev. **95**, 1329 (1956)
Levitan, L., Yahalom, A., Horwitz, L., Lewkowicz, M.: Chaos **23**, 023122 (2014)
Lewkowics, M., Levitan, J., Ben Zion, Y., Horwitz, L.P.: Handbook of Chemical Physics (CMSM), Chap. 15, pp. 231–252. Chapman and Hall, CRC Press (2016)
Maassen, H.: Prob. Theor. Rel. Fields **83**, 489 (1989); Attal, S., Lindsay, J.M. (eds.): Quantum Probability Communications vol. XII. World Scientific, Singapore (2003)
Mackey, G.W.: The Theory of Unitary Group Representations. University of Chicago Press, Chicago (1976)
Milgrom, M.: Astrophys. Jour. **270**, 365, 371, 384 (1983)
Milnor, J.: Morse Theory, Annals of Mathematics Studies 51. Princeton University Press, Princeton (1969)
Misra, B., Prigogine, I., Courbage, M.: Math. Ann. **207** 315 (1974), see also, (Howland 1979)
Moiseyev, N.: Phys. Rep. **302**, 212 (1998)
Moser, J., Zehnder, E.J.: Notes on Dynamical Systems. American Mathematical Society, Providence (2005)
Newton, I.: Philosophia Naturalis Principia Mathematica, London (1687)
Newton, R.G.: Scattering Theory of Waves and Particles. McGraw Hill, New York (1967)
Nikol'skii, N.K.: Treatise on the Shift Operator: Spectral Function Theory. Springer, New York (1986)
Nikol'skii, N.K.: Operators, Functions and Systems, an Easy Reading, vol. I. American Mathematical Society, Providence (2002)
Oloumi, A., Teychenne, D.: Phys. Rev. E **60**, R6279 (1999)
Paley, R.E.A.C., Wiener, N.: Fourier Transforms in the Complex Domain. American Mathematical Society, Providence (1934)
Parravincini, G., Gorinim, V., Sudrashan, E.C.G.: J. Math. Phys. **21**, 2208–2226 (1980)
Parthasarathy, K.R.: An Introduction to Quantum Stochastic Calculus. Birkhauser-Verlag, Basel (1992)
Patrignani, C., et al. (Particle Data Booklet, from Review of Particle Physics): Chin. Phys. C. **40**, 100001 (2016)
Pauli, W.E.: Handbuch der Physik. Springer, Berlin (1926)
Pavlov, B.S.: J. Math. Sci. **77**(3), 3232–3235 (1995)
Pavlov, B.S.: Self-adjoint dilation of the dissipative Schrödinger operator and its resolution in terms of eigenfunctions. Math. USSR Sb. **31**, 457 (1977)
Pettini, M.: Geometry and Topology in Hamiltonian Dynamics and Statistical Mechanics. Springer, New York (2006)
Pettini, M.: Geometry and Topology in Hamiltonian Dynamics. Springer, New York (2007)
Rosenblum, M., Rovnyak, J.: Hardy Classes and Operator Theory. Oxford University Press, New York (1985)
Ruelle, D.: Chaotic Evolution and Strange Attractors. Cambridge University Press, Cambridge (1989)
Saad, D., Horwitz, L.P., Arshansky, R.I.: Found. Phys. **19**, 1125 (1989)
Safaai, H., Hasan, M., Saadat, G.: Understanding Complex Systems, p. 369. Springer, Berlin (2006)
Schwinger, J.: Phys. Rev. **82**, 664 (1951)
Simon, B.: Ann. Math. **97**, 247 (1973)
Simon, B.: Phys. Lett. A **71**, 211 (1979)
Sjöstrand, J., Zworski, M.J.: Amer. Math. Soc. **4**, 729 (1991)
Strauss Y., Horwitz L.P., Yahalom A., Levitan J.: arXiv:1708.00609 (2017)

Strauss, Y., Horwitz, L.P., Volovick, A.: J. Math. Phys. **47**, 123505(1–19) (2006)
Strauss, Y., Horwitz, L.P.: J. Math. Phys. **43**, 2394 (2002)
Strauss, Y., Silman, J., Machness, S., Horwitz, L.P.: Comptes Rendus **349**(19–20), 1117–1122 (2011)
Strauss Y.: J. Math. Phys. **46**(10), 102109(1–12) (2005)
Strauss Y.: J. Math. Phys. **46**(2), 032104(1–25) (2005)
Strauss, Y.: J. Math. Phys. **51**, 022104 (2010)
Strauss, Y.: J. Math. Phys. **52**, 032106 (2011)
Strauss, Y.: J. Math. Phys. **56**, 073501 (2015)
Strauss, Y., Horwitz, L.P.: Found. Phys. **30**, 653 (2000)
Strauss, Y., Horwitz, L.P., Eisneberg, E.: J. Math. Phys. **41**, 8050 (2000)
Strauss, Y., Silman, J., Machness, S., Horwitz, L.P.: Int. J. Theor. Phys. **50**(7), 2179–2190 (2011)
Strauss, Y., Horwitz, L.P., Levitan, J., Yahalom, A.: J. Math. Phys. **56**, 072701 (2015)
Stueckelberg, E.C.G.: Helv. Phys. Acta **14**(372), 585 (1941)
Stueckelberg, E.C.G.: Helv. Phys. Acta **15**, 23 (1942)
Sudarshan, E.C.G., Misra, B.: J. Math. Phys. **18**, 756 (1977)
Sz.-Nagy, B., Foias, C.: Harmonic Analysis of Operators on Hilbert Space. North Holland, Amsterdam (1970)
Szydlowski, M., Krawiec, A.: Phys. Rev. D **53**, 6893 (1996), who have studied a somewhat generalized Gutzwiller form (with the addition of a scalar potential) which accommodates the Jacobi metric
Szydlowski, M., Szczesny, J.: Phys. Rev. D **50**, 819 (1994)
Taylor, J.R.: Scattering Theory: The Quantum Theory of Nonrelativistic Collisions. Wiley, New York (1972)
Titchmarsh, E.C.: The Theory of Functions. Oxford University Press, London (1939)
Unruh, W., Wald, R.M.: Phys. Rev. D **40**, 25982614 (1989)
Van-Winter, C.: Trans. Amer. Math. Soc. **162**, 103 (1971)
Weisberger, W.I.: Phys. Rev. Lett. **14**, 1047 (1965)
Weisskopf, V.F., Wigner, E.P.: Zeits. f. Physik **63**, 54 (1930)
Wesson, P.S.: Five Dimensional Physics. Classical and Quantum Consequences of Kaluza-Klein Cosmology. World Scientific, Singapore (2006)
Wigner, E.P.: Ann. Math. **62**, 548 (1955)
Wu, T.T., Yang, C.N.: Phys. Rev. Lett. **13**, 380 (1964)
Yahalom, A., Lewkowicz, M., Levitan, J., Elgressy, G., Horwitz, L., Ben Zion, Y.: Int. J. Geom. Methods Modern Phys. **12**, 1550093 (2015)
Yahalom, A., Levitan, J., Lewkowicz, M.: Phys. Lett. A **375**, 2111 (2011)
Yoshida, H.: Commun. Math. Phys. **116**, 529 (1988)
Zaslavsky, G.M.: Chaos in Dynamic Systems. Harwood Academic Publishers, New York (1985). See also, Physics of Chaos in Hamiltonian Dynamics. Imperial College Press, London (1998); Hamiltonian Chaos and Fractal Dynamics. Oxford University Press, Oxford (2005)
Zerzion, D., Horwitz, L.P., Arshansky, R.: J. Math. Phys. **32**, 1788 (1991)

Index

L. Horwitz and Y. Strauss, *Unstable Systems*, Mathematical Physics Studies,
https://doi.org/10.1007/978-3-030-31570-2

Printed in the United States
by Baker & Taylor Publisher Services